Science, culture and popular belief
in Renaissance Europe

Science, culture and popular belief in Renaissance Europe

edited by Stephen Pumfrey,
Paolo L. Rossi and Maurice Slawinski

Manchester University Press
Manchester and New York

distributed exclusively in the USA and Canada by St. Martin's Press

Published by Manchester University Press
Oxford Road, Manchester M13 9PL, UK
and Room 400, 175 Fifth Avenue,
New York, NY 10010, USA

Distributed exclusively in the USA and Canada
by St. Martin's Press, Inc.,
175 Fifth Avenue, New York, NY 10010, USA

British Library cataloguing in publication data
Science, culture and popular belief in Renaissance Europe.
 1. Europe. Social. Life, 1453–1650
 I. Pumfrey, Stephen II. Rossi, Paolo L III. Slawinski, Maurice
 940.21

Library of Congress cataloging in publication data
 Science, culture and popular belief in Renaissance Europe / edited by
 Stephen Pumfrey, Paolo L. Rossi, Maurice Slawinski.
    p.   cm.
   ISBN 0-7190-2925-2 (hardback)
   1. Science, Renaissance, 2. Europe – Intellectual life – 17th
 century. I. Pumfrey, Stephen II. Rossi, Paolo L. 1946–
 III. Slawinski, Maurice.
 Q125.2.S36   1991
 509.4'09032—dc20   90-25569

Paperback edition published 1994

ISBN 0 7190 4322 0 *paperback*

Typeset in Monotype Bell
by Koinonia Ltd, Manchester
Printed in Great Britain,
by Bell and Bain Ltd, Glasgow

Contents

List of figures

The figures are reproduced with the permission of the following: the Director and University Librarian, the John Rylands University Library of Manchester (1 and 4); the British Library (2, 5 and 6); the Trustees of the British Museum (3); The Swedish National Art Museums (7); and the Archivio di Stato, Rome (8).

Notes on the contributors

Jim Bennett is Curator of the Whipple Museum of the History of Science in the University of Cambridge, and a Senior Research Fellow at Churchill College, Cambridge. Among his recent publications in this field is *The Divided Circle: a history of instruments for astronomy, navigation and surveying* (Phaidon, 1987). He has just completed a history of Armagh Observatory.

Stuart Clark is Lecturer in early modern history at University College, Swansea (University of Wales). He is author of several articles on the intellectual and cultural history of witchcraft beliefs, and on recent styles of history-writing in France. He is completing a study of the place of demonology in early modern thought.

Patrick Curry is a Canadian, now living in London. He held a research fellowship from the Research Council of Canada for 1987–89, after obtaining his PhD from University College London. The author of *Prophecy and Power: Astrology in early Modern England,* he is presently completing a book on Victorian and Edwardian astrology in London.

Peter Dear is Assistant Professor of History at Cornell University. He is the author of *Mersenne and the Learning of the Schools* (Cornell University Press, 1988), as well as articles including 'Miracles, experiments, and the ordinary course of nature', forthcoming in *Isis.*

Germana Ernst teaches in the Department of Philosophy of the University of Florence. She has produced critical editions of Tommaso Campanella's *Articuli Prophetales* (Florence, 1977) and *Monarchia di Spagna* (Naples, 1989) and is currently preparing an edition of his *Atheismus Triumphatus.* Her wider interest in the history of ideas in the Renaissance is represented in her collection of essays *Religion, Reason and Nature: studies on Campanella and the late Renaissance* (Angeli, 1991).

Luce Giard is Associate Director of the 'Laboratory for the History of Science' of the Centre National de la Recherche Scientifique in Paris, and is Visiting Professor at the University of California at San Diego. She is the author of numerous essays on the history of science and philosophy in the Middle Ages and the Renaissance, including a series of articles on ' The history of the universities and the history of knowledge: Padua (XIVth-XVIth centuries)'.

John Henry is Lecturer in the history of science and medicine in the Science Studies Unit at Edinburgh University. He has published a number of articles on sixteenth- and seventeenth-century science and medicine and is joint editor (with Sarah Hutton) of *New Perspectives on Renaissance Thought* (Duckworth, 1990).

Julian Martin is developing a science and technology studies programme at the University of Alberta, Canada. Previously he held Research Fellowships at the Wellcome Institute and at Clare Hall, Cambridge. He has published on early modern scholarly medicine, and a book on Francis Bacon will shortly appear with Cambridge University Press. He is currently preparing an edition of Bacon's Elizabethan works.

Roy Porter is Senior Lecturer in the social history of medicine at the Wellcome Institute for the History of Medicine. He is currently working on the history of hysteria. His recent books include *Mind Forg'd Manacles. Madness in England from the Restoration to the Regency* (Athlone, 1987); *Patient's Progress* (Polity, 1989) – co-authored with Dorothy Porter; and *Health for Sale. Quackery in England 1660-1850* (Manchester University Press, 1989).

Stephen Pumfrey is Lecturer in the history of science at Lancaster University, and was previously a Research Fellow of Corpus Christi College, Cambridge. He has published a number of articles on seventeenth-century natural philosophy, and is currently writing on the Scientific Revolution for the *Longman Studies in Modern History* series.

Paolo L. Rossi is Lecturer in Italian studies at Lancaster University, and was previously Donaldson Bye-fellow at Magdalene College, Cambridge. He has also been visiting lecturer at the Universities of Glasgow and Liverpool. As well as the cultural history of the Renaissance his interests include the history of architecture and the relationship between patronage and the arts. He is completing a biography of the artist Benvenuto Cellini, and a three-volume edition of his writings.

Maurice Slawinski is Lecturer in Italian studies at Lancaster University, and has also taught at the Universities of Cambridge and Milan. He is the author of a number of essays on Italian Renaissance and Baroque literature, and on the sociology of intellectual life during the period, as well as on comparative literature. He is currently writing a study of the Italian Baroque poet G. B. Marino.

Foreword

The contributions to this volume are evidence of the great developments which have taken place in recent years in our understanding of the place of science in society. Natural knowledge is now profitably interpreted as a cultural product, made out of and subject to the same social, political and cultural resources which appear in historical accounts of, for example, works of art, political ideologies or popular beliefs. Such interconnections were particularly evident in Renaissance Europe. Consequently, the history of Renaissance science (or, to use the category of the time, 'natural philosophy') has moved away from an analysis of ideas strictly, and wrongly, demarcated as 'scientific', and now addresses the full richness of Renaissance culture and society.

In our teaching we find that the new approach to Renaissance science is of broad interest, reaching students and interested readers of history and other human sciences. However, much of the best new work has yet to move beyond the scholarly journals. To bridge the gap, the authors in this volume, all of whom have advanced research in this more contextual understanding of Renaissance science, have synthesized recent work in fields of their own expertise, whilst advancing their own particular interpretations. The resulting collection of commissioned essays adds up, we hope, to an accessible, stimulating and scholarly account of important connections between science, culture and popular belief in the Renaissance, useful to the serious student and to lay readers alike. To this end, scholarly footnotes have been kept to a minimum, but an annotated bibliography of more specialist literature concludes the volume.

The editors would like to thank John Banks, late of MUP, and our original commissioning editor, for his encouragement and forbearance through many delays, Steve Jenkins of Lancaster University's Computer Centre, for his invaluable help in getting our contributors' word-processors to talk to each other, and the contributors themselves for their co-operation and patience through the editing process.

Stephen Pumfrey
Paolo L. Rossi
Maurice Slawinski
Lancaster, 14 December 1990

Introduction

Roy Porter

It is now just over forty years since Herbert Butterfield declared that the Scientific Revolution 'outshines everything since the rise of Christianity and reduces the Renaissance and Reformation to the rank of mere episodes, mere internal displacements within the system of medieval Christendom'.[1] The claim was bold. It was rare for a humanist historian to award science so prominent a place in the making of modernity. Protestant Europe had long celebrated the key progressive role of the Reformation; over the course of a century, Burckhardt's followers had established the Renaissance as epochal to the creation of modern secular individualism. Living in the land of the 'two cultures', British historians had somehow been especially shielded against the significance of science. It took the events of the 1940s – Nazi science, Lysenkoism, Stalin's war against the intelligentsia, and above all the atom bomb – to bring home, to historians and the common man alike, the extraordinary power science had given mankind, for good or ill, since the early modern era.

In two respects, however, Butterfield was utterly conventional. First, he had not the slightest doubt that the Renaissance, the Reformation, and the Scientific Revolution were all good things. No surprise that a Methodist historian ensconced in a Cambridge college should come down on the side of High culture, conscience, and the advancement of knowledge. Only surprising, perhaps, that the thunderer against *The Whig Interpretation of History* should, when addressing science, have been so unselfconsciously Whiggish himself.[2]

Second: although, like Paris, he made a judgement upon his 'three R's', Butterfield showed no special interest in probing their interconnections. This was typical of the times. There was the history of art, the history of religion, and the history of science; and if it would be unfair to imply that ne'er the three did meet, they were assuredly studied by different scholarly constituencies, with different interests, housed in different academic departments. True, attempts had been made to argue their inner affinities – as, for instance, in the 'Merton thesis', which claimed to show, in a somewhat Weberian manner, how the Puritan ethic promoted the value system of science in the seventeenth century.[3] But specialist scholars peered down at history through their individual microscopes; and, especially in the 1950s and early 60s, historians of science militantly denied, faced with the claims of Marxism and sociology, that 'external influences' shaped the intellectual course of science, which they tended to see as Objective Rational Truths, discovered by Great Minds.[4]

The present volume, incorporating and advancing the scholarly developments of the last couple of decades, is ample proof of radical reorientations in outlooks since Butterfield's day – proof of a veritable 'revolution in the revolution'. For one thing, the 'Berlin Walls' separating academic disciplines have been undermined if not utterly dismantled. It would be difficult to categorize this volume exclusively as 'history of religion', 'history of culture', 'history of science', 'history of medicine', or any such. It embraces and subsumes all of these, and much more besides. If label be needed at all for the 'thick description' pursued in the following pages, some such very broad name as 'cultural anthropology of thought' might be called for. It may not be the perfect term, but it has the virtue of underlining more clearly than some such phrase as 'history of ideas' or 'history of science', the ways in which terms such as nature, order, matter, spirit, have meanings that transcend the technical, and link up with wider models of social order and boundaries. It better indicates the sense of inter-connectedness between different fields and levels of discourse which the contributors convey.[5] But pigeon-holing may, in any case, be unnecessary. For nowadays, as Patrick Curry emphasizes in his contribution below, scholars aim to avoid the teleological ('Whiggish') pitfalls involved in diachronic, 'tunnel history' (history of physics, history of scientific method, etc.). Learning the history of mentalities, as pursued by the *Annaliste* school, they find it more rewarding to operate structurally, tracing the patternings of veins along synchronous strata, laying bare the common assumptions, language, and rationalities informing such diverse enterprises as astrology, magic, witchcraft, and philosophies of nature, and emphasizing how bodies of ideas can simultaneously be speaking to religious, technical, political and even gender issues.[6]

Putting hindsight behind them, today's historical scholars have sought to abandon time-honoured but question-begging 'before/after' explanatory models: from darkness to light, from error to truth, from superstition to religion, from magic to science. Instructed by cultural anthropology and by new trends in the sociology of knowledge, the accent nowadays is not upon plotting the triumphal progress of truth, but upon interpreting structures of belief.

This involves various strategies. For one thing, reversing old assumptions, scholars are newly eager to reveal the 'rationality' in the 'before'. Several essays below explore such 'traditional' frameworks of belief. Stuart Clark lays bare the internal logic governing notions of diabolical powers; Peter Dear recovers the coherence of those Aristotelian natural philosophies dominant before the rise of the mechanical philosophy; Germana Ernst and Patrick Curry explore the historical plausibility of astrology's belief in astral influences over human destiny.[7]

At the same time, and in a complementary gesture, the question is put to the 'after'. Did the science of the Scientific Revolution really make a clean break with occult causes? Did the new medicine of the emergent profession of physicians truly differ from old Galenism and the old wives' wisdom of kitchen physic, asks John Henry, or was it basically old wine in new semantic bottles? Did it, indeed, work more cures? Was the 'new philosophy', which promised to free people from dread of demons, devils, and witches, authentically liberating? Did rationality replace rhetoric? Was the new rationality any more rational?[8]

If the essence of such 'paradigm switches' thus lies not in breakthroughs but in 'difference', a changed conception of reality but not necessarily a more truthful picture, how are we to explain the transitions from 'before' to 'after'? Final causes ('truth is great and will prevail') can no longer be part of the historian's armoury. No longer, for instance, is it acceptable to cast the demise of astrology or sorcery in terms of ignorance and irrationality bowing before the triumphal car of observation, experiment, and scientific method. Historians today feel the need to defenestrate science, or at least take it off its pedestal.[9] Knowledge is no transcendental force for progress. Historically understood, it is local, it is plural, it embodies interests, it mobilizes the claims of groups and classes, and, above all, it is recruited, willy-nilly, on all sides in wars of truth. As has been variously suggested from

Bacon to Foucault, knowledge cannot be understood independently of power, knowing from doing, *verum* from *factum*.[10] Thus, to understand the fate of folk healing, of judicial astrology, of Galileo's championship of Copernicanism, or of an obscure Friulian miller's belief that the world arose, like cheese, by fermentation, we must recognize that early modern times were embroiled in a Hundred Years' War between all manner of rival knowledge-claims, and multiple models of epistemological legitimacy; and, behind these, it goes without saying, rival ideologies, promoted by disparate interests, such as religious, national and factional.[11]

And these are complicated indeed! Older and often self-serving historiographies habitually conceptualized changes in terms of mind wars between intellectual superpowers: Magic *versus* Science, Ancients *versus* Moderns, Genesis *versus* Geology – and many other sorts of *Kulturkämpfe* fought to the death.[12] The reality was always far more complex, fluid and confused. As Peter Dear emphasizes below, we should probably not even be talking in respect of such contending abstractions as 'theology' and 'science', but rather in terms of ephemeral, localized and tangible ecclesiastical factions, grouping in shifting alliances and patronage networks in respect to individual scientific challenges. There were leading Italian churchmen who found uses for Campanella and Galileo (or at least thought they could use them), no less than there were others who shunned or silenced them. The attitudes of individual theologians to Copernicanism themselves changed. Faced with the new science, Protestant ecclesiastics sometimes reacted in ways analogous to the Catholic hierarchy, and sometimes not.[13]

In many conventional accounts, the Protestant churches in the post-Galileo age are represented as less antagonistic towards the mechanical philosophy than was Rome. This should not be misinterpreted. It was, perhaps, not so much the product of a policy actively friendly towards the new science, but rather a symptom of having no coherent policy at all. For, Dear suggests, the real character of the Protestant churches was not that they were particularly liberal. Rather, they tended to be small, weak, and divided. Unlike Jesuit-dominated Rome, with its vast philosophical and pedagogic apparatus, they were hardly in a position to formulate intellectually coherent programmes towards natural philosophy.

Other features of the making of modern science, other victories and defeats, hinged upon local circumstances, temporary alliances, situational logics. The growing prestige and pretensions of learned, humanist 'physic' in the early modern period owed little, and contributed little, to genuine medical advance, to anatomy and physiology, to any real increase in healing power. It had most to do, suggests John Henry, with the patronage policies of the new monarchies and, as Luce Giard properly emphasizes, with the advent of the printed book, sometimes lavishly illustrated, conferring the permanent authority of print.[14] Likewise, it will not do simply to speak of 'mathematics', as if it were some homogeneous whole. There were many different sorts of mathematics in the early modern world, emphasizes Jim Bennett, and the branch most often neglected by historians – practical mathematics – in many ways proved the most progressive, thanks to urgent new contexts of use, such as navigation.

Similarly, as Stuart Clark particularly emphasizes, throughout the seventeenth century the struggles concerning the reality both of miracles and of demonic forces rarely hinged upon anything so direct as their assertion and denial (still less their convincing proof or disproof). What was at stake, rather, were the circumstances under which such manifestations as supernatural or preternatural occurrences would occur, and in what discourses and categories they would be explained, authorized, or disconfirmed. Niceties of terminology were crucial: thus 'wonders' were not necessarily 'miracles'. And the orientations of royal, ecclesiastical, and academic politics weighed heavily. In early Stuart England, both Puritans and Papists felt they had all to gain by arguing for the real presence of the preternatural; the Anglican establishment, by contrast, sought to encompass such mira within the order of nature, as it were as part of the kingdom of law.[15]

In an exemplary demonstration of the way in which changes in natural beliefs should be seen as instances of a wider ideological warfare, Patrick Curry handles the decline and fall of astrology in post-Restoration England as a rather direct product of an emergent class hegemony. Astrology had become tainted with Civil War radicalism. It was ever after represented as 'vulgar' by those who saw in the espousal of Newtonian cosmology a further, galactic, championship of a general ideology supporting law-'n'-order.[16]

The contributors to this volume have thus moved beyond the old vision of the 'Scientific Revolution' as the progressive victory of truth or scientific method; they emphasize, instead, the prominence of cultural conflicts respecting natural knowledge. One of the great strengths of this shift in emphasis is that it enables us to gain a much fuller understanding of what to the old judgemental vision seemed puzzling or perverse. The use by many early fellows of the Royal Society of the canons of the new science to clinch their convictions in witchcraft or magic, or Newton's enduring preoccupation with alchemy – these no longer need appear anomalous, scandalizing our image of the 'new science' and its 'new society'. Scholars can nowadays dispassionately debate the likelihood that alchemical pursuits may have had some part in leading Newton to notions of attraction, rather as his Neoplatonic, Neopythagorean (and even Hermetic) commitments probably contributed to his advocacy of action at a distance. It no longer seems bizarre or anomalous that experimental philosophers such as Robert Hooke used the language of astrology.[17]

Above all, abandoning the dictionary definitions inscribed in modern 'presentism' enables us to grasp why early modern debates within natural philosophy and about the order of things were so preoccupied with, as Maurice Slawinski discusses, the true status and rhetorical power of language,[18] and with the negotiation and establishment of proper conceptual boundaries. What were the right ways of defining the relations – theological, metaphysical, empirical – between body and soul, nature and human nature, the living and the dead, or male and female?[19] These were crucial questions. The making of modern science was not merely a matter of the increase of positive knowledge or the development of adequate conceptual abstractions: it was about radical transformations in the capacity to think about nature and man's place in it.

Today's scholarship, amply reflected in this volume, insists upon the coexistent flourishing of a plurality of discourses about natural knowledge in the early modern period. This recognition expands our canvas, and forces us to rethink the 'canon'. Histories of the Scientific Revolution cast in the mould of a 'great tradition', running linearly from Copernicus to Newton, must be replaced. Today we are confronted with a far more complex panorama, one perhaps even to a Brueghelesque degree, in which numerous factions and their spokesmen – academics, churchmen, courtiers, wandering scholars, 'wise women' and witches, religious sectaries, alehouse philosophers, hack pamphleteers, self-styled prophets, and many others, including not a few *sui generis* intellectual loners – wander in clusters across the stage disputing the truths of creator and creation, force and matter, law and miracle, body and soul, mostly deeply aware of the fact that upon the determination of such natural truths was hanging the fate of more directly personal, human issues: questions of government, authority, freedom, salvation.[20]

Today's scholarship is showing that early modern debates about nature were rarely conducted with the sophisticated charm of the Symposium, that noble meeting of minds, often resembling by contrast a heady, heckling, public shouting match, a Babel even. Babel? A strong term indeed, but it seems to me that a real strength of modern scholarship lies in its demonstration of the enormous heterogeneity, diversity, and fragmentation, in the early modern intellectual world, wreaked by the rise of printing, of confessional differentiation, and the sheer proliferation of the sites of intellectual endeavour. In the long run, as I suggest below, there were countervailing forces which finally brought new shapes of order out of disorder.

What else should we expect of the wars of words of early modern intellectual debate? With their destabilizingly rapid population expansion, massive persecutions and migrations, wars, and vertiginous urbanization, the sixteenth and seventeenth centuries saw sociocultural chaos and fantastic changes. Many regions, many peoples, became intellectually vocal for the first time. Heresy, witchcraft, and radicalism proved powerfully contagious. The wildfire spread of literacy rendered even the poor argumentative and assertive. The first century of the widespread circulation of printed books led to unparalleled and seemingly inexhaustible battles for the mind. The sudden ferocity of early modern intellectual conflict could be, to some degree, an optical illusion – with the arrival of print, more evidence survives. But who would deny that the decibel level also soared, or that cacophony was often the result?[21] The essays below give a flavour of ferocious intellectual warfare: within astrology, as Germana Ernst reveals, scores of rival schools staked out their own distinctive positions.

But it is also the historian's job to stand back, take the broader chronological perspective, and evaluate some of the consequences, many of them unintended. This is not easy. We are faced, for one thing, with an outmoded triumphalist historiography, which has liked to celebrate the Renaissance, Reformation, and Scientific Revolution as, between them, ushering in a new era of man, free, powerful, progressive. At the very least, this vision must be qualified. After all, were that the case, how do we explain why the *philosophes* needed to campaign so violently for basic freedoms in the Enlightenment politics of *écrasez l'infâme*?[22]

Today's historians mainly paint a darker landscape. Seventeenth-century elites, we are told, be they those of science, or Counter-Reformation, were engaged in a desperate struggle to jam the lid back upon the Pandora's box of popular intellectual debate, to restrict the authorization to speak, to police beliefs. The 'intellectual consequences of the English revolution', Christopher Hill tells us, were essentially reactionary, a betrayal. Such consequences perhaps went far wider in the second half of the seventeenth century, the age of rising absolutism, and of what Foucault has termed the 'great confinement'. Enlightenment rationality and political absolutism (argue Adorno and Horkheimer, and, more recently, Foucault) together defined a new hegemonic order which, far from being liberal or liberating, was exclusionary and repressive.[23] Erstwhile lively cultures were marginalized as 'vulgar' or as 'pseudo-sciences'; hermetic arts became 'forbidden' or 'occult'; popular culture was 'reformed' from above; gender hierarchies were reinforced, thereby silencing women's voices and protests; and the tyranny of 'reason' dealt a blow against intellectual and cultural pluralism. If literal 'witch-hunts' were brought to an end, metaphorical ones grew more prominent, targeted against such emblems of the Other and Unreason as the poor, the mad, Jews, and homosexuals.[24]

This almost Rousseauvian verdict upon the rise of rationalism as a new fall, the product of biting the tainted Newtonian apple, deserves serious consideration. Yet it may be stated in overly melodramatic terms, it may itself provide a false perspective. Education, the printing press, the book trade, the unstoppable enrichment of material life with all the proliferating meaning-systems attached to it, to say nothing of the division of Christendom into a multitude of states and confessions – all these collectively ensured the survival, the spread, of cultural diversification.[25]

But the fact of the matter is that if reason and science were in their own ways exclusive and authoritarian, their effect was not to eliminate alternative and counter discourses but to marginalise them. Many – not least astrology – are still going strong today: witness the power of the counter-culture of the 1960s.[26] Recent events in Eastern Europe should give us pause before we overestimate the capacity of monolithic state apparatuses, past or present, to enforce orthodoxies and permanently proscribe dissent. The new technologies and disciplines of the early modern age – above all, print – aided the forces of centralization and uniformity, but they also proved, as Paolo Rossi shows for a variety of contexts, eternally double-edged.

The 'new' social histories tend to regard the poor, deviants, and, above all, women as the real victims of the politics of mind control explored below.[27] Perhaps. But over the historical long haul, the developments, discoveries, and debates surveyed in this book were more decisive in another way. They broke the predominant grip (*monopoly* would be somewhat too strong a term) exerted for many centuries by the Church and ecclesiastical institutions upon intellectual matters and the directing of life. To point this out is not necessarily to argue for an irresistible tide of unbelief. It is, however, to insist that, by the eighteenth century, churches and clerics had ceased to dominate the intellectual life of Europe, and that culture was growing rapidly more secular in both form and content.[28] The long-term significance of Renaissance humanism was the creation of the writer; the impact of the Scientific Revolution was the establishment of new techniques of inquiry, norms of truth, and sites of intellectual authority. The invention of historicity, as is demonstrated below in Stephen Pumfrey's discussion of early modern views of truth and time, helped to liberate science from the hidebound domain of the text, and relocate it in new contexts with far greater potential for future development, above all, the site of the laboratory and the medium of the experiment. Partly thereby, as Julian Martin emphasizes in his evaluation of the career of Francis Bacon, science and politics became inextricably linked from around the close of the sixteenth century in a prospect of future power simultaneously managerial and utopian. Neither Rome nor the Protestant communions had answers to these challenges; eventually they lost their intellectual dominion. The author, the public sphere, public opinion, the fourth estate, Grub Street even, and certainly (as Luce Giard stresses in her discussion of early modern academe) the scientific community – these may be seen as the most momentous consequences of the developments surveyed in this volume.[29]

Recent scholarship has been addressing the shifting historical meanings of natural knowledge. These essays concentrate upon such issues. They go beyond the old Whiggism. Yet, inevitably, in so doing they inadvertently constitute a new Whiggism: like all other new historiographies, this one too thinks it is better than those it is superseding. So the cause for congratulation is also a cause for humble reflection; a reminder that we are all children of our time, a warning that the task of criticising previous outlooks entails no less a powerful duty of being self-critical about one's own stances. It is in this spirit that the following book is offered.

Notes

1 Herbert Butterfield, *The Origins of Modern Science, 1300-1800*, London, 1950; for recent assessments of the concept of the Scientific Revolution see John A. Schuster, 'The Scientific Revolution' in *Companion to the History of Modern Science*, edited by R. C. Olby, G. N. Cantor, J. R. R. Christie, and M. J. S. Hodge, London, 1990, pp. 217-43; I. B. Cohen, *Revolution in Science*, Cambridge, Mass., 1985; Roy Porter, 'The scientific revolution: a spoke in the wheel?', in *Revolution and History*, edited by Roy Porter and Mikulas Teich, Cambridge, 1986, pp. 297-316.

2 Herbert Butterfield, *The Whig Interpretation of History*, London, 1931; for a discussion of whiggishness and the interpretation of science, see A. R. Hall, 'On Whiggism', *History of Science*, 21, 1983, pp. 45-59.

3 Robert K. Merton, *Science, Technology and Society in Seventeenth Century England*, New York, 1970; for later attempts, see R. Hooykaas, *Religion and the Rise of Modern Science*, Edinburgh, 1973; S. Jaki, *The Road of Science and the Ways to God*, Edinburgh, 1978; Charles Webster, *The Great Instauration: Science, Medicine and Reform, 1626-1660*, London, 1975. For the theological underpinnings see Amos Funkenstein, *Theology and the Scientific Imagination from the Middle Ages to the Seventeenth Century*, Princeton, N.J., 1986; and, for a survey, John Hedley Brooke, 'Science and religion', in *Companion to the History of Modern Science*, edited by R. C. Olby, G. N. Cantor, J. R. R. Christie, and M. J. S. Hodge, London, 1990, pp. 763-82. A pioneering cross-disciplinary work showing the links between art and science was R. Klibansky, E. Panofsky and F. Saxl, *Saturn and Melancholy: Studies in the History of Natural Philosophy, Religion and Art*, London, 1964.

4 For entrée into the once-ferocious 'internalist/externalist' debate see H. Kragh, *An Introduction to the Historiography of Science*, Cambridge, 1987; Steven Shapin, 'Social uses of science', in *The Ferment of knowledge: Perspectives on scholarship of eighteenth century science*, edited by G. S. Rousseau and Roy Porter, Cambridge, 1980, pp. 93-142. On the characteristic 1950s denial of social influences, see Roy Porter, 'The history of science and the history of society', in *Companion to the History of Modern Science*, cit. (n. 3), pp. 32-46.

5 See C. Geertz, *The Interpretation of Cultures*, New York, 1973.

6 Structuralist and post-structuralist notions are obviously important here. See Stuart Clark, 'French historians and early modern popular culture', *Past and Present*, 100, 1983, pp. 62-99; Peter Burke, 'Popular culture between history and ethnology', *Ethnologi Europaea*, 14, 1984, pp. 5-13; *idem*, 'Revolution in popular culture', in *Revolution in History*, edited by Roy Porter and M. Teich, Cambridge, 1986, pp. 206-25; *New Perspectives on Historical Writing*, edited by Roy Porter and M. Teich, Cambridge, 1991.

7 For studies emphasizing the 'rationality' of former belief systems, see

Frances Yates, *Giordano Bruno and the Hermetic Tradition*, London, 1964; I. Couliano, *Eros and Magic in the Renaissance*, trans. by Margaret Cook, Chicago and London, 1987; Michael MacDonald, 'The inner side of wisdom: suicide in early modern England', *Psychological Medicine*, 7, 1977, pp. 565-82; *idem*, 'Insanity and the realities of history in early modern England', *Psychological Medicine*, 11, 1981, pp. 11-25; *idem*, 'Science, magic and folklore', in *William Shakespeare: His World, His Work, His Influence*, edited by J. F. Andrews, I, New York, 1985, pp. 175-94. Important here have been the insights of anthropology. See Michael MacDonald, 'Anthropological perspectives on the history of science and medicine', in *Information Sources in the History of Science and Medicine*, edited by P. Corsi and P. Weindling, London, 1983, pp. 61-96; Mary Douglas, *Purity and Danger: An Analysis of Concepts of Pollution and Taboo*, Harmondsworth, 1966; *idem*, *Natural Symbols: Explorations in Cosmology*, Harmondsworth, 1973; *Rules and Meanings: The Anthropology of Everyday Knowledge*, edited by Mary Douglas, Harmondsworth, 1973; *idem*, *Implicit Meanings: Essays in Anthropology*, London, 1975; *idem*, 'Cultural bias', *Occasional Paper* No. 35, London: Royal Anthropological Institute, 1978. For the latest assessment of the contribution of the sociology of knowledge, see Barry Barnes, 'Sociological theories of scientific knowledge', in *Companion to the History of Modern Science*, edited by R. C. Olby, G. N. Cantor, J. R. R. Christie, and M. J. S. Hodge, London, 1990, pp. 60-76.

8 For scholarship underlining the ambiguities of the new 'scientific' view, see Keith Hutchinson, 'What happened to occult qualities in the Scientific Revolution?', *Isis*, 73, 1982, pp. 233-53; M. MacDonald, *Mystical Bedlam: Madness, Anxiety and Healing in Seventeenth Century England*, Cambridge, 1981; Larry Briskman, 'Rationality, science and history', in *Companion to the History of Modern Science*, edited by R. C. Olby *et al.*, London, 1990, pp. 160-80; J. Ravetz, 'Orthodoxies, critiques and alternatives', *ibid.*, pp. 898-908. There is a defence of the 'rationality' of the new science in the editorial introduction to *Occult and Scientific Mentalities in the Renaisssance*, edited by Brian Vickers, Cambridge, 1984.

9 Especially valuable here is the insistence of the so-called 'Edinburgh school' upon 'explanatory symmetry' in treating different (and putatively scientific and non-scientific or pseudo-scientific) thought systems. The very same structures must govern the explanation of what perhaps seem to us true, or at least, progressive and successful beliefs as govern the explanation of those that appear irrational or which fell by the wayside. See D. Bloor, *Knowledge and Social Imagery*, London, 1976; H. Collins, *Changing Order. Replication and Induction in Scientific Practice*, Beverly Hills and London, 1985; B. Barnes, *Interests and the Growth of Knowledge*, London, 1977.

10 For Foucault on the interpenetration of knowledge and power see M. Foucault, *The Order of Things: An Archaeology of the Human Sciences*, London, 1970; *idem*, *The Archaeology of Knowledge*, London, 1972; *idem*, *Discipline and Punish: The Birth of the Prison*, Harmondsworth, 1976; Michel Foucault, *Power/Knowledge: Selected Interviews and Other Writings 1972-1977*, Brighton, 1977; for knowledge as a power, see Bruno Latour, *Science in Action*, Milton Keynes, 1987.

11 See R. Redondi, *Galileo Heretic*, trans. by R. Rosenthal, Princeton, 1987; C. Ginzburg, *The Cheese and the Worms: the Cosmos of the Sixteenth Century*, London, 1980; on the logic of folk-healing, see Françoise Loux, 'Popular culture and knowledge of the body: infancy and the medical anthropologists', in *Problems and Methods in the History of Medicine*, edited by Roy Porter and Andrew Wear, London, 1987, pp. 81-98.

12 For a classic account of such a struggle, see J. W. Draper, *History of the Conflict between Religion and Science*, reprint, Farnborough, Hants, 1970.

13 For the politics of knowledge, patronage and faction, see Mario Biagioli, 'Galileo and patronage', *History of Science*, 28, 1990, pp. 1-62. For the interaction between natural philosophy and theology, see Pietro Corsi, 'History of science, history of philosophy, and history of theology', in *Information Sources in the History of Science and Medicine*, edited by P. Corsi and P. Weindling, London, 1983, pp. 3-28.

14 For histories of medicine confirming this view, see L. M. Beier, *Sufferers and Healers: The Experience of Illness in Seventeenth-Century England*, London, 1987; Barbara Duden, *Geschichte unter der Haut*, Stuttgart, 1987; *Patients and Practitioners: Lay Perceptions of Medicine in Pre-Industrial Society*, edited by Roy Porter, Cambridge, 1985; Roy Porter and Dorothy Porter, *In Sickness and in Health: The British Experience 1650-1850*, London, 1988; Dorothy Porter and Roy Porter, *Patient's Progress: Doctors and Doctoring in Eighteenth-Century England*, Stanford, Ca., 1989. In *The Medical Revolution of the Seventeenth Century*, edited by R. French and A. Wear, Cambridge, 1989, the supposed revolution pointed to is in medico-scientific explanations, not in healing power.

15 On the politics of witchcraft, see C. Larner, *Witchcraft and Religion: The Politics of Popular Belief*, Oxford, 1984; John Putnam Demos, *Entertaining Satan: Witchcraft and the Culture of Early New England*, Oxford and New York, 1982; for magic, see John Henry, 'Magic and science in the sixteenth and seventeenth centuries', in *Companion to the History of Modern Science*, edited by R. C. Olby *et al.*, London, 1990, pp. 583-96.

16 See *Astrology, Science and Society: Historical Essays*, edited by Patrick Curry, Woodbridge, Suffolk, 1987; Patrick Curry, *Prophecy and Power: Astrology in Early Modern England*, Cambridge, 1989; for Newtonianism as ideologically conservative, see M. C. Jacob, *The Newtonians and the English Revolution, 1689-1720*, Ithaca, 1976; *idem, The Radical Enlightenment: Pantheists, Freemasons and Republicans*, London, 1981.

17 On Newton's alchemy, see M. J. T. Dobbs, *The Foundations of Newton's Alchemy*, Cambridge, 1975; and, more broadly, Simon Schaffer, 'Newtonianism', in *Companion to the History of Modern Science*, edited by R. C. Olby *et al.*, London, 1990, pp. 610-26; *idem*, 'States of mind: enlightenment and natural philosophy', in *Mind and Body in the Enlightenment*, edited by G. S. Rousseau, Los Angeles, 1990. On the diverse intellectual outlooks of early fellows of the Royal Society see Michael Hunter, *Establishing the New Science: The Experience of the Early Royal Society*, Woodbridge, 1989.

18 J. V. Golinski, 'Language, discourse and science', in *Companion to the History of Modern Science*, edited by R. C. Olby *et al.*, London, 1990, pp. 110-26. For contemporary theories of language, see Hans Aarsleff, *From Locke to*

Saussure: Essays on the Study of Language and Intellectual History, Minneapolis, Minn., 1982; Brian Vickers and Nancy S. Struever, *Rhetoric and the Pursuit of Truth: Language Change in the Seventeenth and Eighteenth Centuries: Papers read at a Clark Library Seminar 8 March 1980*, Los Angeles, Calif.: William Andrews Clark Memorial Library, University of California, Los Angeles, 1985.

19 For concepts of the body see *Fragments for a History of the Human Body*, edited by Michel Feher, 3 vols., New York, 1989; Leonard Barkan, *Nature's Work of Art: The Human Body as Image of the World*, New Haven, Conn., 1975; Piero Camporesi, *The Incorruptible Flesh: Bodily Mutation and Mortification in Religion and Folklore*, Cambridge, 1988; *idem, Bread of Dreams: Food and Fantasy in Early Modern Europe*, Cambridge, 1989; *idem, The Body in the Cosmos: Natural Symbols in Medieval and Early Modern Italy*, Cambridge, 1989; Natalie Zemon Davis, 'The sacred and the body social in sixteenth-century Lyon', *Past and Present*, 90, 1981, pp. 40-70; *idem*, 'Boundaries and the sense of self in sixteenth century France', in *Reconstructing Individualism: Autonomy, Individuality and the Self in Western Thought*, edited by T. C. Heller, *et al.*, Stanford, Calif., 1986; G. Vigarello, *Le Corps Redressé: Histoire d'un Pouvoir Pédagogique*, Paris, 1978; P. Barber, *Vampires, Burial and Death: Folklore and Reality*, New Haven, Conn., 1988.

Nature and human nature are discussed in Keith Thomas, *Man and the Natural World*, Harmondsworth, 1984; *Languages of Nature: Critical Essays on Science and Literature*, edited by L. Jordanova, London, 1986.

On death, see Philippe Ariès, *Western Attitudes Towards Death: From the Middle Ages to the Present*, Baltimore, 1974; London: Marion Boyars, 1976; *idem, Images of Man and Death*, trans. by J. Lloyd, Cambridge, 1985; John McManners, *Death and the Enlightenment*, Oxford, 1985.

Notions of sexual differences are discussed in I. MacLean, *The Renaissance Notion of Woman*, Cambridge, 1980; Carolyn Merchant, *The Death of Nature: Women, Ecology and the Scientific Revolution*, San Francisco, 1980; J. R. R. Christie, 'Feminism and the history of science', in *Companion to the History of Modern Science*, edited by R. C. Olby *et al.*, London, 1990, pp. 100-9.

20 A good illustration of the utter intertwining of natural knowledge with social knowledge is offered by David Warren Sabean, *Power in the Blood: Popular Culture and Village Discourse in Early Modern Germany*, Cambridge, 1984.

21 For social transformations, see F. Braudel, *Civilization and Capitalism, 15th-18th Century*, I, *The Structures of Everyday Life*; II, *The Wheels of Commerce*; III, *The Perspective of the World*, New York, 1985; J. De Vries, *European Urbanization, 1500-1800*, Cambridge: Mass., 1984; E. Eisenstein, *The Printing Press as an Agent of Change*, 2 vols., Cambridge, 1979, and for the ambivalences of attitudes towards books see David Cressy, 'Books as totems in seventeenth-century England and New England', *Journal of Library History*, 21, 1986, pp. 92-106. The geographical spread of the Renaissance is explored in *The Renaissance in National Context*, edited by Roy Porter and Mikuláš Teich, Cambridge: Cambridge University Press, 1991.

22 Michael Adas, *Machines as the Measure of Men: Science, Technology, and Ideologies of Western Dominance*, Ithaca, N.Y., 1989; on the relationship

between the 'three R's' and the Enlightenment, see Peter Gay, *The Enlightenment: An Interpretation*, 2 vols., New York, 1967-1969; and especially Peter Gay, 'The Enlightenment as medicine and as cure', in *The Age of the Enlightenment: Studies Presented to Theodore Besterman*, edited by W. H. Barber, Edinburgh, 1967, pp. 375-86.

23 Christopher Hill, *Intellectual Consequences of the English Revolution*, London, 1980; for Foucault, see above, note 11. See also Max Horkheimer and Theodor W. Adorno, *Dialectic of Enlightenment*, trans. by John Cumming, New York, 1989.

24 On modern scapegoatings and silencings, see Sander L. Gilman, *Seeing the Insane*, New York, 1982; idem, *Difference and Pathology*, Ithaca and London, 1985; H. Mayer, *Outsiders: A Study in Life and Letters*, Cambridge: Mass., 1984; Thomas S. Szasz, *The Manufacture of Madness*, New York, 1970; London: Paladin, 1972; idem, *The Myth of Mental Illness: Foundations of a Theory of Personal Conduct*, London, 1972; rev. edn, New York, 1974; idem, *The Age of Madness: The History of Involuntary Hospitalization Presented in Selected Texts*, London, 1975. On stigmatizing women, see Brian Easlea, *Science and Sexual Oppression: Patriarchy's Confrontation with Woman and Nature*, London, 1981, and on the simultaneous stigmatizing of women and nature, see Carolyn Merchant, *The Death of Nature: Women, Ecology and the Scientific Revolution*, New York, 1980. On the 'reform' of popular culture, see Peter Burke, *Popular Culture in Early Modern Europe*, London, 1978.

25 Edward Gibbon offered the common Enlightenment reflection that the political pluralism of modern Europe happily prevented the reinstatement of the imperial and intellectual tyrannies of Rome (imperial and Christian). See Roy Porter, *Edward Gibbon: Making History*, London, 1988, Conclusion. On the development of material life, see *The Social Life of Things: Commodities in Cultural Perspective*, edited by Arjun Appadurai, Cambridge, 1986; Jean Baudrillard, *Le Système des Objets*, Paris, 1968; Mihaly Csikszentmihalyi and Eugene Rochberg-Halton, *The Meaning of Things: Domestic Symbols and the Self*, New York, 1981; Chandra Mukerji, *From Graven Images: Patterns of Modern Materialism*, New York, 1983; Hiram Caton, *The Politics of Progress*, Gainesville, 1988; *Consumption and the World of Goods*, edited by John Brewer and Roy Porter, London, 1991. One of the themes of Simon Schama's *The Embarrassment of Riches: An Interpretation of Dutch Culture in the Golden Age*, London, 1988 is the mutually reinforcing nature of material life, the pursuit of knowledge, and social pluralism.

26 A good instance – though one which preserves Whiggish evaluations in the title – is J. Devlin, *The Superstitious Mind: French Peasants and the Supernatural in the Nineteenth Century*, New Haven, Conn., 1987; H. Leventhal, *In the Shadow of the Enlightenment: Occultism and Renaissance Science in Eighteenth-Century America*, New York, 1976; Robert Darnton, *Mesmerism and the End of the Enlightenment in France*, Cambridge, Mass., 1968.

27 The 'new' social, cultural and intellectual histories valuably demonstrate the squeezing of popular cultures. See *Modern European Intellectual History: Reappraisals and New Perspectives*, edited by Dominick LaCapra and Steven L. Kaplan, Ithaca, N.Y., 1982; K. Park and L. J. Daston, 'Unnatural conceptions: the study of monsters', *Past and Present*, 92, 1981, pp. 20-54; P.

Stallybrass and A. White, *The Politics and Poetics of Transgression*, Ithaca, N.Y., 1986; Robert Muchembled, *Popular Culture and Elite Culture in France, 1400-1750*, trans. by L. Cochrane, Baton Rouge, La., 1978.

28 On the 'sociologization' of religion and the waning of enthusiasm, see Frank E. Manuel, *The Eighteenth Century Confronts the Gods*, Cambridge, Mass., 1959; *idem, The Changing of the Gods*, Hanover and London, 1983; Colleen McDannell and Bernhard Lang, *Heaven: A History*, New Haven, Conn., 1988; Jean Delumeau, *La Peur en Occident XIVe-XVIIIe Siècles*, Paris, 1978; *idem, Le Péché et la Peur: La Culpabilisation en Occident XIII-XVIIIe Siècles*, Paris, 1983; D. P. Walker, *The Decline of Hell: Seventeenth-Century Discussions of Eternal Torment*, London, 1964; Michael MacDonald, 'The secularization of suicide in England, 1660-1800', *Past and Present*, cxi, 1986, pp. 50-100; R. D. Stock, *The Holy and the Daemonic from Sir Thomas Browne to William Blake*, Princeton, N.J., 1982; Keith Thomas, *Religion and the Decline of Magic: Studies in Popular Beliefs in Sixteenth and Seventeenth Century England*, London, 1971; reprinted, Harmondsworth, 1978.

29 For this 'long revolution', see Raymond Williams, *The Long Revolution*, New York, 1984. See also J. Habermas, *The Structural Transformation of the Public Sphere. An Inquiry into a Category of Bourgeois Society*, Cambridge, Mass., 1989; Robert Darnton, 'The High Enlightenment and the low life of literature in pre-revolutionary France', *Past and Present*, 51, 1971, pp. 81-115; *idem*, 'In search of the Enlightenment: recent attempts to create a social history of ideas', *Journal of Modern History*, 43, 1971, pp. 113-32; *idem, The Business of Enlightenment. A Publishing History of the Encyclopédie, 1775-1800*, Cambridge, Mass., 1979; *idem, The Literary Underground of the Old Regime*, Cambridge, Mass., 1982.

PART I

Reconstructing authority

2

Remapping knowledge, reshaping institutions*

Luce Giard

Renaissance philosophy came in many forms. All however depended to some extent on the dominant Aristotelian tradition, variously interpreted in the light of other philosophical sources which might be associated, compared or preferred to it, depending on the issue discussed, the theologico-political situation, and the social context where teaching and research took place. Renaissance universities continued to provide a stable framework for institutionalized learning, but were themselves undergoing radical change. Furthermore, they were losing their intellectual monopoly, as philosophy (which also concerned itself with questions now addressed by natural scientists) quit the cloistered world of academic debate and the religious orders, returning to civil society and developing in a variety of situations which together formed a disparate, ever-changeable mosaic of micro-contexts, informal intellectual gatherings whose lively debates were the very essence of the Renaissance ideal of civilization. The return of intellectual debate to the community, the spread of printing and the concomitant expansion of the educated class all contributed to the development of a lay intelligentsia much larger than that of the medieval universities and Church. This expansion, which benefited philosophy but also had implications for matters of faith, was made possible by a conjunction of economic, technological, political and social developments, which together determined the emergence of a new intellectual regime characterised by a change of scale, a widening of the social base of culture, a transformation in the means of access to texts and the channels of circulation of knowledge.

* Translated by Maurice Slawinski.

1   Old universities and new colleges

The change of scale first manifested itself in the growth of teaching establishments. The number of universities increased dramatically. The greatest expansion took place in the territories of the Holy Roman Empire, which possessed no university before 1385, and could boast of seventeen by 1506, when France, previously much better endowed than the Empire, had fourteen. Italy continued to be well-provided, as its universities 'co-opted' the humanist movement, which had earlier been highly critical of scholastic teaching: by the sixteenth century *humanae litterae* had become an integral part of the curriculum. Even in England, where Oxford and Cambridge remained the only universities, the colleges there grew in number and size, and the grammar school system was greatly enlarged.

Expansion had the positive effect of increasing the number of teachers and graduates, but could also reduce the catchment area of single institutions. This shift from an international to a national, and later a regional perspective was accentuated by conflicts between emerging nation states and religious divisions, with unfortunate long-term consequences. Local academic dynasties developed, which harboured mediocrity and showed increasing disinterest in learning. Absenteeism and the worst excesses of patronage, the use of university chairs as sinecures, the granting of degrees as personal favours, gave rise to unhealthy rivalries and meant that some universities existed in little but name. Many of the new foundations were short-lived: some had grown out of conflicts between neighbouring cities or principalities and declined when those conflicts subsided; in others student recruitment was disrupted by wars of religion; many suffered because of protectionist measures which obliged students to pursue their studies in the state of their birth. These effects taken together explain why in the course of the Renaissance even ancient and prestigious foundations forfeited their earlier renown. As their institutional activities shrank to little more than issuing (academically worthless) degree certificates, serious scholars came to despise and actively sought to abandon them.

The consequences of this alienation would become apparent in the career itineraries of Galileo, Descartes or Pascal in the seventeenth century. Its root causes, however, go back to the preceding century, when men of learning foresook the medieval model of *universitas*, and questioned traditional forms of teaching and debate, turning to other social milieux more open and conducive to intellectual innovation. A vast relocation of intellectual energies took place which resulted in a new relationship between culture, political power, the Church and the dominant class. Gradually, from the end of the fifteenth century, it was the princely courts, the public and private academies, the schools run by religious orders, which attracted the most original minds, drawn there by more favourable conditions. Their resources (libraries, scientific instruments, cabinets of curiosities) were all the more attractive because they were accompanied by more worldly inducements: jobs and career prospects; powerful patrons whose support could bring fame, stimulating greater intellectual competition; hopes of converting the ruling class to the 'true faith'. All these factors conspired to make these institutions, removed from the universities, the new pole of attraction of philosophers, scientists and men of letters.

Only the great Italian universities, chief among them Padua, avoided decline, thanks to far-sighted government policies (particularly on the part of the Republic of Venice), and thanks also to the continued influx of foreigners drawn by the fame of humanist scholarship, the libraries and printing houses, the interaction between 'town' and 'gown', between old and new poles of intellectual attraction. The fact that Italian universities were the first to feel the full force of humanist critiques encouraged them to reform before it was too late. They maintained the lead thus gained until the early seventeenth century, when they were overcome by the dead weight of the Counter-Reformation and the consequent exodus of intellectuals: the final say would go to authority, repetition and tradition. But long before that humanist criticism, the impatience of the 'moderns', the extent and diversity of the new demands which society made, had already forced the universities to give up their earlier monopoly on the transmission of knowledge and compete with other institutions. The most advanced research was increasingly conducted elsewhere, in the Collège Royal founded by François I (Paris, 1530), for example, which the University of Paris vainly sought to oppose. At an intermediate level, a network of new institutions made up for the deficiencies of the universities, or addressed those audiences which knew no Latin. Thus in London institutions like the Inns of Court and Gresham College together constituted what a contemporary called 'the third university of England'.[1]

The fiercest competition, however, related to the lower rungs of the University curriculum: to provide teaching at this level a large number of new institutions were developed, such as the 'collèges de plein exercice' in France. These were autonomous of university authorities, often founded and staffed by the religious orders (especially after the Counter-Reformation). They taught courses in Latin grammar and literature, as required for university entrance, to which might be added Greek, and some or all of the subjects traditionally covered in the universities by the Faculty of Arts. And they did so in a novel way which was structured, effective and methodical. Since the Middle Ages the Faculty of Arts had taught logic and philosophy (including natural philosophy) as the compulsory first stage leading on to study in one of the higher faculties (law, theology or medicine). By virtue of its role as an obligatory introduction to university study (which many students did not go beyond) it brought together large numbers of young and volatile students. It supplied ardent combatants for all manner of battles, from the philosophical and theological to those merely concerned with the students' own living conditions and privileges, providing a fertile breeding-ground for student protests against rival teachers, local bishops, law-enforcement officers and civic burghers, to the detriment of academic discipline and serious study.

The new colleges were the major educational innovation of the Renaissance. They first developed in Paris betwen 1490 and 1510 out of the movement for intellectual and spiritual reform. They aimed to instil discipline in their pupils, and the model, satisfying as it did the new social demand for education, was rapidly imitated elsewhere. They served to break up, control and discipline the often rowdy army of 'arts' students and raise academic standards. At the same time they assisted in the diffusion of that ideal of *civilitas* praised by Erasmus and so many other humanists, becoming an essential weapon in the struggle for minds and souls between Reformation and Counter-Reformation Churches. Parisian practices inspired both Johannes Sturm's *gymnasium* in the Protestant city of Strasbourg, and the great network of schools established by the Jesuits in the service of Rome. After 1550 burghers, princes and churchmen alike contributed to the establishment of such institutions, arguing doggedly over their location, funding, endowments, and teaching staff. Religious schisms meant that education was now a counter in the theologico-political game, and to rivalries between churches were added those between the newly constituted nation states, with their expanding, ever-more complex bureaucracies. The state demanded that its servants be educated and competent, and the administrators, counsellors, advisers and diplomats with whom the Prince increasingly surrounded himself were often men of lowly birth but good education, because the great nobles were still attached to the feudal model of a military education: 'Noblemen born / to learn they have scorn, / but hunt and blow a horn, / leap over lakes and dikes, / set nothing by politics'.[2] In his treatise *Concerning the Benefits of a Liberal Education* Richard Pace, a former student at Padua who served Henry VIII as a diplomat, was among the first to spell out the advantages which low-born graduates could expect to derive from their superior education.[3] Scholars everywhere were conscious of the prize at stake and turned their minds to the education of princes: Guillaume Budé, Juan Luis Vives and Roger Ascham all devoted treatises to the subject, while the likes of Jean Calvin busied themselves with the Christian education of the common people.

In the Constitutions which ruled their Order the Jesuits identified the education of their brethren as a key distinguishing feature of the Order, even before making the education of laymen (and members of the ruling class in particular) their chief mission, the main weapon of the Counter-Reformation, as set down in the *ratio studiorum* of 1586, the curriculum and pedagogical system to be adopted by all Jesuit schools throughout Europe and in their missions 'to the pagans'. Their methods would soon be taken up by others, as each camp drew on the same sources and shared similar aims, preoccupations and techniques. All remained dependent on the 'architecture of knowledge' constructed by Aristotle, which helps to explain the varieties of Renaissance Aristotelianism. Doctrinal choices may have differed, but intentions and aims were the same: to make sure that each confession could rely on an army of publicists well versed in the techniques of public debate, both oral and written. Everywhere the 'true faith', whichever it might have been, required trusty foot-soldiers: propagandists, polemicists, preachers, educators and pastors. The art of dialectic, beloved of medieval universities, thus quit the examination hall, and set to tackling questions of value to the community at large. In the Jesuit schools and Calvin's Geneva alike, in Lutheran Germany and the English College at Douai, which dreamed of returning England to the Roman Faith, pride of place was given to the study of rhetoric and logic. And if both were now attributed equal importance, that was because the perceived need was no longer to master the logical subtleties of a semi-artificial language (scholastic Latin), but to know how to win an audience over to the principles of one's faith. In its purest form the medieval *disputatio* had been a key element in a highly specialized investigation of logical processes. In the Renaissance, this kind of exercise was transformed into a virtuoso display of the art of communication. As its objectives changed, so did its tools: the art of disputation had been replaced by the 'art of conferring'.[4]

The multiplication of new colleges also had important consequences for provincial intellectual elites. Their establishment in cities without a university brought a faculty of teachers who made friends there, owned and lent out books, or made copies and disseminated extracts from them. They would organize solemn public assemblies (prize-days, school plays and the like), establish bonds with their pupils' families and the civic worthies, become leading lights in informal discussion groups, spread ideas and theories, establish collections of curiosities or astronomical observatories. There developed between the colleges and their immediate social milieu a dense network of exchanges, while thanks to the visits and letters of other members of the same religious order who travelled from one House to another in the course of their educational and pastoral activities, the colleges also acted as a channel for novelties from the capital or from other countries: travellers' reports, scholarly news and gossip, echoes of debates in progress. The numerical expansion of the colleges filled the provinces with a thousand little-known routes for the circulation and discussion of texts and ideas. Clusters of micro-contexts thus developed, held together by affinity, or simply geographical proximity, around forgotten heroes who may have left little or no mark on cultural history, but whose hidden role as 'intellectual mediators'[5] then possessed a significance which has left a trace in innumerable letter collections.

Such correspondence was private only in its outward form: it was meant for public reading and group discussion, and written accordingly. Letter-writing was transformed into a distinct literary genre which would play an essential role in cultural life and the communication of knowledge. For their authors these learned, stylistically refined epistles were also a means of self-advertisement, as with due (but largely false) modesty they set about displaying their erudition and communicative skills. It was a tool used with great ability by Erasmus who, *noblesse oblige*, composed a treatise *On letter-writing*, as did Vives.[6] Many humanists, Erasmus and Ficino among them, took care to publish an anthology of their best correspondence, sometimes revised and amended in keeping with intervening political and religious changes. Alongside such distinguished practitioners were many others, long forgotten, who reviewed books, offered critical judgements, reported snippets of conversation, playing their part in intellectual renewal. They quoted, summarized, reproduced, compared and contrasted texts and news, contributing to the constitution of a public opinion around the great intellectual issues of the day, and disseminating the 'new philosophy', even when themselves hostile to it. In short, they served some of the functions that would be performed in the seventeenth and eighteenth centuries by literary salons, which in the nineteenth century would be taken over by periodicals (and in our day by research conferences and seminars).

2   Words and images: writing, reading and printing

There is one further factor – last in this account but pre-eminent in its importance – which altered the modes of intellectual activity in the Renaissance, giving rise to a new 'configuration' of knowledge, a new 'episteme', and breaching the intellectual monopoly which the medieval Church had hitherto enjoyed in all disciplines, except medicine and the law. This was of course the invention of printing. The material effects of printing are inseparable from its intellectual consequences, combining the 'novelty of economic exchange, social relations and intellectual networks'.[7] Large print runs, the earliest form of mass production, reduced costs and thus multiplied buyers and readers, altering for ever the relationship between reader and written word. The personal ownership of a text made both 'continuous reading' and 'occasional consultation' possible.[8] It accelerated the circulation of ideas and extended their geographical spread by shortening the time necessary to produce texts and the time-lapse between writing and large-scale distribution. This circulation, in turn, allowed large numbers of readers to assimilate a work simultaneously and in its entirety. Each owner of a copy could now read, annotate and discuss it again at his leisure. Reading thus came to combine the motility of impressions on the reader's mind with the fixity of the text on the page. Printing protected the text from the corruption that went with manuscript reproduction, where the copyists tend to introduce errors, variants, emendations, cuts.

Textual editing became the object of a concerted effort whose importance was widely recognized. Beginning in the fifteenth century, even before they could draw on printed texts, the humanists laid the foundations of a philological method, which was further refined in the course of the sixteenth century, stimulating the development of a new discipline of historical linguistics. The classical corpus, from Aristotle to Quintilian, from Euclid to Galen, and not excluding the Old and New Testaments, gained by this enterprise: the same philologists often applied themselves equally to philosophy, science and the scriptures. Whatever the object of their study, they all pursued the impossible dream of reconstituting the *ur-text*, the original error-free version, at the same time as they reconciled themselves to the pragmatism of a historical discipline which could only reconstitute meaning through all the doubts of the hermeneutical process.

Thanks to this early philology the letter of the texts changed, as did the ways in which they were read. Henceforward it would be possible to collate editions of the major authors and to draw on auxiliary critical tools such as chronological tables of their works, anthologies of 'choice extracts', indices and dictionaries. New editorial practices also served to facilitate understanding: page numeration and italic type were introduced; quotations were separated from the main body of the text and references to their sources provided; systems of cross-references from one chapter to another and to other works were evolved. The appearance of the first complete critical editions of the classics of antiquity dramatically altered their understanding. The new-found wealth of printed sources opened up a fruitful era of comparative, pluralist readings of the classical sources. All this was of immense benefit to Aristotle, who continued to be the object of a vast labour of editing, translation and commentary. But other philosophical traditions were moving out of the shadows. The philosophical scene, like the religious, became fragmented, and for similar reasons. As the reader's mode of appropriation of the text was 'privatized' and 'internalized', so the ways in which he related to the text and constructed its meanings were diversified, the procedures of critical reading were refined and radicalized, the reader shook off the weight of received opinion and determined for himself the value of authority, opening the way to a new, eclectic pluralism. The philologically-based criticism of sources thus led to the criticism of interpretations.

With the growth in the circulation of information brought about by printing a new kind of author emerged, reader-writers stirred by the restless ambition to master the printed sign in all its manifestations. Their vast erudition served to transfer rather than produce knowledge, sifting, quoting, compiling and combining the writings of others for the benefit of their readers. In their anthologies and dictionaries all things are set down, apparently mixed at random, oblivious to contradictions. Only the shrewdest among them eluded this quagmire by inventing new fields of research: national antiquities in the case of Etienne Pasquier's *Recherches de la France*; huge bibliographical repertoires such as the naturalist Conrad Gesner's *Bibliotheca Universalis*.[9]

The new scholars combined the passion for collecting the least fragment of wisdom and the encyclopaedic compulsion to re-assemble the totality of knowledge, with a taste for rare words, borrowings from the classical languages, ostentatious quotations, mythological allusions and anecdotes where legend had the better of history. Sixteenth-century scholars were not concerned to draw clear boundaries between the past and the present, reality and fantasy, the arts and sciences, because in their time these categories were neither firmly established, nor yet deemed the necessary prerequisite of meaning. Debates took place concerning the classification of the sciences, as disciplines previously deemed subordinate attempted to shake off their dependance on others (as in the case of mathematics), but we are still far from the modern separation of knowledge into distinct self-contained disciplines. Everything is worth collecting, because everything may one day serve to make a show of learning, which is the same as being learned. In preparation for the vicissitudes of a public career, in order to acquire fortune and renown, the new scholars must amass a vast store of learning, to be copiously exhibited at the appropriate moment: what started off as the pursuit of knowledge for its own sake, ended up as a means of letting people know how knowledgeable they were. Whence the dual function of scholarly compilations: on the one hand the service rendered to the reader, on the other the celebration of the compiler's erudition and, further along the line, that of the reader who will in turn be able to display his learning. The probity, curiosity and devotion of the author-as-reader are combined with the vanity of the author-as-writer, informed about everything, so attentive to 'the vicissitude or variety of the things of this world'[10] as to lapse, more often than not, into a complete suspension of critical judgement.

The rolling back of the frontiers of the known world, the sense of intoxication which this immense field of observation and knowledge engendered and the dizzy realization of the consequent dissolution of the medieval order all stimulated the collection of this 'prose of the world'.[11] To read the resulting texts is sometimes to admire the effort to develop new methods of observation and experiment, and just as often to be dismayed by their lack of critical judgement: judicious hypotheses, acute observations and sound arguments jostle with absurd beliefs, improbable accounts of 'prodigies' and 'marvels' and the painstaking description of chimerae, in a promiscuous medley of ancient authorities, magicians and occultists, modern astronomers and long-distance travellers, whose reports often owed more to imagination than information or observation. The preconditions of such erudition would appear to have been boundless credulity, the worship of the written word, and a limitless admiration for nature, making everything seem possible, and therefore credible. The result was a style of writing alien to our contemporary criteria of scientificity:

> 'When one is faced with the task of writing an animal's *history*, it is useless and impossible to choose between the profession of naturalist and that of compiler: one has to collect together into one and the same form of knowledge all that has been *seen* and *heard*, all that has been *recounted*, either by nature or by men, by the language of the world, by tradition, or by the poets. To know an animal or a plant, or any terrestrial thing whatever, is to gather together the whole dense layer of signs with which it or they may have been covered; it is to rediscover also all the constellations of forms from which they derive their value as heraldic signs'.[12]

Whence the contrasting affinities we have already noted: the tangled teratology and the inventive yet thoughtful medicine of Ambroise Paré; the rigid demonology and the first-rate philosophy of Jean Bodin; the impressionable credulity combined with great observational accuracy of the naturalist Ulisse Aldrovandi; the intellectual itineraries of John Dee or Cornelius Agrippa, whose progression from Euclid and rationalism to magic and the occult betrayed not the least sense of discontinuity.

Thanks to its combination with wood engravings, and later copper-plate, printing brought a second decisive change, by giving a cognitive role to illustrations. From being secondary and an embellishment, the image became significant and primary because it now guaranteed its own invariability, fixed by the engraved plate on to every copy of the work, overturning the previous method of transmission. In the manuscript age copies had been produced individually by hand, and from one copy to the other of a given work, the content of a given image was corrupted, altered, confused. Often the medieval illustrator mis-read or failed to understand the Latin text he was illustrating, relying on a familiar repertoire of iconological stereotypes which bore no direct relation to the text in question.

The invention of printing reversed this situation, to the image's advantage. Stabilized by the very process of their material transmission-reproduction, illustrations could now fulfil with some degree of accuracy their informative role in relation to the text. What is more, the image itself was becoming an object of research, leading to the codification of perspective geometry, whose definition of planes and vanishing points ensured a readability, precision and complexity previously unattainable. Henceforth there would be a vocabulary of images as signs, a grammar of rules concerning their use, a system of scales governing the relationship between a body in space and its representation on the printed page. The language of graphics and representation was thus legitimized as a language of knowledge, alongside the language of words.

Advances in engraving techniques, and in particular the detail made possible by copper-plate, meant that illustrations could match the most disparate subjects. Maps, plans, structural and logical diagrams, mathematical figures, drawings of machines and cogwheels, reproductions of animal or plant species, and synoptic tables invaded the printed page, clarifying, qualifying and completing it. Learning was transformed by 'the new interplay between literate, figurative and numerate forms of expression', the beneficial consequences of which were felt by every discipline: 'Logarithms and slide-rules served Latin-reading astronomers as well as vernacular-reading artisan-engineers. Charts and diagrams helped professors as well as surgeons, surveyors and merchant adventurers. Increased freedom from "slavish copying" and fruitless book-hunting was particularly helpful to teachers and students. Until the recent advent of computers, has there been any invention which saved as many man-hours for learned men?'[13]

The advent of high-quality, precise and faithful reproductions first affected those disciplines where the object demanded to be accompanied by figures and diagrams: mathematics, astronomy, optics, anatomy, botany, zoology, geography. The addiction spread. Very soon Thomas Murner attempted to put logic into map-form, with his *Logical Chart Games*.[14] After 1560 Ramus and his followers established a veritable industry, producing manuals whose distinctive features were the recapitulatory diagrams and the tables of dichotomies, which in their view constituted the ultimate tool, the methodological key by

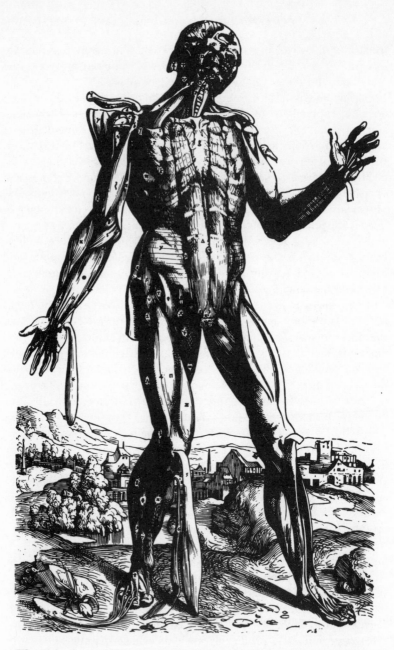

1 The Vth plate of the muscles (Vesalius, *De Humani Corporis Fabrica*, Basle, 1543).

means of which the encyclopaedia of knowledge might be structured and mastered. The image acquired a philosophical role, and the ensuing redefinition 'in figures and signs' of the totality of knowledge would play its part in the development of a new conception of man and the cosmos, as exemplified in Vesalius' anatomical treatise, *The Structure of the Human Body* (whose commercial and intellectual success was at least partly due to its superb wood-engraved plates).[15] Here, on page after page the reader marvels at the noble proportions of skeletons and flayed bodies presiding over vast landscapes where town and country harmoniously combine to create an atmosphere of startling serenity which silently testifies to man's domination of nature (fig. 1).

Nowhere is the key signifying role attributed to the image more apparent than in the emblem book, whose large-scale production and dissemination were again made possible by printing. Each page or double page of the collection associates a figure or an epigram with an often sibylline title or motto. The jurist Andrea Alciati initiated the genre, though we owe its definitive form (as described above) to the Parisian editor Christian Wechel, who adopted it in his re-edition of Alciati's work.[16] Beginning in Italy and especially France, such collections became hugely popular throughout Europe, and more than 2,000 editions appeared before 1700.

Here the image is king. It speaks and makes sense, while the secondary, illustrative role now goes to the text which accompanies it. Eye and spirit engage in extraordinary mental gymnastics, an incessant coming and going between two registers of signs: figures and words. The emblem's compounding of levels of abstraction within the order of representation is something on which all sides drew: Counter-Reformers made use of it, and so did the Reformers, despite their hostility towards the cult of images. The Jesuits did not hesitate to turn emblems into a pedagogical and apologetical tool and taught their pupils how to compose them.

The success of emblem books was only one of the routes by which the printed image established a hold on intellectual life. Images of all sorts abounded in Renaissance publications: their readers had grown accustomed to study texts which were inseparable from their illustrations. The discourse of words finds an answer in that of the image, in a series of visual prompts which participate in the construction of meaning. The eye learns to move between two systems of signs, images are now an integral part of a knowledge which language alone can no longer contain. A visual translation into charts, tables and illustrations had seemingly become indispensable in order to memorize and order knowledge. Whence that profusion of pictorial classifications of which the followers of Ramus were the most vocal advocates, but which they were far from alone in practicing. Images invaded even those most austere and medieval of texts, the logical treatises. All these elements of a new semiotic bespeak an urge to generalize, to attain a higher level of abstraction in the construction of intelligibles; they constitute a milestone along the road to that language of symbols which will transform seventeenth-century mathematics.

3 The languages of philosophy

These transformations also reverberated upon the form and language of philosophy. Medieval texts had been cast in a conventional mould, organized according to an ever-present logical framework which guaranteed a single method of investigation and exposition, consisting in precise, detailed analysis of the syntax and semantics of the Latin statements which formed the structure of the argument. Statements were considered from a formal viewpoint. This formalism conditioned the language employed, a particular variety of Latin, which was neither the elegant literary language of imperial Rome, nor a living popular language constantly transformed by the intervention of those who speak it. Scholastic Latin was a technical language, remote from everyday life. What it gained in precision, abstraction, singularity of meaning and formal perfection it lost in variety of registers, ambiguity, metaphoric or concrete expression. It was not yet a formal language, but in so far as it was no longer the language of a heterogeneous living community it had already ceased to be a natural language. In all Europe, scholastic Latin had the role of a second language, to be acquired only by the tiny minority who had access to academic instruction. Learning Latin in the Middle Ages was not simply an indication of culture: it was also a marker of class, separating and distinguishing its students from the rest of society. But above all its peculiar status and semi-artificiality determined or re-enforced certain features of medieval culture.

Since all intellectual life centred on Christianity and the task of reading and interpreting Scripture, its mode of inquiry derived from exegesis. In theology, philosophy and the natural sciences alike investigation began with the close reading of a canonical work, whose primary aim was to resolve seeming contradictions while respecting the letter of the text. But contradictions abounded, because the Middle Ages relied on middling translations based on no consistent principles, had no access to the Greek originals and – in the absence of adequate philological tools – accepted numerous apocryphal writings as genuine. The result was a 'dialectical exegesis' and countless commentaries aiming to 'insert a speculation in the fabric of the text'.[17] The procedures of this exegesis are always the same. First the commentator gives a detailed paraphrase of the canonical text, whose *auctor* is also an *auctoritas*, and is always treated as such, with no reference to either an intellectual or a historical tradition: 'Medieval philosophers rarely treated the ancients as historical individuals working in a distinct time and place,' viewing them 'as impersonal *auctoritates* rather than fallible individuals'.[18] This is followed by a stringent logical analysis of arguments for and against, gathered from a collection of parallel texts by other authorities and the occasional modern commentator. Having thus developed the 'question', examined 'objections' and contributed 'answers' one can move on to the closing 'determination', subtly reconciling the theses present and discreetly voicing a personal preference. The dual imprint of logical method and exegetic model were accentuated by the institutionalized context in which this research took place, whether at the – often mediocre – level of the faculty of arts, where the teaching of logic was dominant, or at the higher level of the faculty of theology. As for the mode of transmission, for all the celebration of the written word, in the first instance this was oral. The master 'read' the source-text before his students who, in the majority of cases, did not possess a copy. The effect was to fragment the object of study, examined in small blocks which were paraphrased and glossed according to a fixed, repetitive procedure marked by clear divisions. This facilitated understanding, but it also emphasized the static character of the exposition and the logical formality of the analysis, which needs must fit the pre-determined format of the 'question'. This mode of composition did not disappear completely in the Renaissance, but was increasingly viewed as boring, heavy-handed and artificial.

Where the formal structure of the 'question' was retained, the logical method lost its exclusive role, to the benefit of other modes of argument and presentation. At the same time, scholastic Latin retreated before the harsh criticism of the humanists who sought to restore to the language its suppleness, richness and appeal. Other modes of investigation (philological, historical, mathematical or experimental) joined the fray. Though Latin remained the chief language of learning, the sphere wherein the vernaculars could be employed was considerably enlarged. But above all the structure of Latin was no longer deemed to reflect the order and immutable truth of the universe, or to define and circumscribe the ways in which sense might be produced, as in medieval 'speculative grammar', which under cover of studying the structure of Latin had proposed at one and the same time a theory of knowledge, a logic, a philosophy of language, and even, in part, a metaphysics. In the Renaissance, this pan-logicism turned to the problem of method: where to find it, how to organize it? Aristotelians, Ramists, mathematicians, Hermeticists, all put forward their solutions. None of them proved satisfactory, but such insistent questioning points very clearly to the fact that enlightenment was no longer expected to result from logic alone.

Once the old faith was lost, the demise of scholastic logic was swift. Though it retained some strongholds (up to the 1530s in Paris, later still in Spain) in practice it was everywhere losing ground to another logic, the *nova dialectica* of the humanists. The old logic did not, however, vanish completely: those parts of it which were deemed most useful were preserved, but logic was no longer expected to provide a comprehensive method. What was now sought was more a tool of communication than understanding, an instrument that would prove agile rather than constricting, versatile rather than rigorous. It was now a question of having at one's command a 'rhetoricized logic', respectful of the obscurities, twists and byways of natural language. The new dialectic was a return to dialogue, and much emphasis was placed on the significance of its etymology (*dia*-lectic: 'speaking between'). We should not fall into the error of supposing, however, that it was devoid of philosophical significance or logical content. Rather, it was no longer blind to the pleasures of language and style. Following Cicero, Valla insisted on 'grace' and 'elegance' of expression, and these two qualities would come to count for as much as the formal validity of the argument. With the 'new logic' the Renaissance abandoned the medieval realm of necessary truths and a-historical authorities, immersing itself in the historicity of the human condition and the contingency of public debate.

It was not the attitude to logic alone which changed. So did the language of philosophy. By returning to the grammatical traditions of antiquity, the fifteenth century not only acquired a greater mastery of Latin and Greek, but also lay the foundations of philology and the linguistics of natural language. The classical corpus was extended, critical editions, translations and commentaries could now be compared with each other. Aristotle was read in his entirety and alongside his ancient Greek commentators. Plato was read together with the neo-Platonists, no longer artificially distanced from them, and due admiration was paid to the metaphorical and poetic style of both, so far removed from the rigorous simplicity of Aristotle or the asperities of scholastic Latin. The sceptics, stoics and Epicureans were rediscovered, and so were Euclid, Archimedes, Galen, Herodotus and Polybius. The new-found variety of sources and schools of thought, as well as their philological study, meant readers were more sensitive to the nuances of language, newly conscious of the semantic evolution of words in history and the different ways in which they were used by different authors. A new, methodical lexicography devoted itself to the study of the conceptual language of philosophers, learned to interpret words in their context (contrary to medieval practice) and resulted in the remarkable *Lexicon Philosophicum* of Rudolphus Goclenius (Frankfurt and Marburg, 1613-1615).

The exploration of the history of ideas, concepts and words which accompanied the birth of history of philosophy drew attention to the literary genres deployed in philosophy and science. Soon the uniform language, style and form which characterized scholasticism became a mark of 'barbarism' and 'ignorance' (an accusation endlessly bandied about by the humanists, from Valla to Vives). Its rejection opened up new spaces for the use of the vernacular, particularly among those philosophical circles concerned with civil society, impatient of the intellectual monopoly of Latin (whether scholastic or humanist).

Under the influence of the Platonic model and thanks to the diffusion of classical literary culture (whereas the Middle Ages had learned its Latin chiefly from Christian literature and late-classical anthologies) a new desire for eloquence and elegance made itself felt. In Italy Aristotelians and Platonists engaged in a series of famous controversies: the debate in the circle of the Greek exiles Argyropoulos, George of Trebizond and Gemistos Pletho was concerned both with the doctrinal content and the style and form of philosophical texts. Equally significant was the polemic concerning whether eloquence is legitimate in philosophy, which divided Italian humanists, with Ermolao Barbaro pleading in favour, and the great

Pico della Mirandola against. The issue was important because philosophy had come to understand its role in the community differently: henceforth it would seek to seduce civil society into right conduct, counsel the prince, partake in government, seek as its final purpose to reform human affairs. In place of imposing itself through the rigour and irrefutable logic of its arguments, a whole current of philosophical writings now sought first to be admired and loved.

Philosophy continued to be taught in the universities and the schools of the religious orders, in a mixture of old and new which varied according to place, topic and lecturer. The significant change was that these institutions no longer held a monopoly on philosophical research. Philosophers now lived in princely courts, in the papal *Curia*, in the palaces of cardinals and bishops, wherever their talents might prosper, gain recognition and attract the protection of some powerful patron. Having returned to the *polis*, philosophy paid much greater attention to rhetoric, ethics, politics. The decline of logic brought with it the decline of the form of writing which had been specific to philosophy in the Middle Ages, its organization into 'questions', 'objections' and 'responses'. These survived for didactic purposes, but gave some ground, even in teaching, to 'theses' (to be propounded), courses and manuals, which sometimes took the form of questionnaires. This decline was understandable, since the 'question' had originated in oral transmission by a master who 'read' the work of an 'authority'. Printing freed culture from the constraints of oral transmission. Everyone could now own or consult a copy of the object of study, the teacher was able to distance himself from his sources and develop a more personal point of view, no longer obliged to follow the letter of a text so as to explain it, but free to arrange the subject matter of his course as best suited him, according to themes. Around 1500 the first philosophical manuals appeared in Parisian circles, initially as an 'introduction' to specific philosophical topics, later as a systematic and more or less detailed presentation of the different parts of philosophy. Easy to consult and well suited to private study, readily available and relatively cheap thanks to printing, this new kind of textbook became very popular, answering as it did the growing demand for education and serving the new needs of college students outside the Faculty of Arts.

In the great theological and political debate which mobilized intellectual energies and talents in the Reformation, each side would thus have its writers of manuals, whose works served its educational objectives and were employed in its educational institutions. But all of them continued to be dependent on Aristotle, while neither the other philosophical schools of antiquity, nor the new natural philosophies (from Pomponazzi to Bruno) were attributed any legitimacy in the schools. If Aristotle remained a massive presence, it was because of the weight of tradition, because Christian theology had developed out of his conceptual language in the Middle Ages, but also because Aristotelianism constituted a framework encompassing and supporting the totality of different sciences, such as no other philosophy could yet provide. But Aristotle's centrality did not prevent efforts to free his works of medieval additions, or add to it elements of contemporary natural philosophy, anatomy, astronomy. Melanchthon for the Lutherans, the Jesuits of the Collegio Romano and the University of Coimbra, popularized complementary Latin expositions of a renewed Aristotelianism: more elementary that of the former, intended for beginners; more subtle those of the latter, which included theological issues given to advanced students. Calvinists strove to construct, without Aristotle, a 'Christian philosophy' purged at last of pagan impurities, while the Ramists in turn flooded the market with their best-selling, but often superficial didactic treatises, which drew on all the resources of typography to produce highly readable texts, tabulating or otherwise schematizing the material to be committed to memory. And in due course the philosophical manual spawned more specialized handbooks for each branch of learning, as these gradually became distinct and autonomous and the practice gained ground of distinguishing betwen metaphysics and natural philosophy, a distinction which contributed to the division of arguments, methods and objects of study between philosophy and science.

Beyond the walls of colleges and universities, we find a more varied production, freer in tone, which attributed paramount importance to debate, hence rhetoric and the plurality of views. The dialogue form, modelled on Plato, on the humanists' beloved Cicero, on Lucian (whose satires were a favourite set-text for students of Greek) suited these aims perfectly. The dialogue combined all the techniques of elegant style and rhetoricized logic, captivating the reader and dramatizing the conflict between various schools of opinion and the search for truth, as embodied in the various interlocutors. This combination of the new dialectic with humanism, originated in fifteenth-century Italy. Over the following two centuries some of the greatest intellectual figures of their day, from Valla to Bruno to Galileo, would contribute to this form (first in Latin, later in the vernacular) guaranteeing its respectability. Other forms of writing were equally practiced by philosophers and scientists, from the letter (fictional or polemical, and often both together) to the travelogue (real or imaginary, and again as likely as not both), to poetry; towards the end of the sixteenth century the 'essay', destined for such greatness, was invented. Major authors associate themselves with each new form, from Thomas More (*Utopia*, 1516), to Maurice Scève (*Microcosme*, 1562) to Montaigne (*Essais*, 1580-81). These forms had in common the fact that they no longer privileged a clearly displayed, omnipresent logical structure, leaving room for rhetoric as well as logic in their argumentation. They implied a multiplicity of possible authorial strategies, depending on the effects intended, the language adopted and the kind of readership envisaged.

4   The Academy: serving new needs

No institution associated with the new learning better typifies its plurality of interests, or its preoccupation with the 'art of conferring' than the Renaissance Academy, a term which covered a constantly changing phenomenon in the course of the fifteenth and sixteenth centuries. Beginning in Italy with small informal gatherings based on ties of friendship or patronage between a magnate and his clients, a group of intellectuals, professional and amateur, gathered to share debates, ideas, readings. Discussion could be oral or epistolary, mutual criticism and encouragement were the order of the day. It was a model of male sociality probably inspired by the professional guilds and religious fraternities of the Middle Ages. In the fifteenth century academies had no fixed timetable of events or objectives: in the absence of detailed records their activities, with a few exceptions, are only known in the most general outlines. Meetings were held in the house or gardens of the presiding figure, who often lent his name and impressed his outlook on the group. His departure from town, or his death, was likely to bring their meetings to an end. After 1500 the practice gained a firm footing in France, was installed in the countless petty courts of Germany. As it expanded and became less transitory, so it was institutionalized.

Academicians grew proud and jealous of their affiliation, gave their gathering a mutually agreed name founded on some metaphor, witticism or scholarly allusion (like the *Infiammati* around Sperone Speroni in Padua, 'aflame' with the love of learning, or the later *Lincei* in Rome, 'lynx-eyed' in their observation of natural phenomena, whose membership was to include Galileo). They chose a motto, an emblem, and often also a specific purpose and a field of endeavour. They established a constitution: the first such to have come down to us belongs to the Accademia Veneziana, founded by Aldo Manuzio, whose prime objective was the study of Greek. After the Council of Trent religious concerns became more marked. Soon the nature of academies changed, developing either in the direction of the gentlemen's club, an association of local worthies concerned primarily with cultivating their class identity and strengthening their mutual social ties, or else becoming closely associated with the Prince, and serving as official guardians and propagandists of a 'national' language or culture.

But before that happened, for a century and more, the academies functioned as catalysts and points of reference for the humanists and their studies, offering them a network independent of the universities and monasteries. They were thus a factor in the growing autonomy of intellectual life and contributed to the formation of a sector of the bourgeoisie. The academy was a training ground where young people came into contact with their elders, were introduced to the exchange of ideas and initiated into the ways of the world, established useful contacts with powerful and influential individuals who might later help them to a job, a benefice, a pension or some other form of patronage. Albeit on a limited scale, the academy achieved a degree of social mixing, bringing together men of learning of different social status: lawyers with an interest in rhetoric and history, physicians versed in philosophy, aristocrats disenchanted with the military profession, princes of the Church attracted by the new culture, publishers and translators of the classics.

In their earliest phase the academies were dominated by the taste for rhetoric, Latin style and what Paul Kristeller has described as literary humanism.[19] Later some at least moved on to more philosophical, and some to natural-philosophical concerns, while others devoted themselves to the question of vernacular languages, literary and artistic theory. Their growing specialization was accompanied by changes in recruitment and a shift to more technical debates. In learned academies (as distinct from those which remained simply gatherings of local notables) content mattered at least as much as the form of discourse. By engaging in philosophical or scientific speculation, this kind of academy served as a sounding-board for minority views like those of the Platonists and the sceptics, or heterodox ideas such as the new astronomy. They also witnessed lively debates concerning the hierarchy of the sciences and the nature of method.

Though largely concerned with intellectual abstractions, their activities were not entirely free of risk. From their very inception academies were frequently the object of suspicions and denunciations. They were accused of heresy in some instances, of sedition, immorality or magical practices in others.

How are we to interpret such 'trials by opinion', which often led to real trials, imprisonment, torture, condemnation? On occasion they were no doubt due to ill will or jealousy, and aimed to discredit an intellectual rival or a prominent public figure. Certain suspicions found fertile ground in the minds of a political or religious establishment mistrustful of any group or organization which it did not control. In other cases, the verbal incontinence of intellectuals whose theorizing assumed increasingly radical positions (without ever intending to move from words to deeds) must itself have excited retribution. Closed groups may well have drifted into occultism, subversion or libertinism, such temptations are common to all ages. But more generally, and whatever the truth of particular suspicions, it must be clear that in a society undergoing rapid transformation suspicions of this kind are a natural outlet, a spontaneous defence against the anxiety that comes with change. The danger was particularly acute in fifteenth-century Italy. Legal proceedings were initiated against several academies, followed more occasionally by their disbanding or even persecution. After the outbreak of the wars of religion in particular, fear of heresy was everywhere, and like all other institutions academies were forced to render pledges to the authorities of the day, and make an active show of conforming to religious and political orthodoxy. These setbacks notwithstanding, for almost two hundred years from the mid-fifteenth century, the academies remained a driving force, a proving ground for intellectual innovations. They undermined the old division of knowledge, acted as catalysts in the spread of a new culture, raised the general awareness of what was at stake in the debate concerning the status and use of the vernaculars, and began the great task of codifying the modern national languages.

Already in the fifteenth century Rome, Naples and Florence, where the humanist movement was established among the governing elites, hosted influential academies. In Rome, under the protection of the princes of the Church, wealthy literary amateurs and gifted young provincials aspiring to a clerical career frequented a number of such societies, most notably that which grew around Valla's Roman disciple, Pomponius Laetus, around 1460. It was chiefly concerned with the classics, Roman archaeology and Latin style, held debates on history, ethics, rhetoric and poetics, and placed great stress on the composition and recitation of elegant Latin orations. Suspected of republican leanings, a dangerous heresy in the eyes of the Pope, the academy was dissolved in 1468. It was reconstituted in 1478, by which time its interests had shifted to religion and the vernacular, but its recruitment base gradually dried up with the decline of Roman humanism, until the Sack of Rome in 1527 put an end to its increasingly mediocre and conventional existence. Similarly in Naples, at the court of Alfonso I a brilliant academy was instituted in 1447 by Panormita (Antonio Beccadelli). Its early interests were primarily literary, but after 1471, under the leadership of Giovanni Pontano, it turned to philosophy. Among its most active members at this time was Agostino Nifo, well-versed in Aristotelian philosophy, but equally imbued with neo-Platonism, whose interests extended to logic, natural philosophy, medicine, astrology and magic. Whether mere coincidence, or a sign that this tradition continued after him, the academy finally closed in 1542, under suspicion of heresy. Similar fates befell the Accademia Veneziana, founded by the humanist publisher Aldo Manuzio with an ambitious programme of editing and printing the Greek classics, disbanded by order of the Venetian government in 1562 under suspicion of political subversion, and the Neapolitan Accademia degli Svegliati, established in mid-sixteenth century to promote literary and scientific discussions (its members included the young Campanella and were heavily influenced by the doctrines of the natural philosopher Bernardino Telesio) which was closed by order of the Spanish Viceroy in 1593.

Of all philosophical academies the most celebrated was without doubt the Accademia Platonica which gathered around Marsilio Ficino in his home outside Florence, under the active patronage of the Medici. To his vast labour of translation and commentary we owe the revival of Platonism and neo-Platonism[20] to which must be added original works in which he developed his own philosophical synthesis: a metaphysics founded on Plato and his heirs, centred on the doctrine of the immortality of the soul, and the theory of love as a means of ascending to God. To this was appended the outline of a natural philosophy according to which magic and astrology would enable man to read in the fabric of creation the secret of the universe. In the process Ficino erased the Aristotelian distinction between sublunary and super-lunary worlds, and placed the Christian revelation within a continuum of fragmentary revelations, beginning with the dawn of human society. Since then, Ficino believed, it had been the task of philosophy to reconstitute these scattered fragments into a single meaningful whole, what he called the *prisca theologia* ('first', or 'original' theology).

The quality of Ficino's translations, the breadth of his learning, the impact of his personality, the abundance of his correspondence, had a profound influence in Italy and throughout Europe, from Erasmus to Symphorien Champier and Lefèvre d'Etaples in France, John Colet and Sir Thomas More in England (and right through to the Cambridge Platonists in the seventeenth century). Ficino typified the new philosophy, though in one respect at least he did not follow the humanist norm, since he lived a retired contemplative existence devoted to study among a few friends, and removed from public life. He did however conform to another humanist practice, remaining aloof from the universities (he never held a formal teaching position) and the Church. With his vast erudition, the multiplicity of his interests, he mobilized for the benefit of the Platonist enterprise all available resources, from the field of medicine to that of philosophy, from astrology to music, drawing on the scholastic tradition as well as on the work of the humanists. On the one hand, he continued the enterprise of the early Church Fathers, seeking to reconcile Platonism with Christianity; on the other he extended the boundaries of medieval Christian thought, reconciling man with nature by placing the ascent towards God within the grandiose frame of a mystical cosmic vision. By combining the ancient and the new (the intelligible beauty of the world, man's dignity and creative power, religious toleration) Ficino's Christian Platonism broke Aristotelianism's monopoly as the only philosophy compatible with Christianity, and developed a modern philosophy where a new individualism could emerge.

5 The endurance of Aristotelianism

Aristotelianism would however remain the dominant philosophy for many years to come. In Ficino's lifetime, and that of his disciples, the doors of the universities remained closed to Platonism. Until the seventeenth century it remained confined to the academies, to small circles of intellectuals, and owed its diffusion chiefly to private reading of the printed word. A minority struggled to introduce it into the university curriculum, but neither Paulus Niavis (Leipzig, c. 1490) nor Niccolò Leonico Tomeo (professor at Padua c. 1500) achieved any durable influence in this. The same was true of the Collège Royal in Paris, where Ramus, Omer Talon and Adrien Turnèbe in turn attempted to establish Platonist teaching. In Germany Melanchthon, while admiring the nobility of Plato's concepts, deemed him too difficult and too sceptical for his students. In Padua the Aristotelians had numbers and tradition on their side, particularly as they had successfully assimilated humanist teaching. In Paris the best Greek scholars, those most attached to Plato, were estranged from the university, and soon their links with the Reformation led to their death or to exile. In any case the Collège Royal was only a small institution, young and fragile, by comparison to the centuries-old University of Paris, which drew prestige from its theological tradition. Thus we have to wait until the last thirty years of the century before the first university chairs reserved exclusively to Platonist philosophy appear: in Ferrara in 1570, in Rome in 1592 (both held by Francesco Patrizi, and not filled again after his death). Even then they merely offered students a secondary option. It was a start, but other, less direct routes were of more influence in the diffusion of Platonism, such as Francesco Barozzi's mathematics courses in Padua, around 1562, or through the new astronomy. By all manner of back doors Platonism penetrated the philosophy of nature and contributed to the ruin of the Aristotelian hegemony.

Nevertheless, throughout Renaissance Europe Aristotelianism continued to determine the overall structure of thought, the division of academic disciplines and their frame of reference, and succeeded in turning humanism to its own use. It succesfully modified itself in accordance with intellectual location, philosophical issues and hermeneutic context. In the sixteenth century it was still impossible to identify it exclusively with any one confession. All the Churches (Catholic, Anglican, Calvinist or Lutheran) continued to use it as the basic institution of knowledge. The various schools of Aristotelianism crossed confessional divides. If Luther had little love for Aristotle, whom he knew chiefly in his medieval guise, his lieutenants, chief among them Melanchthon, 'teacher of all Germany'[21], came to believe that the very foundations of education were at stake, and how necessary it was to preserve a strong framework of Aristotelianism upon which to build both their theology and their propaganda. The Catholic side came to the same conclusions, for the same reasons. Suarez in Spain, Zabarella in Padua, the Jesuits of the Collegio Romano and Coimbra all revitalized their Church with Aristotelian works of the highest quality. Despite confessional divisions such texts circulated freely, because all were serious discussions of Aristotle, whose thought mattered in and for itself.

Aristotelianism survived because it served everyone, not least in the range of learning that it encompassed. This very breadth was what made it capable of being adopted and adapted to different needs, leading to a plurality of eclectic Aristotelianisms, each of them characterized by the way it drew selectively on the Aristotelian corpus, and in relation to a particular field of interest and a specific problematic, and by its combination with other intellectual traditions. A second factor, originating in the Aristotelian corpus itself, favoured this process of differentiation. The humanists had taken the study of Aristotle back to the original Greek, seeking to purge the text of errors and accretions and locate it in its historical context. Placed alongside his classical commentators and opponents, his texts came to be read and understood differently. A critical distance was re-established: even the most faithful and rigorous Aristotelians learned to discuss the master, weigh all the evidence, and sometimes disagree with him. Through this multiple process, local varieties of Aristotelianism developed. Thus Aristotelianism as a whole continued to be a living philosophy, with all that this entailed in terms of incoherence, contradiction and heterodoxy. Witness the abundance of printed texts, almost 4,000 editions, translations, etc. before 1600 (as opposed to less than 500 for Plato), to which one must add the innumerable manuscript lecture courses, theses and student notes still dormant under the dust of the archives.

But even revitalized, restored to its original form or enriched by other traditions, Aristotle's work did not give an equally satisfactory answer to all questions. Perceptions as to what the study of nature involved had sharpened, a sense had grown of the need for natural philosophy to become more concrete. In the face of the new astronomy Aristotle's physics lost plausibility, although progress in natural history and anatomy actually meant that his biology gained ground. Cracks opened up in the system and destroyed it little by little, leaving on the one hand a metaphysics rooted in Aristotelianism, which served as the privileged instrument for translating Christian thought into the language of natural reason, and on the other a philosophy of nature stimulated by cosmology or the technological desire to transform the world. This philosophy of nature engendered impatience and dissidence, beginning in the Italian university milieu, then outside these institutions, and beyond Italy itself, as the rigidity and censorship imposed by the Counter-Reformation forced the most adventurous thinkers to cross the Alps and take to the highways of Europe.

By the beginning of the seventeenth century the evidence is irrefutable: a complete break has developed between orthodox Aristotelianism, which is slowly atrophying despite its textual erudition and the intelligence of some of its defenders, and the heterodox natural philosophies which are opening the way to the emancipation of science. Henceforward, in many countries the most fruitful work would be done at some distance from Aristotelianism and the Schools. Leading medieval universities like Paris and Padua lost their leadership and their drawing power. Most original thinkers looked to positions outside University or Church. In an expanding civil society state administrations, royal academies, the Court or (in the case of the well-off) an enlightened rural retreat, were more conducive locations for philosophy and science. The divorce between science, philosophy and religion became operative. The new intelligentsia, no longer identified with schoolmen or clerics (two categories which in the Middle Ages had more often than not been one and the same) was loosely related by informal networks through innumerable letters. All over Europe, corresponding was a core intellectual task and an appropriate way to share information, publicize scientific challenges, sustain long-lived quarrels, dispute on questions of intellectual property and priority. In their way, these informal intellectual networks nourished and regulated the seventeenth-century European community of scholars and scientists, assuming all the functions of medieval universities, except for teaching. But this exception marked the beginning of a great divide between teaching and original research. In this sense, it was the birthdate of a new configuration of knowledge whence would come our modern cast of mind. A new configuration in which there would be no place for one major philosopher ruling over the whole range of knowledge as exclusively as Aristotle had from the Middle Ages to the end of the Renaissance.

Notes

1 Sir George Buck, *The Thirde Universitie of England* ... (1612), quoted in Kenneth Charlton, *Education in Renaissance England,* London, 1965, p. 169.

2 John Skelton, *Poetical Works,* ed. A. Dyce, London, 1843, I, p. 334, quoted by J. H. Hexter, 'The education of the aristocracy in the Renaissance', *Journal of Modern History,* 22, 1950, p. 2.

3 Richard Pace, *De Fructu Qui Ex Doctrina Percipitur,* Basle, 1517.

4 Michel de Montaigne, *Essais,* III, VIII.

5 P. Dibon, 'Communication épistolaire et mouvement des idées au XVIIe siècle', in *Le edizioni dei testi filosofici e scientifici del '500 e del '600,* edited by G. Canziani and G. Paganini, Milan, 1986. p. 86.

6 Erasmus of Rotterdam, *De Conscribendis Epistolis,* Basle, 1522; Juan de Vives, *De Conscribendis Epistolis,* Basle, 1536.

7 Roger Laufer, 'L'espace visuel du livre ancien', in *Histoire de l'édition française,* edited by R. Chartier and H.-J. Martin, Paris, 1982, I, p. 497; for the concept of 'configuration' see Norbert Elias, *What is Sociology?,* London, 1978, and *Norbert Elias and Figurational Sociology,* special issue of *Theory, Culture and Society,* 4, 1987, nos. 2-3; for the concept of 'episteme' see Michel Foucault, *The Order of Things,* London, 1970, pp. xx-xxii.

8 Laufer, *art. cit.* (n. 7), p. 497.

9 *Bibliotheca Universalis, sive Catalogus omnium scriptorum locupletissimus, in tribus linguis, latina, graeca, et hebraica* ('The universal library: being a comprehensive catalogue of all authors in the three languages, Latin, Greek and Hebrew'), Zurich, 1545-55; Etienne Pasquier, *Des Recherches de la France,* Paris, 5 Vols., 1560-1611.

10 The title of a compilation by Louis Le Roy, *De la Vicissitude ou variété des choses de l'univers,* Paris, 1575.

11 M. Foucault, *Order of Things, cit.* ch. 2, 'The prose of the world', pp. 17-45.

12 *Ibid.,* p. 40.

13 Elizabeth L. Eisenstein, *The Printing Press as an Agent of Change,* Cambridge, 1979, II, pp. 520-21.

14 *Chartiludium logicae, seu logica poetica et memorativa* ('Logical chart-games, being a system of poetic and mnemonic logic'), Cracow, 1507.

15 Andreas Vesalius, *De Humani Corporis Fabrica,* Basle, 1543.

16 D. Russell, 'Alciati's emblems in Renaissance France', *Renaissance Quarterly,* 34, 1981, pp. 534-54; H. Diehl, 'Graven images: Protestant emblem books in England', *ibid.,* 39, 1986, pp. 49-66; M. Bath, 'Recent developments in emblem studies', *Bulletin of the Society for Renaissance Studies,* 6, 1988, No. 1, pp. 15-20.

17 M.-D. Chenu, *Introduction à l'étude de Saint Thomas d'Aquin,* 2nd ed., Paris, 1954, p. 133.

18 Anthony Grafton, 'The availability of ancient works', in *The Cambridge History of Renaissance Philosophy,* edited by C. B. Schmitt and Q. Skinner, Cambridge, 1988, p. 771.

19 Paul Oskar Kristeller, *Renaissance Thought. The Classic, Scholastic and Humanist Strains,* rev. ed., New York, 1961, ch. 1; 'The humanist movement', pp. 3-23.

20 Ficino completed the translation of Plato's dialogues in 1470, and these were followed by the works of Plotinus (1484–92), Proclus, Porphyrius and several others.

21 The appellation 'Praeceptor Germaniae' was already being used by German historians at the end of Melanchthon's life, see 'Melanchthon' in *Contemporaries of Erasmus. A Biographical Register of the Renaissance and Reformation*, edited by P. G. Bietenholz, II, Toronto, 1986, pp. 424–29.

3

The history of science
and the Renaissance science of history

Stephen Pumfrey

In this essay, I shall develop a general argument of the volume (that the dramatic changes in the practice of natural philosophy or science during the Renaissance can be closely related to other, broader changes in Renaissance culture and beliefs) through one specific but central relationship: Renaissance philosophers legitimated their chosen sources of natural knowledge with arguments based on philosophies of history. In particular, the seventeenth century's preference for empirical sources depended for its plausibility upon a new 'sense of the past', a broader change in the way late Renaissance writers compared their own times with antiquity.

Scholars in 1450, and still some in 1650, believed that knowledge of nature was most authoritatively established through the analysis, often critical, of ancient texts, especially those of Aristotle in philosophy, Ptolemy in astronomy, and Galen in medicine. By 1650 self-styled 'new philosophy' had gained ground, and its more empiricist proponents claimed to be replacing the false 'opinions' of the ancients with the authority of certain deductions based on first-hand observations and experiments on nature. For many historians, this methodological shift is a central element of the 'revolution' which created modern empirical science. In traditional 'internalist' historiography, the causes of the shift were primarily new developments within the study of nature itself – new discoveries, new methods, new mathematics and new technologies which established modern science as an independent discipline. From this viewpoint, contemporary attitudes to humanist studies such as history and statecraft are of little relevance.

However, the shift from a textually to an experimentally based natural philosophy can be seen as part of a broader dispute about where to locate authority. We shall explore the ways in which the ideological determination of authority in the understanding of nature paralleled the general ideological determination of authority in other important areas such as the understanding of justice, beauty and social organisation: in short, Renaissance science evolved as part of Renaissance culture. We shall follow such similarities and dependencies from the fifteenth century, when the process of restoring classical culture and wisdom was undertaken, through the late fifteenth and early sixteenth centuries, when it was increasingly suspected that the classical tradition would not easily yield answers to contemporary questions, to the later sixteenth and seventeenth centuries, when for some areas, such as law and, of course, science, the relevance of Greek and Roman practices was rejected. We shall particularly follow the parallels between law and science for, as Kelley, Huppert and others have shown, it was legal humanists first who developed a radically new sense of the past.[1] This sense of the past was needed and used by some advocates of empirical science to justify their own revolution. Let us examine the nature of history and the nature of empirical science, particularly as they were understood in the Renaissance in order to see why the link was necessary.

1   The necessity of history

'Every society', writes J. G. A. Pocock, 'possesses a philosophy of history – a set of ideas about what happens, what can be known and what done, in time considered as a dimension of society – which is intimately a part of its consciousness and functioning.' He argues that such philosophies are central to all political ideologies, conservative and radical.[2] The conservative will defend the status quo by appeal to tradition, deriving the authority of current practice from its continuity with past practice. This is clearly an historical argument. The radical (such as the anti-Aristotelian natural philosopher), in seeking support to institute new practices, is forced to engage with, and demolish, the traditionalist historical construction of authority. He must propose a new interpretation of history, a different relationship between past and present that simultaneously renders current tradition as false, and his innovation as true. We might say that Renaissance radicals developed low risk (which could be called 'reactionary') and high risk ('rational historicist') strategies to solve this difficult historiographical problem. The reactionary will accept the premiss that authority resides in

49

C

tradition, but then claims to have identified an older, purer and more authoritative tradition, from which current 'tradition' deviated in the past. Rational historicists will argue that future institutions should be, or can be, different from all past models, but this requires them to develop a philosophy of history which removes authority not just from the conservative's accepted tradition, but from all constructible traditions, thus leaving their utopian projection as the only guide to action. An effective way to block recourse to some authoritative past is to use historicism, in its original relativistic sense of interpretation of the past in terms of its own local, particular and transient causes and contexts. The radical thus frees himself for judgement by ahistorical, rational criteria.

Since all these strategies rely upon philosophies of history, interpretations of the past are always implicated, intentionally or by appropriation, in ideological conflicts. The pace of change in the Renaissance, leading to conflicts between church and state, and between churches and between states, helped to create new, critical histories written to advance or destroy the authority of a major institution, together with its legitimating history. Faced with such change and conflict, certain French legal humanists concluded that simple appeals to the past were no longer valid, and they historicized it. This is not to say that such historiography was, or is nothing but disguised ideology. Indeed, Renaissance historians tried to develop standards for their discipline precisely to lift it above propaganda. Furthermore, once history had acquired some disciplinary autonomy, it led to ideological change in other areas such as natural philosophy.

Why should new natural philosophers have drawn on the new historiography to argue for the superiority of modern over ancient science? Historians of science have usually seen progress in science as the major, or at the very least the decisive, battle in the 'quarrel' between supporters of modern over ancient learning, as a self-evident cause rather than the consequence of the decline of Renaissance ideology. In the last section of the essay we shall examine evidence that the new science did indeed draw on the new history. But, to carry through Pocock's suggestion that all activities require historical legitimation, we need to examine the mixed fortunes of empiricism.

Modern empirical science is the latest of a line of disciplines that claim to discover, more truly than their predecessors, 'the nature of things'. Francis Bacon is often cast as the heroic founder of modern empirical science, but he (and following him, some twentieth-century historians) gave the title 'the first of the moderns' to the late Renaissance Italian Bernardino Telesio. In 1565 Telesio had published his own account *Of The Nature of Things*. Like many empiricist philosophers of nature, he considered that the current ignorance of nature's processes was due to an uncritical acceptance of traditional authorities – in his time, the ancient Greeks Aristotle and Plato. A true knowledge of nature, he argued, would accrue once people laid aside erroneous theories and observed impartially, so that nature could 'announce herself'. He advanced the conclusions from his own cautious observations. The first was that 'the nature of things' was, fundamentally, the conflict of hot versus cold!

Telesio was merely one of the first, Descartes in the 1640s perhaps the last, of an unprecedented flurry of natural philosophers who claimed (unsuccessfully in the long run) to have broken with the past and to have established a true, new science of nature, in accordance with experience and experiment. In between, the better known names include the English Queen Elizabeth's doctor William Gilbert, Galileo Galilei and, above all, Bacon himself. All, of course, played crucial roles in the reshaping of natural knowledge called 'The Scientific Revolution', and yet many of their conclusions, sanctioned by the power of observation and experiment, are as unacceptable as Telesio's. Gilbert's chief claim in *De Magnete* (London, 1600) was to have proved experimentally the existence of an Earth soul, something rarely accepted even in the seventeenth century.[3]

The historical graveyard for casualties of naive empiricism has many representatives from the late Renaissance and Early Modern periods (while philosophers of the preceding centuries died instead from a surfeit of respect for antiquity). In one sense they suffered the heavy losses of all advance parties establishing bridgeheads. We should recall how recently empirical science has become the dominant, most trusted form of knowledge. The claims were first made during 'the Scientific Revolution' of the Early Modern period, and then secured and advanced by the *philosophes* of the eighteenth-century Enlightenment, for whom Newtonian natural philosophy provided a methodological template for the study of man and society. The nineteenth century saw the development of sociological science and scientific history.

Today it is popularly believed that the success of science is owing to its rigorous empirical character. The popular ideology is, of course, legitimated by a popular, pro-empiricist 'Whig' history of science in which the above graveyard represents instead a pantheon for heroic contributors of objective observation. The view persists despite the consensus of contemporary analysts of science that the regularities we find in nature depend upon the pre-existing concepts and interests which structure our perceptions. This post-modern response is but the latest phase of a long dialectic between the empiricist's reliance on, and the sceptic's concern about, inductive knowledge. Hume and Kant undermined the naive empiricism fashionable in the eighteenth century. In the late Renaissance itself, the arguments of the classical sceptics Pyrrho and Sextus Empiricus were all the rage. Their chief target had been Aristotle, whose inductive logic had answered Plato's own idealist attack on the empiricism of the Pre-Socratics! In short, many late Renaissance intellectuals knew well the pitfalls besetting the empirical path to natural knowledge, into which Telesio and Gilbert had fallen.

How, then, was empiricism made plausible in the late Renaissance – a time particularly unsuited to it given the preference for classical over modern authorities in general and, increasingly for some, a preference for classical scepticism over all authority? Without a constructed historical tradition of successes of the kind popularly invoked today, how could advocates of the new empirical method justify novelty in an emulative culture? This volume suggests several answers, but here we shall look for one in the close relationship between science and history in the Renaissance.

2   Natural philosophy and the authority of the classical tradition

Medieval philosophy, especially natural philosophy, in the Latin West, was based closely on Latin versions of Aristotle and his Islamic commentators, in a growing dialogue with contemporary Aristotelians such as Aquinas. What distinguished the medieval privileging of Aristotle (and in natural philosophy other ancients such as Galen and Ptolemy) from a similar emphasis in the Renaissance was, as Burke argues, an absence of a sense of the past. It was not imagined that significant cultural change had occurred during the one and a half millennia which separated medieval Europe from classical Greece, and consequently there were no *historical* reasons, negative or positive, for the study of the Greeks. Their texts were allowed to set the agenda because they were considered to be incomparably superior. Aristotle was 'The Philosopher' because he seemed the best, and not initially because he was ancient. Indeed, his position as a classical Greek (and hence outside the Judaeo-Christian tradition) provided an original and recurring ground for hostility from ecclesiastical authorities. Lacking Burke's criteria of a sense of anachronism and an awareness of evidence, medieval clerks interrogated Aristotle (and, when they misattributed authorship, 'pseudo-Aristotle') about natural verities as though he were a living master.

Some parts of his natural philosophy fitted their static society – for example, his principle that all bodies seek rest as their natural state, a principle rejected in the seventeenth century. But the lack of concern with context precluded this kind of criticism, even of a more doubtful generalisation (as Aristotle's splendidly Grecian characterisation of air as hot and wet might have seemed to northern Europeans). However, for most medieval and Renaissance thinkers, the philosophy of nature dealt with the unchanging everyday experience of natural processes of growth and decay, such as the heavenly motions, the growth of plants, the functioning of the human body. The real, natural existence of experimentally (one could say artificially) created phenomena, such as Galileo's frictionless motion, or Boyle's vacuum would later be disputed, but the phenomena of pre-experimental science required no special techniques to be made visible. The difficulty lay in attributing causes to these well-known phenomena – a problem of reason, not experiment. Although the optical researches of Theodoric of Freibourg or Roger Bacon show that an experimental approach was not unknown, medieval natural philosophers saw greater merit in logical analysis of the concepts of 'the Master of those who know'.

While the medieval ignorance of an historical dimension to scientific truth claims may not strike us as strange (and many contemporary philosophers ignore it too), we should note that the same ahistoricity pervaded the other faculties of medieval universities. In numerous law faculties jurisprudence was similarly based on two ahistorical assumptions. First, Roman law was held to represent a permanently excellent system for policing human affairs. The *Corpus Juris* was treated as the unique foundation of civil law, with no sense of the anachronistic late Imperial institutions and problems with which it dealt. Secondly, and by the same token, medieval jurists did not conceive that the glossed versions available to them, let alone Tribonian's original code, were corrupted, contradictory and partial versions of a legal 'system' which had been constantly adapted to changing social needs. Scholars consequently answered legal questions using precedents torn from their historical context. With the identity of past and present taken for granted, true meanings required logical, not contextual analysis.

Renaissance ideology, in disdaining the products of medieval scholasticism, constructed a sense of the past. More concerned with the civic and literary learning of the *studiae humanitates*, it distanced itself from medieval preoccupations with logic and theology, in part by an historical device. The familiar periodisation was created of a classical Graeco-Roman age, a debased Middle Age, and a glorious Modern era in which the achievements of Antiquity would be restored, imitated and emulated. In practice, Renaissance learning owed more to scholasticism than it admitted. In natural philosophy especially, the problems set by Aristotelianism were still dominant in the seventeenth century. Renaissance humanists did not dispute the centrality or value accorded to classical learning. Rather they charged that the schoolmen had failed to appreciate the pricelessness of classical achievement. Having no knowledge of Greek or good Latin, of the secular culture of the city state, republic or empire, the clerks had culpably misconstrued the texts and instruments of that culture into their own decadent and barbarous terms. Thus the early humanists claimed to be purging an ancient wisdom of its barbarous medieval accretions, re-discovering lost treasures, and restoring the classical tradition to a condition and use fit for a new age capable of using it's 'exemplary ways of thinking and living' to the full.[4]

If we grant the Renaissance its belief in the authoritative genius of antiquity, and the belief that nature does not change, the project of a text-based natural philosophy makes sense. However, early Renaissance scholars similarly held less familiar beliefs in unchanging universal principles of aesthetics, rhetoric, jurisprudence, statecraft and so on, an understanding of which was available through a similar reconstruction of the classical legacy. By 1550, Vasari judged that art was so well restored that it might only decline. In law, Pocock imagines that the early legal humanists 'set about their task in the belief that the true principles of Roman jurisprudence, when found, would prove of such surpassing wisdom that they need only be directly imitated and applied to the present day.'[5] This early Renaissance 'sense of the past' is not the progressive, proto-historicist 'sense of history' of the sixteenth-century humanists. The claim that Ptolemaic astronomy, Ciceronian rhetoric and Justinian law were of even greater contemporary value than the medievals had realised, despite the parallel claim that social structures and cultural achievement had changed dramatically, reveals the ahistoricity of the early Renaissance. Renaissance ideology incorporated a model of cultural change which assumed decline or, at best, cyclicism. It was not historicist, for it saw greater similarities with the distant than with the immediate past. And it was not progressivist, because both Christian and humanist teaching conspired to predict and describe an imminent completion of human endeavours. It was, of course, fundamentally retrospective, and therefore fundamentally different to both the medieval assumption of uniformity, and to our own of open ended development. Thus, while medieval thinkers looked to develop atemporal truths which just happened, as it were, to have been well perceived by classical writers, the Renaissance elevated Antiquity to a unique age of heroic achievement, the original *locus classicus* of truth, and searched for the *ur-text*.

In the dynamic years of the High Renaissance many classical texts were indeed located, properly established and edited, made increasingly available through printing and increasingly intelligible through new critical apparatus. The surest way for a scholar to reward his patron or impress the republic of letters with valuable learning was to assist this restoration of the *prisca sapientia* (ancient wisdom), rather than to run conspicuously after novelty. Much work was genuinely novel but was presented, in accordance with Renaissance ideology, as an outgrowth of classicism. This is well known for the humanities, but the same controlling ideology operated in the discipline of natural philosophy. For the more conservative natural philosophers, the first pressing task was the application of new linguistic and philological skills to produce faithful editions of Aristotelian works, whence medieval errors could be expunged to leave the pure *sapientia* of the classical tradition. As Giard notes, the new skills combined with the new printing technology to make the ancients available to be read in a new way. A retrospective and textual humanism prepared the ground for the simultaneous publications in 1543 of, in astronomy, Copernicus' *On the Revolutions of the Heavenly Spheres* and, in anatomy, Vesalius' *On the Structure of the Human Body*, works which have traditionally marked the beginning of the Scientific Revolution, but which more recent historians regard as conservative.

The 'scandalous' ignorance of medieval astronomers was signified by the inaccuracy of the Julian calendar. Assuming that this arose from errors in transmission of Ptolemy's *Almagest* through Arabic manuscripts into poor medieval translations, which were then compounded by crude textbooks such as Sacrobosco's *Sphere*, late fifteenth-century astronomers strove to reconstruct the pure Ptolemy. Peurbach and Regiomontanus collaborated to translate and interpret an early Greek manuscript from Constantinople, of which an *Epitome* was published in 1496. They and subsequent scholars revitalised astronomy by publishing several attempts to fit the demands of recent observation and contemporary cosmology around a restored Ptolemy. By 1515, when Copernicus first circulated his heliocentric hypothesis, it was clear to him, from his privileged multiple reading position, that the fit was imperfect, but his critical strategy is pure Renaissance. It is text-centred, pointing to the incompatibility of different systems (including the restored Ptolemy) to each other, not to Ptolemy's data – which he himself relied on. Furthermore, Ptolemy is criticised for having encouraged a decline from the pure classical principles of seeking true hypotheses based on uniform circular motion. Copernicus introduces no significant observational or theoretical novelties, except the central position of the Sun – a position which he claimed, with apparent sincerity, was sanctioned by several classical mathematicians. Epistemologically and historiographically, Copernicus' work is typically humanist, and it made no modernist challenge to Renaissance ideology.

A similar, though modified account can be given of the emergence of a 'new anatomy'. Again the real stimulus came from the Renaissance concern to reject the corrupt texts and oversimplified pedagogy of the scholastics, based on the work of Mondino (*c.*1275 – 1326), who read lectures based on Arabic versions of Galen, while a surgeon-demonstrator dissected a body. By the early sixteenth century most anatomists, including Vesalius, sought improved knowledge primarily in the translation and editing of Greek Galenic manuscripts, including some previously unavailable, such as the *Anatomical Procedures*, Galen's account of his own method of dissection, which Vesalius published in 1541. We should not deny that Vesalius' own dissection led him to criticize Galen heavily, especially for his reliance upon animal cadavers, but in his method and particularly in his theory, Vesalius did not depart from his contemporaries' renewed confidence in a purified Galenic tradition. 'There was no powerful methodological drive beyond this [textual study],' comments Hall, 'rather it was the case that the endeavour to emulate Galen's practice led to the abandonment of Galen's precepts.[6]

Hall's comment can be read more widely as a description of the inherent self-deconstructing fallacy of Renaissance ideology as a whole, for it was not primarily in natural philosophy that reconstruction work of the kind undertaken on Ptolemy or Galen began to fail. The more humanists like Lorenzo Valla succeeded in excavating and restoring the ruins of classical civilization, the more they realized that they were indeed ruins. Yet it was not until the early sixteenth century that what had become a conservative strategy of seeking answers from the classical tradition became problematic, and radical constructions of authority gained ground. Some natural philosophers pursued the reactionary strategy of seeking a purer, more ancient wisdom, while legal philosophers developed rational historicism into a 'powerful methodological drive' away from the ruined sites altogether, perhaps because the need for workable solutions to their socio-political problems was more pressing. We shall begin with the reactionaries.

3   Reactionary strategies and an ancient wisdom of nature

One of the most distinctive features of the study of nature in the sixteenth century was the growth of the Platonic and Hermetic 'traditions', both of which owed much to the work of the Florentine humanist Marsilio Ficino. The excitement engendered by the restoration of Platonic natural philosophy in the Renaissance can easily be understood. Quite apart from Plato's 'intrinsic' merits, humanists compared favourably the polished rhetorical style of Plato's dialogues with the dense didacticism of his pupil Aristotle. Moreover, Plato's nature was a divine and animate creation, lending itself more easily to Christianized interpretation. Finally, the philosophy advanced by Aristotle when he left Plato's Academy to found his Lyceum, explicitly opposed many aspects of Platonism. He denied both that a god had created man and the cosmos around the same geometrical structure of a 'World Soul', and that mathematics was a key to natural knowledge. Above all, he denied that empirical observation was unreliable because matter was corrupt, and hence that nature was best known through the abstract contemplation of Ideas. Aristotle's break with Plato was itself a reason for the appeal of Renaissance Platonism.

To radicals, a scholastic education, however much restored, humanized and tempered, was still Aristotelian and still redolent of the Middle Ages. Younger humanists like Francesco Patrizi, impatient of fame, set themselves up as dissident Platonist lecturers, or offered this more elegant learning to receptive patrons. A Platonist emphasis could also be used, as it was by Leonardo and Galileo, to dignify the work of mathematicians and artists, and by natural magicians for whom Plato's creation myth underwrote the existence of hidden correspondences between microcosm and macrocosm which they claimed to control. In short, a sympathy for Platonism was adduced by most sixteenth-century anti-Aristotelians but, of course, the sympathy required historiographical legitimation.[7]

Reactionary natural philosophers, who assented to the authority of antiquity, found Plato's position as a classical philosopher necessary, but not sufficient to justify departure from a tradition begun by his pupil and virtual contemporary. That permitted only comparison of their respective merits, and the common conservative activity of reconciling the two traditions into one. Following Ficino, the crucial differentiation was Plato's 'theology'. 'The divine Plato', it was argued, had somehow been sensitive to the revealed truths of Judaeo-Christianity, while Aristotle had not. Plato, then, was a superior source of a more ancient wisdom, either because he had a superior mind, or because he had learned it, or even because he had been divinely inspired (Kepler called him 'Moses speaking Attic Greek'!). On this reading, Plato's slightly greater antiquity to Aristotle becomes important, providing the reactionary with an older tradition, obscured by the dominance of Aristotelianism, but surfacing occasionally with Neoplatonists like Plotinus and Proclus.

Renaissance Hermeticism was a more extreme version of the reactionary Platonic strategy, developed in tandem with it. It is well known that Ficino suspended his work on Plato when he received texts purportedly written by Hermes Trismegistus, believed to be an Egyptian priest contemporary with Moses himself. In fact they were pious 'forgeries' composed by Neoplatonists in the late Roman period as part of a campaign to advance their mystico-magical worldview against secularism. The Hermetic texts are thus examples in themselves of the reactionary strategy. They emphasized the power of human knowledge, and the importance of spiritual magic and number-based harmonies in nature. They therefore held a considerable interest for heterodox natural philosophers. To neutralize theological objections to Hermetic magic, Ficino constructed 'The Hermetic Tradition', perhaps the most extreme possible example of reactionary historical legitimation. In this view, God had revealed a true knowledge of His creation and its secret processes to the Jews. The 'ancient theology' had passed from Moses to Hermes during their exile in Egypt. Passed on only to initiated *magi*, in cryptic form, it had been learned by Pythagoras, who took it back to Greece, where Plato had been its last exponent before its submergence under Aristotelianism. Other ancient thinkers, such as Zoroaster and Orpheus, all of whose ideas were (un-)fortunately only known through obscure fragments, could be fitted into the tradition. Within the historical strategy membership and sequence of this new old tradition varied according to need. Sixteenth-century investigators satisfied an envious emerging nationalism with other inclusions, the Druids, for example, in England and France, and the extreme radical Hermeticist Giordano Bruno legitimated his heretical religious beliefs by placing Hermes before Moses, and he was executed.

The Hermeticist's historical strategy is thus a radical development of the Renaissance Platonist's. Historians see more Neoplatonists and fewer Hermeticists than they did twenty years ago, but it remains that the legitimating history was similar enough to be accepted, and allowed thinkers to place themselves within a spectrum of radical belief, and to pick and choose their sources. The pre-classical authority of the *Corpus Hermeticum* was not refuted until 1614, although the philological and historical skills employed by Isaac Casaubon to date it correctly were those famously deployed by Lorenzo Valla some 200 years earlier to expose the Donation of Constantine. In any case, the nugatory size of the extant work of many of the newly respectable and newly expounded ancient philosophers indicates that Renaissance thinkers effectively used the sources as (somewhat unroadworthy) vehicles for their own philosophies of nature. The Hermetic tradition served to bring within the Pale a mass of nonclassical, unorthodox or ambiguous conceptions of nature. Its texts could be made to speak with ancient, even divine sanction of new or hitherto dismissed ideas, such as communication with spirits, the immense powers of magic and mathematics and a Sun-centred or infinite universe.

Just as it would be wrong to claim that rational historicism *inspired* a later empiricism, so these heterodox natural philosophies were not inspired by reactionary historiography. However, we can claim, in both cases, that the new attitudes required the new histories in order to find space to develop within the ideological boundaries of Renaissance culture. Before the late sixteenth century, 'new' principles were clothed in ultra-antique authority: Copernicus dubiously asserted that his heliocentric astronomy had been anticipated by a succession of pre-Ptolemaic figures, including Hermes, Bruno asserted an ancient knowledge of infinity. Isaac Newton would later hold that the Pythagoreans had had a secret knowledge of his law of gravity. The power of reactionary ideology is clearly shown in the development of Paracelsian chemical philosophy and medicine. Paracelsus (d. 1541) ridiculed all elite knowledge of nature, especially Galenic medicine, and insisted upon a novel 'chymical' approach to health. His confused writings were systematized after his death, for example by Richard Bostocke in *The Difference Between The Auncient Physicke... And The Latter Physicke...* But Bostocke's title hermetically transforms the new Paracelsianism into 'the auncient physicke, first taught by the Godly forefathers, consisting in unitie, peace and concord' and hence superior to 'the latter Physicke proceeding from Idolators, Ethnicks and Heathen: as Gallen [*sic*], and such other consisting in dualitie, discord and contrarietie'.[8]

The 'genealogical' history of the Hermetic Platonists is, then, a prime example of reactionary radicalism. It produces radical, deconstructive consequences for the authority of the mainstream classical tradition (for peripatetic philosophy began with apostasy), using a grotesque exaggeration of the appeal to antiquity, decadence and restoration. A continuing authority of (very) antique textuality suggests that the new experimentalism required a different 'sense of the past' in which to appear plausible. As before, we should note that the reactionary escape backwards was not a strategy peculiar to natural philosophers escaping Aristotelianism, but common to many groups whose interests were not furthered by the Italo-centric character of the High Renaissance. While it is easy to draw erroneous parallels, two withstand comparison.

First, there is emergent nationalism, and history in its service. 'What is all history but the praise of Rome?' asked Petrarch, but once non-Italian scholars had acquired the culture of Renaissance human-ism, their national sensibilities were not promoted by such incessant Romanity. Nor could the French and Germans share Italian reasons for deploying a pejorative concept of mediaevalism, whose purpose had been to undermine the legitimating doctrine of *translatio imperii*. By attacking the traditional and documentary authority which the Holy Roman Empire claimed to have inherited, Italian civilization, past and present, was left unrivalled. Opponents constructed new histories which, by inventing complex Frankish and Gallic societies before the Roman conquest, lessened the cultural debt to Rome, and allowed pride in a continuous indigenous culture, frequently traced back to pre-Roman ancestors such as the Trojans.[9]

Consideration of Northern Europe's escape from Rome brings us to the Reformation, the second more general example of reactionary strategy, which historians consider to be dependent, in part, upon new Renaissance attitudes to the past. Indeed the problems, and solutions, worked on by Reformation ideologues were those of the Hermeticists writ large. Distrustful of the possibility of eliminating 'medieval corruptions' from the institutionalized Church, they proposed to re-establish the earlier true church of the apostles. This required an extraordinarily powerful historical relocation of authority, for what was at stake was a body of not merely natural but divine truth, and sanctified Patristic tradition. Encouraged by Luther and Melanchthon, Protestant historians employed reactionary strategies. They excluded all but the most antique texts, Scripture, in a 'return to the sources'; they exposed erroneous traditions with the techniques of critical humanism; and they constructed a revised tradition of saints ('witnesses to the True Church') while debunking the hagiography of the old church, now a merely human 'Kingdom of the Antichrist'. The historical battle culminated in rival histories of the world, with the Reformers' so-called 'Magdeburg Centuries' provoking the magisterial and scholarly Catholic response, the *Ecclesiastical Annals* of Cardinal Baronius. Interestingly, there were similar attempts in early Protestant universities to reshape the antique authority of natural philosophy. Briefly, and disastrously, in the mid-sixteenth century, Melanchthon based the curriculum upon Pliny's *Natural Histories,* and the Dutch divine Lambert Daneau extracted a natural philosophy from the Bible itself. It is clear that, while the dominant ideologies of both radical Protestant and traditional Catholic thought privileged the past, the experimental philosopher's call to tear up the past and observe anew would convince few.[10]

In the light of the above examples we can agree with Burke and others that the conflicting political and religious interest groups had rested too much of their legitimacy in the ideology of tradition to give up lightly their conservative and reactionary strategies. This is why Renaissance scholars took so long to acknowledge that their classical tradition was flawed and irrelevant. In sixteenth-century France, however, some scholars finally found value in a historized past.

4   Rational historicism and the rise of empiricism

In France, several ideological conflicts conjoined to produce a climate
in which attitudes to the past became highly politicized and dangerous
indicators. France was evolving rapidly into a centralized state, for
which French apologists sought a cultural history which revolved
neither around Rome nor the Roman conquest. Secular pressure to
invent a Gallican culture was compounded by pressures for a Gallican
Catholicism with greater autonomy from Rome. And these conflicts
were minor compared with the tensions building in France between
sympathizers and opponents of the German Reformers. Finally, the
Wars of Religion were incited by a feuding nobility, whose arguments
over monarchical succession and the balance of power between king
and nobility were naturally backed up by competing histories of the
'true' French constitution. Ideological clashes produced incompatible
historical interpretations of the authority of central institutions such
as the church, the crown and the learned universities. Since the
collapse of religious and political certainties had, in any case, fostered
scepticism, the reliability, indeed the possibility of historical knowledge, was a pressing question. The concern of some of the French
noblesse de robe with a proper, critical practice of history reflects their
attempts to find new, sure and useful ways to legitimate French
institutions. A major focus for this issue was the debate over the
history and theory of law in France. The debate has recently received
thorough scholarly attention, and we need only summarize it before
applying it to natural philosophy.[11]

The problematic relationship of ancient to modern learning came
under scrutiny early in the sixteenth century from the first great
French humanist, Guillaume Budé. Employed by the French court,
Budé knew the practical and symbolic importance of French custom-
ary law. He studied Roman law, and Romanity in general, with
scholarly respect but, controversially, treated its remains rather like
valuable museum pieces. His successors developed a new 'French way'
(*mos gallicus*) of teaching law, based on the axiom that legal systems
were the products (indeed useful historical records) of particular
societies. Thus Roman law was not a stable codification of a perfect
jurisprudence, nor could it be relevant to a France very different from
classical Rome. French jurists, in their capacity as civil servants,
thereby freed themselves to evolve contemporary solutions to contemporary problems.

Here was a new use of philology, and a new sense of the past. The French used the philological techniques pioneered by Valla to plot the historical distance of classical from contemporary discourse. Valla's generation had urged a return to the past but the French humanists, while agreeing that the past was excellent, asserted that it was also past. To know one's noble history and traditions remained important but while the past initiated the present, it did not authorise or resemble it. It can be disputed whether certain legal humanists had arrived at a progressive historiography, or whether it was basically cyclical, but the differences are unimportant compared with the fundamental rejection of both traditional and reactionary modes of historical legitimation, and the re-presentation of human history as a secular *process* of civilization. Different members of the circle attacked different aspects of antique legitimation. We shall find these attacks redeployed by advocates of empiricism.

The comparatively conservative Etienne Pasquier, applied the legal historicism he had learned from Baudouin and Hotman to produce a revolutionary history of France. The first of his group publicly to reject the legitimating device of tracing a nation's ancestors back to the Trojans, he emphasized the changing historical shape of France and French culture, for which he gave secular, social explanations. He distinguished a still progressing French culture from a dead Roman one. Jean Bodin advanced the superiority of modern over ancient more promiscuously, and was particularly savage in his attack on the decadent myth of 'the age which they call "golden", which if it be compared with ours, would seem but iron', because of its uncivilised violence. He enlisted Thucydides to show that the Greeks too had once been barbarous.

> There you have your famous centuries of Gold and Silver! Men lived dispersed in the fields and forests like wild beasts...: it has taken a long time to pull them away from this savage and barbarous life and to accustom them to civilized behaviour and to a well-regulated society such as we now have everywhere.[12]

Even if Bodin's thinking was still cyclical, his circle's full development of secular rationalist historicism into a theory of historical evolution, from primitive to advanced culture, was soon advertised by the contemporary publications of Louis Le Roy and Lancelot de La Popelinière. Indeed it is in Le Roy's *Of the Interchangeable Course, or*

Variety of Things in the Whole World... from the Beginning of Civility, and Memory of Man to this Present, that we first find the concept of 'civilization' as a process. Although Le Roy's philosophy of history is derivative, his account is interesting for its selection of civilized culture, especially the progress of the practical arts and sciences, as the yardstick of history, and for its display of the great variety of civilizations, both in time and in geography. It also shows the double importance of the discovery of America in the development of historicist ideology. On the one hand, Amerindian society is used as proof that primitive societies are uncivilized, from which he inferred that Europeans were once barbarous, as Bodin held, and 'remained for a long time without letters'. On the other hand, the discovery of a whole New World is proof of our superiority over Antiquity. Thus Le Roy can eulogize recent progress: 'new lands, new customs, new kinds of men, mores, laws, folkways: new plants, trees, roots, liquors, fruits; new diseases and new remedies: new routes in the sky and on the ocean never tried before, new stars seen'.[13]

La Popelinière 'summed up the historical thought of an entire generation', in three treatises on the new historiography published in 1599. His ideology centred on the rejection of notions of antique authority, golden ages, unique cultural flowerings, and privileged traditions with heroic founders. In place of such myths, he volunteered to write a pre-history of Europe based on observation of American savages. His interest is in the contingent process of how and why societies become *civilisé*. This must be explained 'according to the times, the places, their causes, progress and results.' The concept of civilization transforms the Renaissance 'sense of the past' into a 'sense of history'. The present becomes the most informed and most privileged vantage point. La Popelinière is therefore openly dismissive of antiquity, condemning those 'fonder of antiquity than of reason' who had 'an excessive respect for their predecessors'. Kelley's assessment of this representative of French historiography is not, I think, extravagant: 'Like Bacon he had tried to open up a new road to knowledge.'[14]

The connection with Bacon is an important one. His legal thinking is very different and very English, but Bacon read the French jurists. And while their historiography found little official support in seventeenth-century France, it was developed in England. Some historians have been tempted to suggest methodological similarities between the French-led historical revolution and the Baconian scientific revolution in England. The methodological link is unconvincing but, in this last section, I wish to suggest that aspects of the 'new history' survived long enough to provide not the method but a legitimating strategy for an emerging empirical science in England. Since the ancients were accorded authority in the human sciences in just the same way as in natural philosophy, so the late Renaissance claim that knowledge of nature could best be advanced by replacing classical models with contemporary observations was no different and no easier a claim to advance than were the similar claims made about knowledge of society by the French legal humanists.

We can argue that modern empirical science required, as a condition of its possibility, plausible arguments that the ancients knew only imperfectly. Otherwise, while new, independent and direct investigation of nature could have taken place (as of course it had), it would have had to continue to locate itself within the traditional Renaissance ideologies of ancient wisdom. We have already seen how rarely Renaissance natural philosophers represented their thought as anti-traditional novelty and, in the case of Paracelsianism, how quickly and easily new thinking was re-appropriated by traditional ideology. Advocates of a new empiricism, therefore, had to establish not merely that a new approach would work, but that old approaches would not.

The extremely radical claim that the ancients knew imperfectly not merely human history but also natural history does not seem to have been made explicitly by the legal humanists. For Bodin, in fact, natural history was different from human history precisely because 'in nature there is nothing uncertain.' Both he and Le Roy discuss natural philosophy in very conventional Aristotelian terms, unsurprising perhaps for writers whose expertise lay elsewhere. They did, as we saw, place great emphasis as an argument for progress on developments in the practical arts, discussing at length printing, gunpowder and the compass, and their immense social consequences and benefits. However, these examples were well worn *topoi*, used by traditional humanists to signify the vibrancy of their age. They had no implications for ancient philosophy because Renaissance ideology preserved the classical distinction between *scientia* (philosophical truth, a patrician pursuit) and *ars* (technical skill, with plebeian overtones of manual labour). Renaissance ideology could celebrate the haphazard discoveries of natural facts, which they permitted to be cumulative and unknown to Antiquity, without conceding that the same was true of natural philosophical causes. For the same reason, the increasingly vocal claims of the technically skilled to be making progress in the exploitation of nature by no means proved that traditional philosophy of nature had been superseded. Nevertheless, that conclusion was advanced by the propagandists of empiricism.

The two chief early propagandists of empirical and experimental science, certainly in England, were William Gilbert and Francis Bacon. Their influence, where it was felt, was complementary. On the one hand, Gilbert's book *On the Magnet* is recognized as virtually the first textbook of experimental science, demonstrating the power of experimentalism by example, but it propounded no coherent methodological principles. On the other hand Bacon, in his voluminous publications, charted in detail the inductive method and practical aims of experimental science, without doing any.[15] Despite this great difference, and despite Bacon's criticism of the narrowness of Gilbert's research, the two men shared remarkably similar attitudes to the history of natural philosophy. Both rejected antique authority, and considered erroneous both Aristotle and the Aristotelian natural philosophy they had learned at Cambridge University. They agreed that its particular fault was that it dealt with 'words not things', meaning that it consisted in the logical analysis of concepts which did not, in fact, describe nature at all. They agreed that the fault was best corrected by starting from scratch, relying only upon observational and experimental, not textual, knowledge. Consequently they purported to esteem more highly the artisan's practical experience of nature than abstract philosophy about nature. Their radical iconoclasm aroused considerably more hostility than it did support.

These parallels are well known, but perhaps the most significant parallel has been overlooked. While recent historians have concentrated on the positive accomplishments of Gilbert's experimental practice and Bacon's experimental logic, neither Gilbert nor Bacon expected the virtues of their empiricism to be self-evident. Both, in fact, prefaced and justified it with an historiographical deconstruction of the ideology of tradition, which is indebted to the French legal humanists. They argue for a new experimental foundation only after they have used historicist attacks on all existing foundations. We might say that Gilbert and Bacon practised the critical history of science, for they apply to science the relativistic methods used by the legal humanists to depict legal systems as cultural products. The concept of progressive civilization, of development from ignorance to knowledge, is used to undercut a privileged ancient wisdom. As Bacon observed, 'antiquitas seculae, iuventus mundi' (the time of the ancients was the world's infancy') – the ancients lacked our experience. Having denied a privileged or divine origin to knowledge, their account of its development is that of an unguided, secular evolution. The natural philosophical thought of each age is related to its particular character, interests and personalities. Since both are concerned to present the history of natural philosophy as a history of error, they invoke specific faults of a society to account for specific philosophies. For example, both argue that Aristotle's temperament made him subdue facts to logic, and that his system conquered because he was patronized by Alexander the Great. The ecclesiastical dominance of medieval society produced an erroneous conflation of theology and philosophy. Recent times have been more concerned with style than truth. As Rossi astutely observed for Bacon, this is a primitive sociology of knowledge[16], and that which Rossi has claimed for Bacon, we can claim for the earlier and more vigorous approach of Gilbert, who may even have been a source for Bacon. His interpretation of history is developed most extensively, although still rather scantily, in an unpublished work provocatively entitled *A New Philosophy of the Sublunary World in Opposition to Aristotle*, which Bacon read in manuscript, but a similar strategy is also visible in *On the Magnet*.

Gilbert's problem was that of the extreme radical. Since he wished to claim that truth resided only in his new philosophy, he had to argue that all previous philosophies were groundless, i.e. that no constructible tradition was reliable. He achieved this by combining the French concepts of progressive civilization (to establish that the ancient philosophers were not reliable) and relativism (to account for the ancients' errors). Like Bodin, he began at the beginning with an uncompromising attack on the Golden Age and the Hermetic Tradition. 'There are [he commences] people who think that a precise understanding of all things and sciences, or the advanced arts and sciences, were imparted from on high to the first men in the world. This obvious mistake shows them to be devoid of understanding.' On the contrary, mankind began 'naked, ignorant and nomadic', which is confirmed by that 'mirror of the past' – America. 'From then until now men have gradually gained in wisdom.' Gilbert agreed that theological and scientific thought began with the individuals named in the Hermetic tradition but their 'crude awareness' was only to be expected 'in such ignorance and poverty'. Gilbert approved of the progress in natural knowledge up to the time of Aristotle but, like his patron Alexander the Great, Aristotle and his ancient followers such as Galen enforced conformity to their opinion. Aristotle's theory of the four elements was deemed absurd by the Englishman, who indeed observed that only a parochial Greek could describe air as hot and wet! Gilbert related the survival, and even revival, of this false philosophy since the Middle Ages, to the medieval 'error' of incorporating it into Christian teaching, for this made dissent dangerous. Gilbert also adduced the power of the cult of antiquity, and the successful effort expended by the 'shallow' scholars of the universities in glossing over its difficulties and contradictions, and noted that scholars were required by law to uphold it. Gilbert has no time for proponents of other ancient traditions such as Patrizi, who had 'cobbled together a philosophy from tatty ancient fragments'.[17]

Given his interest in direct knowledge of nature, Gilbert is far more impressed by the progressiveness of the practical arts and technology. The Moderns have now explored the whole surface of the Earth, and sunk mines more deeply into it. Printing, gunpowder and the compass (which is, of course, his chief piece of experimental apparatus in the study of magnetism) all represent the possibility of progress through a new empirical philosophy.

Thus Gilbert's rational historicist plea for a new start in science is, like Bacon's, built around the modern sense of history developed by the French jurists. While we know that Bacon had read some of the new French history, the complete lack of documentation makes any determination of Gilbert's sources impossible. He does however mention Bodin, and some passages are redolent of him. Beyond this, we can only note that Gilbert was at Cambridge University in the 1560s and 70s, a time when, according to Gabriel Harvey, who had an influence on Gilbert, 'You can not steppe into a schollars studye but (ten to one) you shall likely finde open either Bodin de Republica or Le Royes Exposition uppon Aristotles Politiques.'[18]

Neither Bodin nor Le Roy nor any humanist provided the English empiricists with their blueprints for a future science, but they did provide the tools to liberate future science from the dead hand of ancient texts, and to provide naive empiricism with a legitimating philosophy of history. As R. F. Jones argued in 1936, the battle between ancients and moderns, which began as an historical question, was won in seventeenth-century England through the work of experimental scientists who emulated Gilbert and Bacon. We could say that scientists were merely returning the favour done for them by historians, and restoring the mutual dependency of the history of science with the science of history. Bacon's history of error and its use to justify a new start formed the origin of influential Enlightenment histories of science, such as d'Alembert's *Preliminary Discourse* to the *Encyclopédie*. The *philosophes*, however, were able to extend the history with one hundred and fifty years of heroic success, beginning with Lord Chancellor Bacon himself and passing through Descartes and the incomparable Newton. A new practice had been legitimated, a new tradition constructed, a new alliance made between science and history. The surprisingly modern alliance between an empirical science and a historiography of progress has come to seem a natural one and yet, as we have seen, it had its contingent origins and uses in the culture of the Renaissance.

Notes

1 George Huppert, *The Idea of Perfect History: Historical Erudition and Historical Philosophy in Renaissance France*, Chicago, 1970, chs. 9, 10; Donald R. Kelley, *Foundations of Modern Historical Scholarship: Language, Law and History in the French Renaissance*, New York, 1970, part 4. More generally, see Peter Burke, *The Renaissance Sense of the Past*, London, 1969, especially ch. 2.

2 J. G. A. Pocock, *Politics, Language and Time: Essays on Political Thought and History*, London, 1971, pp. 233-72. Quotation from p. 233.

3 For Telesio see Bernardinus Telesius, *De rerum natura iuxta propria principia*, Rome, 1565. See Book 1, Proemium and ch. 1. The work is described in Neil C. Van Deusen, *Telesio – The First of the Moderns*, New York, 1932. For a brief review of seventeenth-century empiricism and the ideas of Bacon, Descartes and Galileo see A. Rupert Hall, *The Revolution in Science 1500-1700*, London, 1983, ch. 7. For Gilbert see note 17 below.

4 For humanist learning see Luce Giard in ch. 2 of this volume; Anthony Grafton and Lisa Jardine, *From Humanism to the Humanities*, London, 1986, esp. ch. 3. For a brief view of humanist historiography see D. R. Kelley in *The Cambridge History of Renaissance Philosophy*, edited by Charles B. Schmitt and Quentin Skinner, Cambridge, 1989, ch. 21.

5 J. G. A. Pocock, *The Ancient Constitution and the Feudal Law*, 1st ed. Cambridge, 1957, p. 9.

6 Hall, *op. cit.* (n. 3), ch. 2, compares the conservative features of Copernicus and Vesalius. Quotation from p. 49. For the influence of new printed sources see Elizabeth L. Eisenstein, *The Printing Press as an Agent of Change*, 2 vols., Cambridge,1979, vol. 2, chs. 6, 7. For anatomy see A. Wear, R. K. French and I. M. Lonie, *The Medical Renaissance of the Sixteenth Century*, Cambridge, 1985.

7 For an introduction to these alternative traditions see Ingegno and Copenhaver in Schmitt and Skinner, *op. cit.* (n. 4), chs. 9, 10.

8 See D. P. Walker, *The Ancient Theology. Studies in Christian Platonism from*

the Fifteenth to the Eighteenth Centuries, London, 1972; Frances A. Yates, *Giordano Bruno and the Hermetic Tradition*, London, 1964; on Bostocke and Hermetic Paracelsianism see Allen G. Debus, *The English Paracelsians*, 2 vols, New York, 1966 and Charles Webster, *From Paracelsus to Newton*, Cambridge, 1982.

9   See Huppert, *op. cit.* (n. 1), ch. 4; Kelley, *op. cit.* (n. 1), ch. 11; S. Kinser, 'Ideas of temporal change and cultural progress in France 1470-1535', in *Renaissance Studies in Honor of Hans Baron*, edited by Anthony Molho and John A. Tedeschi, Florence, 1971, pp. 703-54; Ernst Breisach, *Historiography: Ancient, Medieval, Modern*, Chicago, 1983, ch. 11.

10 Kelley, *op. cit.* (n. 1). Daneau's *Physica Christiana* was translated as Lambertus Danaeus, *The Wonderfull Workmanship of the World, wherein is contained an excellent discourse of Christian naturall Philosophie*, tr. T. Twynne, London, 1578.

11 Huppert and Kelley, *op. cit.* (n. 1). See also the more extensive references to French legal humanism in the bibliography to this essay.

12 Jean Bodin, *Method for the Easy Comprehension of History*, tr. B. Reynolds; New York, 1945, ch. 7, pp. 291ff.

13 Loys le Roy called Regius [Louis le Roy], *Of the Interchangeable Course, or variety of things in the whole world...*, tr. R. Ashley; London, 1594, Book 10, p. 110ᵛ. See also Huppert, *op. cit.* (n. 1), p. 109.

14 D. R. Kelley, 'History as a calling: the case of La Popelinière', in Molho and Tedeschi, *op. cit.* (n. 9), pp. 773-92. Quotation from p. 786.

15 Guilielmus Gilbertus [William Gilbert], *De magnete magneticisque corporibus...*, London, 1600. A readily available translation is William Gilbert, *De Magnete*, tr. P. Fleury Mottelay, 1893; reprint ed., New York, 1958. Bacon's view is clear in many works, most readily in *The Advancement of Learning*, London, 1605.

16 Paolo Rossi, *Francis Bacon: From Magic to Science*, London, 1968. The Italian edition was published in 1957.

17 Guilielmus Gilbertus [William Gilbert], *De Mundo nostro Sublunari Philosophia Nova...*, Amsterdam, 1651; reprint ed., Amsterdam, 1965. The work was printed from partially edited manuscripts. There is no published translation, although a discussion of its contents, including the question of Bacon's readership is Sister Susanne Kelly, *The De Mundo of William Gilbert*, Amsterdam, 1965. It does not discuss Gilbert's historiography. The translation is mine, mainly of passages in Book I, chs. 1, 2. Similar statements can be found in Gilbert's *De Magnete*, *cit.* (n. 15), particularly in Book I.

18 Gabriel Harvey, *The Letter Book of Gabriel Harvey A.D. 1573-1580*, edited by E. J. L. Scott, London, 1884, p. 79.

4

Rhetoric and science/rhetoric of science/ rhetoric as science

Maurice Slawinski

To the modern way of thinking science (the disinterested pursuit of knowledge of the physical world) and rhetoric (self-seeking hot air) are diametrically opposite activities. Nor is this a particularly recent view. In 1663 Robert Hooke described the newly founded Royal Society's aims as 'to improve the knowledge of naturall things, and all useful Arts, Manufactures, Mechanick practices, Engines and Inventions by Experiment – (not meddling with Divinity, Metaphysics, Moralls, Politicks, Grammar, Rhetoric or Logick)'.[1] Save for the absence of any distinction between 'pure' and 'applied' sciences, and the heterogeneous disciplines grouped with rhetoric, to a modern reader the description would seem unexceptionable. Yet if Hooke felt the need to spell out these exclusions (and stress the method which distinguished the new philosophy) it was in part at least because what did or did not constitute a 'science' was by no means as 'obvious' then as it appears to us now. Within the Aristotelian scheme which dominated learning, *all* the disciplines in question could lay claim to scientific status, where a 'science' was any systematic body of knowledge dealing with a discrete field of inquiry and resting upon logic or authority (or a combination of both).

All the disciplines listed by Hooke could claim to be sciences in this sense of the word, with one exception, since there was considerable debate as to the status of rhetoric. Yet rhetoric, the study and application of the techniques of persuasion, was the cornerstone of education up to, and even beyond, the Renaissance. For all the vehement denunciations, it could not help but play a part in the way natural philosophers set about describing and explaining the physical world, at a time when the divisions of subject-matter and method which were to characterize the 'scientific revolution' had yet to evolve.

71

It did so in a number of ways. First, in the Renaissance rhetoric held pride of place in attempts to revitalize learning, transform traditional ways of thinking, provide a key to the understanding of Creation, and claim for intellectuals a leading role in society. In so doing, it played a part in setting the agenda for the work of natural philosophers generally. Secondly, in so far as rhetoric sought to persuade, it became a vital tool in the battle to gain acceptance for their views. But there is a third more intimate link between natural philosophy and rhetoric. Both sought to analyse facts, draw inferences and marshal evidence. Indeed, like all educated people in the Renaissance, natural philosophers received their first grounding in these skills in the rhetoric course which, together with grammar (i.e. Latin) and logic, formed the core of school and university curricula. It can be no surprise therefore if, as I shall argue in the final part of this essay, rhetorical modes of thought contributed to shaping the very logic of scientific enquiry.

1   The 'arts of discourse'

To understand what rhetoric meant to the Renaissance, we must rapidly sketch its development. Its birth as an intellectual discipline is closely associated with the political institutions of classical Greece, and Athenian democracy in particular. Policy was made by assemblies of all the citizens, where the best orator generally carried the day. Teachers soon appeared who claimed to be able to pass on 'wisdom' and 'virtue' in the shape of speaking skills. Hence their name, sophists, from *sophia* meaning wisdom.

The scope for demagoguery was considerable, and in works such as *Protagoras* Plato shows us his teacher Socrates confounding the sophists in debate, and demonstrating the hollowness of their claims. But this neither removed the need for oratory, nor greatly affected the popularity of those who taught it. It was left to Plato's pupil Aristotle to provide a systematic alternative to the sophists, in the conviction that the devil should not have all the best tunes. Give the honest man access to the same 'tricks' as the demagogue, Aristotle implies, and truth will prevail. Particularly if rhetoric could be de-mystifed, ordered into a logical method founded on clear principles. His meditations on the subject were set down in two related treatises: the *Topics*, dealing with dialectics, the art of winning an argument ('topics' were the headings under which arguments could be classified), and the

Rhetoric.[2]

According to Aristotle neither rhetoric nor dialectics is a science (in the sense already encountered). They are enabling skills, what he calls *technai* and the Renaissance was to term the 'arts of discourse'. They deal with no specific field of human activity, but may be applied to any and every subject. They are concerned with practical results rather than first principles or final causes, so that their methods are more empirical, less rigorous, than those of the sciences. Above all, the discourses they produce are addressed to the lay person rather than the specialist, aiming at approximation, sound opinion based on commonly held belief, what is generally true or held to be so, rather than the absolute truth of science.

This is reflected in the kind of demonstration Aristotle deemed adequate for the purpose of public debate. Where science is founded on formally correct deductions (syllogisms) drawing necessary inferences from first or final causes, or on systematic induction (arguing general principles from many particular instances), the rhetorician is to adopt the less stringent proofs of dialectics. His syllogisms will take the simplified form of enthymemes, where one of the stages of deduction is omitted as self-evident (the syllogism 'all men must die, Socrates is a man, therefore Socrates must die' might be re-stated enthymemically as 'all men must die: Socrates too will meet this fate'). Similarly inductive arguments will be reduced to straight-forward examples, where as often as not the audience will be left to draw its own conclusion ('So you want to fly? We all know what happened to Icarus'). Concise arguments of this kind, though less precise and more likely to hide faulty or partial reasoning, are better suited to a popular audience lacking the knowledge or patience to follow complex philosophical proofs. Furthermore brevity, and the fact that the work of interpretation is left to the listener, make them particularly convincing: though guided by the speaker, the listener feels the conclusion is his own.[3]

This does not mean that rhetoric (and dialectics) are forms of studied deception. For Aristotle the most perfect form of truth is that derived from the stringently logical processes of deduction expounded in his *Prior Analytics.* Conversely, in the *Topics* he allows the occasional use of arguments where concision actually serves to conceal faulty or imprecise reasoning. But in distinguishing between the procedures of formal logic and of the 'arts of discourse' he was more concerned to distinguish between arguments which are 'correct' (however complicated, or seemingly absurd) and those which are 'good' (in the sense of effective). That they are good does not exclude that their premises can be validated, that their reasoning is capable of being reformulated syllogistically, or that what is 'proved' by rhetorical sleight of hand is demonstrable 'scientifically' by some other route.[4]

Aristotle was not distinguishing between truth and falsehood, but between different levels of intellectual practice. Dialectic and rhetoric ensured that convincing arguments were put forward where they mattered, in the business of the community. Logic underpinned them by ensuring that they could be validated (which is where philosophers came in). In this sense Renaissance rhetoric, which emphasized the relationship with dialectics while making no clear distinction between this and formal logic, may have come closer to the spirit of Aristotle than is generally supposed. But though rhetoric looks to dialectic for its arguments, and indirectly to formal logic to establish their validity, Aristotle reminds us that persuasion also involves an understanding of human behaviour, both in relation to the subject under debate, and the particular audience addressed, so that it requires a grasp of ethics, psychology and politics.

Aristotle then goes on to distinguish between three types of oratory: *deliberative*, aiming to sway the audience to a particular action by considering its likely effects (as in public assemblies); *forensic*, concerned with whether something has or has not been done, who was responsible for it and if responsibility signifies guilt (as in the lawcourt); *epideictic*, which apportions praise or blame (as in the Greek tradition of public speeches in honour of great men). Between them they encompass every aspect of life, with all its causes and effects, from analysing past events to predicting future ones, from determining what is of use or harm to the community, to offering models of virtue and wickedness.

For the possible modes of arguing the orator may employ in support of his case Aristotle refers the student to the *Topics*. The *Rhetoric* focuses instead on analysing the attitudes, beliefs and expectations on which the orator must found his persuasion. The result is a veritable digest of the opinions (*doxa*) of Greek (particularly Athenian) society. This might appear to confirm the *Rhetoric*'s separateness from the main body of Aristotelian science, but once again the distinction is not as clear-cut as the author suggests. Many *doxa* also appear in 'scientific' works such as the *Ethics* and *Politics*, where Aristotle presents them not as 'common opinion' but as irrefutable axioms (e.g.

the statement 'the end of life is happiness'). The same ambiguity surrounds the basic tenets of Aristotle's natural philosophy. For example, in his discussion whether the earth is at rest at the centre of the universe (in the *De Caelo*), both the arguments against, which he disproves, and his own arguments in favour of the thesis are based on *doxa* owing more to the social practices of Greek society than to physics.[5] This absence of clear boundaries between the *doxa* of rhetoric, and the axioms of moral and natural philosophy will contribute to the conflation of rhetoric, dialectic and logic, and the emergence in the Renaissance of alternative models of rhetoric as a meta-science.

Only the third and final book of Aristotle's treatise deals with what we normally think of as rhetoric: the use of striking verbal embellishments and the techniques of delivery. Of the latter Aristotle has little to say: in so far as he is concerned chiefly with the content and ethics of public speaking it is a marginal issue. There is nothing marginal about style, however. Whereas later rhetoricians tended to see it as mere 'dressing up' of a pre-constituted content, and drew up exhaustive inventories of the figures to be employed, Aristotle's observations suggest that for him style and content are intimately bound up. Thus he states that, as well as commanding attention, vivid description and story-telling add credibility, persuading the audience that they have learned something new and worthwhile. He suggests that simile and metaphor are particularly effective sub-varieties of inductive argument. He implicitly identifies *doxa* with maxims, proverbs and para-logical verbal patterns such as antithesis (the arrangement of successive phrases so that opposite meanings are balanced out by syntactic parallels, as in 'the grievous day, the joyous night') and chiasmus (reversal of word order in successive phrases, as in 'a dark crime, a criminal darkness').[6]

A further clue to the key importance he attributes to style is that Aristotle only goes on to consider the overall structure of a speech and its arguments at the end of the treatise, after establishing these basic points. That this is a 'natural' order will be clear to all who have planned an essay: we may begin with the 'points' to be made, then slip in appropriate quotations and examples, think up memorable phrases to underline key points, and add stylistic embellishments, but the process is seldom as linear as this suggests. A similitude or comparison will spring to mind which opens a new line of argument; a turn of phrase may suggest itself with such force that we alter the original plan to give it greater prominence; we may introduce a narrative excursus to introduce a particular point only to discover, once the story is told, that the point is self-evident, and needs no explanation.

What seems uppermost in Aristotle's mind is not to lay down a series of rigid rules, but a set of guidelines to help find the most appropriate arguments, expressions and order of exposition, for each different subject and audience. This notion of an Aristotelian 'memory' (in the sense of structured thought capable of combining 'things', 'words' and 'ideas' in memorable and persuasive ways) will be of great importance in the Renaissance, when it will re-emerge out of the training of memory in the narrow sense of 'remembering one's speech' as a key element in rhetoric's claim to scientific status.[7]

The immediate concern of the numerous rhetoricians who sought to improve on Aristotle, however, was more mundane. Whether by conflating his teaching with that of other schools, further sub-classifying his division of the subject, or offering alternative divisions of their own, their aim was to produce comprehensive rules for any student to apply. Aristotle's treatise set the terms of the debate, but what had originally been flexible and open-ended was formalized into 'parts', and procedures were set down which involved going through these in a pre-determined way. Within each part, ever-longer lists of options and sub-categories (different modes of arguing, gaining attention and credibility; any number of stylistic ornaments and tricks of delivery) were drawn up. Each teacher had his own scheme, but a broad consensus emerged, concerning the main divisions of the orator's skills, whose roots were in Aristotle, but which almost wholly lost sight of his original aims. It was in this formalized, philosophically disengaged version that rhetoric was passed on from Classical Antiquity to the Middle Ages and the Renaissance.

Particularly influential in this process were two minor works of Cicero, *Topica* and *De Inventione*, dealing with the elaboration of arguments, and the more comprehensive *Rhetorica ad Herennium*, written around 70 bc and long attributed also to Cicero. The respect accorded to its presumed author, the accident of history which made it the only complete classical rhetoric available to the Middle Ages, and the appeal which its schematic nature held for the scholastic frame of mind conspired to make it the 'key' by which all other theorists of rhetoric (not excluding Aristotle himself) were read as they were rediscovered in the course of the Renaissance. *Ad Herennium* divided oratory into five distinct, sequential parts (already mentioned by Aristotle, though with nothing like the same rigid emphasis): *inventio* (working out arguments to support one's thesis); *dispositio* (organizing the arguments into an effective whole); *elocutio* (embellishments of style serving to capture and retain an audience's attention); *memoria* (the methods by which an orator committed his speeches to memory) and *actio* (the manner of delivery). Henceforth the distinction would become canonical.

Ad Herennium's treatment of *inventio* (the greater part of Books I, II and III) is very closely tied to Roman political, and above all legal, practices, and had limited direct relevance for medieval or Renaissance readers. But it did show how the search for arguments and their organization (*dispositio*) might be turned into a quasi-mechanical procedure, complete with a check-list of topics. In the case of deliberative rhetoric, for example, *Ad Herennium* tells us that its purpose is to demonstrate 'advantage' on the grounds of either 'security' or 'honour'. Security is then subdivided into 'might' and 'strategy' and for each the aspiring orator is invited to check off its various parts (armies, fleets, weapons, engines of war, man-power, in the case of 'might') for pertinent debating points. The appropriate place in the speech for arguments relating to each heading is then indicated, and moving on to *actio* we even have an attempt to reduce vocal delivery and accompanying physical gestures to a series of alternatives appropriate to all possible subjects and audiences.

But with this the treatise's logical- schematic compulsion exhausts itself. The final part of Book III dealing with *memoria* gives a rapid and sketchy summary of a mnemonic system based on visual recall (to which we shall have to return). Book IV is dedicated to *elocutio*, an exposition of rhetorical figures where method gives way to inventory. It was precisely this which made it the most popular and influential part of the treatise, a repertory of devices by which the orator's didactic pill could be sweetened and, no less important, a source of ornament that would distinguish its practitioners from the illiterate rabble.

2   Rhetoric as a science

The systematic approach applied to *inventio* and *dispositio* in the first three books of *Ad Herennium* and the inventory of the fourth reinforced certain assumptions of scholastic culture, where the shift from an essentially oral conception of discourse to a written one tended to erase the distinction between logic and dialectic. Aristotle's *Topics* was grouped together with his writings on formal logic in what became known as the *Organon* or 'instrument', and read as a treatise on applied logic (as were the Ciceronian *Topica* and *De Inventione* and the equivalent parts of *Ad Herennium*). Rhetoric, for which medieval culture had few formal occasions, was whittled down to *elocutio* and restricted to such marginal areas as the art of elegant letter-writing (*dictamen*), or identified with poetic style. *Ad Herennium*'s analysis and exemplification of figures made it a perfect textbook for the purpose. Conversely, Aristotle's *Rhetoric*, even where it was known, seemed to excite little interest: it would only be 'rediscovered' when a need emerged for its 'message'.[8]

The Renaissance accepted medieval distinctions between logic and dialectic on the one hand, and rhetoric on the other, but reversed their relative importance. Humanists were concerned not with abstract speculation but with practical social and political issues. They attributed value to logic in so far as this could be made useful through rhetoric, helping to construct persuasive arguments concerning the regulation of human affairs. Where the Middle Ages identified dialectic with logic, and saw rhetoric as subservient to both, a means of 'sugaring the pill' of learning or morality, Humanists viewed formal logic as a specialized part of dialectic, which was equated with *inventio* and absorbed into the greater art of rhetoric. This view was massively endorsed by the re-discovery of Cicero's major works, and the rejection of *Ad Herennium* from the Ciceronian *corpus* on philological grounds.[9]

The rediscovered Ciceronian works were not student handbooks, but theoretical treatises cast in the urbane mould of informal discussions between friends. They presented rhetoric not as a series of recipes for persuasion and embellishment, but as a practical, socially oriented philosophy bridging the gap between right thought and right action through the medium of *bene dicere* (meaning both attractive *and* correct speech). Cicero's orator was no mere technician trained to handle dialectical topics and rhetorical figures, but the ideal philosopher-politician, learned in all the sciences, on which he would draw to promote the good and refute the bad. Humanist scholars, eager to break with scholasticism, saw this as confirmation that rhetoric was no mere instrumental skill, but the science of sciences.

Unfortunately, this view of rhetoric, born out of Cicero's practical experience as lawyer and politician and inspired by the wish to defend the ideals of Roman republicanism, gained currency in the context of growing political absolutism. The problem was not, as in Cicero's time, to define the militant intellectual and train him for political leadership, but to see what constructive subordinate role intellectuals might occupy in relation to an all-powerful prince. A partial answer would came from another Roman rhetoric made newly available by humanist research, Quintilian's *Institutio Oratoria* ('The Teaching of Rhetoric'), written after Cicero's Republic had been replaced by the Empire with the aim of recasting Cicero's ideal in terms compatible with the new regime. The philosopher-politician who played a direct, active part in the running of the state gave way to the philosopher-teacher who exercised influence by first training the young prince, then acting in the background capacity of adviser, and whose prestige and authority were no longer based on the Ciceronian claim to govern through eloquence, but on the ability to endow others (the prince, members of the ruling class) with the power that *bene dicere* conferred.

Quintilian's pedagogical views were absorbed into educational practice only gradually, though they would leave a lasting monument in the Jesuit Colleges. Long before this, however, his views brought about a change of emphasis in the pronouncements of Renaissance theorists of rhetoric, such as the Paduan professor of humanities Sperone Speroni, who shifted attention away from forensic and deliberative rhetoric, proclaiming the greater nobility of the epideictic variety.[10] From an oratory intended directly to affect decision-making we move to one whose function was apparently to heap praise on public figures, but which actually sought to preserve as much as possible of the orator's influence (while acknowledging that in the new order he could no longer aspire to an autonomous political role) by making him the judge of correct behaviour and values.

A related change affects that favoured Renaissance didactic-rhetorical form, the dialogue. Where in the early years of the century it had clearly been modelled on Cicero, and portrayed an exchange of views among intellectual and social equals, this was gradually replaced by a demonstration. The exterior trappings of conversation were preserved, but now portrayed an intellectual 'professional' drawing on all the skills of epideictic rhetoric to criticize the conventional values upheld by his interlocutor (usually a noble amateur) and presenting in their place his own version of the 'right' and 'good' (which the interlocutor will end up accepting as superior). Alongside this the

79

example of the *institutio oratoria* begot a whole new genre, the educational guide in which a 'professional' presented a complete educational programme for the mastery of gentlemanly 'sciences' as disparate as the management of a country estate, hunting and society games.[11]

The emphasis on the orator-humanist's didactic and advisory role reflected a real state of affairs, in so far as princes required a growing number of humanist-trained lawyers and administrators to staff their expanding bureaucracies. But few of these had serious pretensions to educate and 'lead' their sovereign, preferring the (safer) role of self-effacing minion. By the second half of the sixteenth century virtually no leading intellectual figures actually held important offices of state, though as late as 1629 Tommaso Campanella could briefly dream of guiding the policies of Pope Urban VIII. The contrast with earlier humanist scholars, who had combined intellectual originality with substantial political influence, and whose learning had frequently been placed directly at the service of the state, is a telling one.[12] Significantly the one conspicuous late Renaissance success in this field, Francis Bacon, neither owed his high office to his intellectual prominence, nor sought to use it systematically to advance the 'great instauration' of learning (and after he had fallen from favour, the dedication of his natural-philosophical projects to the Crown signally failed to restore him to his lost position).

Yet the attractions of Quintilian's model were great, all the greater because its practicability seemed confirmed by a humanist golden age only one or two generations past: his educational project both heightened, and appeared a solution to, late Renaissance intellectuals' sense of social and political displacement. Natural philosophers were no exception, and their intellectual programmes, however different in assumptions and method, generally placed great stress on the practical applications of scientific knowledge, its usefulness to the state, and the need for a community of learned men who would educate society (or at any rate the ruling classes) in the new philosophy. All had their roots in the humanist rediscovery and defence of oratory, and the practice of a rhetorical education.

Even an idiosyncratic figure like the Neapolitan Giovan Battista Della Porta, somewhat removed from the academic and humanist mainstream and best known among historians of science for his practical work on optics and magnetism (though he also delved into 'sciences' such as physiognomy and cryptography) was not unaffected by this tradition. His 'natural magic' or applied science of the hidden causes of nature was precariously balanced between claims to occult knowledge, the lore of medieval and classical 'natural history', alchemical theories of sympathy and antipathy, a tradition of folk remedies and craft skills and a rudimentary experimental method. When towards the end of his life he sought to organize his disparate and often contradictory interests into a single whole (the never-completed *Taumaturgia*) the chosen form was an 'institutio', a handbook which aspired to teach 'princes' the methods of natural magic, to their greater glory and the prosperity of the state.[13] Conversely, as B. S. Eastwood has shown, in the *Dioptrics,* a handbook written to introduce craftsmen to the principles of optics with a view to improving lens-making techniques, René Descartes relied on a mode of exposition by analogy which is also a transformation, for 'popular' use this time, of the *institutio oratoria*.[14]

But the clearest case of a rhetorically inspired scientific programme is that of Bacon, educated in the law, the discipline most dependent on rhetorical modes of determining causes. Like Quintilian's 'institutio', Bacon's 'great instauration' entailed a reinterpretation of the role of learned men, involving social as much as intellectual reform. The corporation of natural philosophers proposed in *New Atlantis* owed not a little to the example of religious orders, to inns of court and university colleges, but Bacon's insistence on its usefulness to the state and its preoccupation with the *application* of science are very closely akin to the rhetorical tradition; as are its members' role as interpreters of truth and their systematic accumulation and classification of evidence concerning the 'causes, and secret motions of things, and the enlarging of the bounds of Human Empire, to the effecting of all things possible'.[15] The same is true of Campanella's *City of the Sun*, a natural-philosophical utopia whose astrological guiding principles are further removed from modern science, but which like *New Atlantis* was to be guided by an elite who understood and controlled the secrets of nature.[16]

Such hopes were not wholly confined to fictional societies. On a smaller but more practical scale, any number of philosophic and scientific academies were inaugurated, which simultaneously sought to provide an alternative education, to stimulate contacts between intellectuals and prominent politicians, and to attempt to realize something of the utopists' communities of the learned. It was the Jesuit order however which developed most fully the ideals of the *institutio oratoria*. Their pedagogic method, set down in the definitive *Ratio*

81

D

Studiorum of 1599, was the direct product of the principles and aims set down by Quintilian and re-interpreted by the humanists. With its application the Jesuit Colleges moved beyond their early concern with training the order's future cadres. Henceforth they would also educate the ruling class. And for reasons akin to Bacon's they quickly grafted on to the original rhetorical programme a practically inclined mathematical and natural-philosophic education. In the name of Catholic *imperium* a new natural philosophy was to be developed and taught mindful of divine law and theological precepts, wary of dangerous speculations concerning first causes and the order of the cosmos, but able to draw useful lessons from astronomy, hydraulics, mechanics, to the greater glory of God and the Order. And what better way of promoting both than to hold forth to princes the promise of technological innovation?[17]

Even when natural philosophers had no such comprehensive ambitions, however, rhetoric, as an 'applied' philosophy, provided the clearest and most fully articulated model for their enterprise. The modern emphasis on the 'disinterestedness' of science has tended to obscure the fact that for most of its early practitioners and advocates the new natural philosophy was defined as much by its supposed ability to improve material life, as its aspirations to an abstract 'truth'. Hence the absence in Hooke's description of the aims of the Royal Society, of any attempt to determine a hierarchy of 'pure' research and 'useful Arts, Manufactures, Mechanik practices, Engines and Inventions'. This was no doubt in part a tactical consideration. Galileo, for whom the practical application of his work was not the primary concern, was nevertheless acutely aware of how patrons might be attracted by holding out the prospect of such applications. But the fact remains that the new goals of natural philosophy owed much to material changes in Renaissance society (maritime exploration, new farming methods, increased commercial and manufacturing competition) and the attempt to draw on, explain and perfect the empirically derived techniques of artisans, traders and 'wise-men' was an integral part of its efforts.

Yet this ran counter to the most fundamental ideological tenets of elite culture in the *ancien régime*, which tended to attribute greater value to one thing or activity over another, as one theorist of rhetoric so tellingly observed, 'precisely because it is not necessary [for human existence]'.[18] Abstraction, luxury, remoteness from anything related to the mundane needs of the 'vulgar' majority, were the hallmarks of 'virtue'. Rhetoric provided the one example of a discipline which could at the same time be practical, applicable to any specific objective that might benefit the state, and claim an exalted place among the sciences. It was hardly surprising if, consciously or unconsciously, natural philosphers looked to it to provide a model and a point of reference.

3   The key to all knowledge

Most apologists of 'scientific' rhetoric deemed it sufficient to stress the breadth and depth of the orator's culture, the vast *scientia* on which persuasion was based, which made rhetoric the truest and most complete of all disciplines. For a significant minority however it became something more: the key to a radical epistemology which as well as including the social sciences 'rediscovered' by the humanists (moral philosophy, politics, history, linguistics) encompassed the study of the natural world. Platonist endeavours to achieve a synthesis of all philosophical and theological systems, Aristotle's concern with the uses of figurative language, and the codification of rhetoric typical of the *Ad Herennium* tradition were to be combined in the effort to describe and explain the whole 'Book of Creation'.

A detailed account of Renaissance rhetorical-encyclopaedist enterprises and their part in the wider phenomenon of Hermeticism (as the esoteric and magical strands of Renaissance Platonism have collectively come to be known) is beyond the scope of this essay. It is likely that both Hermeticism's consistency and the extent of its influence on the scientific revolution have been exaggerated.[19] There are, however, significant affinities of method and assumption between those who looked to rhetoric to classify, explain and exploit Creation, and exponents of the new natural philosophy. They can best be exemplified by outlining what was at one and the same time the most extreme and the most rationalistic of these rhetorical projects.

Giulio Camillo (*c.*1480-1544) was a 'devout' Ciceronian, whom Erasmus mocked for insisting on slavishly imitating his idol's style and vocabulary. He nevertheless achieved a notable reputation as a rhetorician, persuading a number of princes to finance his research on what is best described as a 'rhetorical computer' which would enable its user to discourse on any subject in Creation with the truth of perfect persuasion. Opinions were divided as to whether he was a genius or a charlatan, but his influence was widespread. The most respected of late Renaissance Platonists, Francesco Patrizi, edited the second volume of his posthumous works. His theories were taken up by Bruno and Campanella in Italy, John Dee and Robert Fludd in England, and had at least an indirect effect on the development of logic from Ramus to Leibniz.[20]

What Camillo proposed was a mnemonic system based on the visual memory, the principle that images and spatial relations act more powerfully on the mind than words or numbers. The technique had supposedly been widely taught and practiced in antiquity, though it came down to the Middle Ages and Renaissance only in brief, tantalizing outlines, best-known of which was the chapter of *Ad Herennium* dedicated to *memoria*. As with rhetoric itself there were many variations, but in essence all consisted of fixing in the mind the topography of an architectural space, real or imaginary (a room, a palace, even an entire city, depending on the amount of information to be memorized). Ideas, words and whole sentences were 'stored' by mentally locating within this 'memory place', in the order in which they were to be recalled, a series of images (mnemonic symbols) capable of jogging the memory. For example, someone wishing to commit to memory Euclidean geometry, beginning with Pythagoras' theorem, might put nearest the door of his memory place a map of Sicily (*Trinacria*, the 'three-angled' island) set in a square frame.

There were numerous Renaissance attempts to reconstruct this lost art, whose greatest practitioners had supposedly been able to memorize books word for word, at a single reading. Camillo went one better: by devising not any arbitrary memory place, but one structured to reflect the essential order of the universe, he promised something much more than simple remembrance. How exactly this could be done, and what kind of 'knowledge' it would produce remain obscure, but clearly Camillo sought to combine and reconcile the contemporary passion for encyclopaedic compendia, and what survived of classical mnemonic systems with the view of memory as structured analytical thought implicit in the Aristotelian 'arts of discourse'.

Camillo resolved the long-standing confusion between logic and dialectic by removing the last vestiges of a distinction between them. Reading the *Organon* through the distorting lens of Ciceronian rhetoric he believed it pointed the way to a simple yet rigorous classification of all predicables, everything that could be said of the nature and attributes of any object or phenomenon. In particular, Aristotle's *Categories* showed how 'all that is in heaven, earth and the abyss might be reduced to just ten principles'.[21] Accordingly Camillo constructed a 'universal' memory place (where earlier practitioners had made up their own, wholly arbitrary ones) whose structure would correspond to the fundamental categories of predication.

The combinatory logic of his system was justified by appealing to Aristotle, but its hermeneutics were Hermeticist. Aristotle had shown the way, but to produce 'true' knowledge his 'ten predicables' were to be replaced by a seven-fold-seven division based on Neo-Platonism and the *Cabbala*. Camillo's memory-place was to be an amphitheatre of seven stepped tiers, split by six gangways into seven sectors, producing an architectural space which was effectively a grid of 49 boxes. The seven tiers, named after classical myths, corresponded to different levels of being, from the basest to the divine. The seven sectors, named after the planets, stood for the principles of creation and divine wisdom, as identified with the seven stones of the Sephiroth of the *Cabbala*.[22]

The Theatre was an analogue of Creation itself, where the sectors represented the fundamental Ideas in the mind of God and the tiers stood for the hierarchy of forms which being took. Its layout did not simply follow logical categories, it reflected the actual structure of the universe. Within it all objects, phenomena and concepts could be located in their proper place. For instance, in the sector of the Moon, on the tier represented by the myth of Pasiphae and the Bull, were to be recorded all things connected with the principle of moisture, that is corruption and generation. In the box would be 'filed' material related to such topics as the descent of the soul into the body, man's physical mutability, cleanness and uncleanness, each under its own sub-symbol, again derived from Greek mythology. Camillo further suggested that each box was to be equipped with a number of drawers, in which to store all the maxims, figures and tropes to be found on each argument in the classical authors (chief among them Cicero), making the Theatre a combination of topical system and encyclopaedia of 'elegant extracts'.

Arguments could be worked out, ordered, embellished with suitable maxims, figures and tropes, and finally committed to memory, all by using this single device of the Theatre. And because of its structure, Camillo argued, the discourses it produced would be perfect representations of the true order of things. The symbols marking each point on the grid were mnemonic devices capable of eliciting immediate recognition (for example the Augean Stables signified uncleanness), but they were also figurative expressions of the mysteries of Creation, as defined by the Platonists, and examples of analogies which according to Aristotle produced both pleasure and understanding in the minds of an audience.

The machine to remember had become a machine to think. How it was to work is not clear (all that remains is two brief essays and a diagram, designed more to whet the appetite of potential patrons than to explain) nor does Camillo say whether the Theatre was to be a physical place which the orator entered or a mental one to be committed to memory, or a combination of both.[23] But the basic idea is clear: the Theatre is a filing cabinet or database, but like a mathematical table it can be read across, up and down to show how its various elements relate. The user can not only find in each box appropriate arguments; he can search along the Theatre's axes for links with every other field of knowledge. And at every point he has at his disposal symbols and figurative expressions which are both striking and rapid concentrates of complex ideas (as suggested by Aristotle) and images charged with the ultimate power of Platonic revelation.

In this respect Camillo's Theatre is in the camp of occult and magical practices. What is significant from our point of view, however, is not its esoteric foundations, but the rationalistic edifice built upon them. It may issue out of an encyclopaedic compulsion to accumulate and classify, its single parts may be drawn from traditional sources, but it also constitutes an attempt to re-organize such encyclopaedias in accordance with a new logic, more productive of meaning.

Two principles presided over encyclopaedic compilations. On the one hand, everything in nature – 'God's book' – had meaning. On the other, that meaning resided not in the single object or phenomenon, but in a chain of associations, similarities and contraries which bespoke pattern, order and harmony. The work of the encyclopaedist consisted in piecing together a part at least of that chain, compiling inventories where each thing was linked with what preceded and what followed. That the criteria by which links were established shifted constantly mattered less than the overall demonstration of unity. Everything had a meaning, but that 'meaning' was an undifferentiated point in a continuous process which had God for its beginning and end. On one level the two principles were mutually supportive, on another they cancelled each other out: the very difference which supposedly made the inventory possible, and significant, was erased.

In the search for a more rigorous topical system Camillo redefines what had previously been a series of arbitrary points on a single undifferentiated continuum, as a network where meaning resides not in things but in the relations between them.[24] Implied in this is an 'epistemological shift' from a view of nature where everything is inherently meaningful, but that meaning is ultimately not of itself but of God (so that, paradoxically, nothing possesses a meaning specific to itself) to its representation as systematic interaction (where by an equal but opposite paradox the denial of inherent meaning leads to the recuperation of each object's specificity). For Camillo these relations remain essentially linguistic, rhetorical in the technical sense of pertaining to the descriptive and persuasive powers of language. The break-through for natural philosophy, its 'epistemological leap' into 'modern science', will come with their experimental and quantitative representation. But the earlier shift contains within it the conceptual seed of that representation, with all its novelty compared to the *mathesis universalis*, the expression of the sovereign principles of order and unity of Creation, pursued by the Platonists.

This does not, of course, make Camillo the 'father' of modern science. His Theatre is only one strand of a tradition which ranged from the quasi mystical 'logic' of friar Ramon Lull in the late fourteenth century to Bishop John Wilkins' attempt to construct a 'universal language' two hundred years later; from the occultist millenerianism of Giordano Bruno, whose *Shadows of the Ideas* supposedly constituted a mnemonic system capable of bringing about a magical reform of society, to the more mundane *Art of Memory* of his contemporary, the Franciscan preacher Francesco Panigarola, simply concerned to teach clerics how to memorize their homilies. Not infrequently we find echoes of mnemonics in other fields. That the walls of Campanella's astrologically governed 'City of the Sun' should be a vast mnemonic device whereon its youth may learn all sciences is not surprising, but the various Chambers and Galleries of Bacon's House of Salomon are also remeniscent of memory places. More precise connections between the art of memory and the new natural philosophy have also been argued, most convincingly in the work of the 'pure' mathematician, Leibniz.[25] Rather than insist on a direct link, however, it seems safer and less contentious to argue that mnemonic logic contributed to that climate of opinion, new ways of thinking old problems, in which the 'new science' developed, and to point to parallels between the 'episteme' of the natural philosophers and that of the rhetoricians.

Mnemonic logic sought to establish rhetoric as truth (rather than mere opinion or probability) by refounding its practice on the

Platonist belief that language is an earthly manifestation of the order and meaning residing in the Divine Mind. But on this metaphysical assumption it proceeded to build an eminently rationalistic structure. The intuitive, visionary revelation of the continuity of God with his creation which, according to Ficino or Pico, could only be achieved through contemplation, by a handful of initiates, has been replaced by a systematic procedure capable of being described and taught to the many.

Something very similar obtains in natural philosophy, where the search for order and relation will be founded on the assumption (no less metaphysical and Platonizing) that numbers hold the key to a higher truth than physical phenomena. The new science is not made possible, as once supposed, by the *divorce* of natural philosophy from metaphysics and theology but, quite the reverse, by pushing back the boundary which Aristotelian scholasticism had drawn between the two orders of discourse. Scholasticism had insisted on the limited applicability of reason or earthly experience, man's inability to grasp the divine order and meaning of Creation except partially, 'through a mirror, darkly', and the dangers of supposing otherwise. In contrast, the new philosophy asserted both the full legitimacy of such study, and the existence of a (Platonist-inspired) system of signs capable of accurately representing creation in all its complexity. As Alexander Pope was to put it: 'Nature, and Nature's Laws lay hid in Night. / God said, *Let Newton be*, and All was *Light*'. For Newton (or Galileo) that system was mathematical, whereas for mnemonic logicians it had been linguistic, but the similarities between the two epistemologies are at least as illuminating as the differences.[26]

4   The rhetoric of scientific method

The parallels between natural philosophy and mnemonic logic suggest that beliefs and practices once dismissed as 'occult', 'irrational' and 'unscientific' were in fact rationalistic endeavours, closely akin to those which characterized the new philosophy. They remind us that the new philosophy itself was built on metaphysical foundations outside the bounds of proof, and may help to explain supposed inconsistencies like Newton's 'Hermeticism'. But rhetoricians, Hermeticists and natural philosophers were also alike (and betrayed a common education) in the way in which they appealed to popular assumptions about, and curiosity in the 'secrets' of nature to gain favour and credibility with an audience, justify their methods in terms accessible to conservative patrons and even hint that they offered more than they could in fact deliver: rhetoric in the derogatory sense of deception and mystification. The pursuit of truth had to be balanced against that of a living, and then as now expensive research was justified in advance by a mixture of plausible-sounding explanations, optimistic claims and inventive descriptions of experiments as yet unperformed.

That this was so, to the point of self-contradiction, is clear in the work of an investigator like Della Porta, who for this reason has been viewed as a transitional figure between old 'superstitions' and new 'science'. His *Natural Magic* opens with a justification of his methods based on the theory of natural sympathies and antipathies, backed up by examples from the lore of natural history ('a wilde Bull being tyed to a Fig-tree, waxeth tame and gentle... the Ape of all other things cannot abide a Snail...')[27] many of which are to be dismissed on experimental grounds in the body of the treatise. Similarly, vagueness and imprecision in describing many of his procedures may be variously attributable to the desire not to give away financially valuable secrets, to preserve something of the aura of mystery and power attaching to the magician (despite claims that his natural magic is lawful, and accessible to all) or even to glide over doubtful assumptions and conclusions.

All these belong to rhetoric in the sense of persuasion. They constitute a parallel discourse to that of experiment and discovery whose greater function, to persuade the reader of the correctness and usefulness of the whole, outweighs any local anomaly and contradiction. Nor are such methods restricted to transitional figures to be reassuringly relegated to the pre- history of modern science. Some of Galileo's pronouncements, especially when patronage was at stake, were no less tinged by hyperbolic claims, and hints of marvellous revelations to come. And when self-interest dictated it Galileo could be extremely economical with explanations (notably with reference to the telescope, where as we shall see he is more than content to borrow Della Porta's laconic account of its workings).[28]

But rhetoric also played a part in the disinterested promotion of new ideas, and again there is no better example than the work of Galileo, from the highly propaganda-conscious manifesto of his astronomical discoveries, *Sidereus Nuncius* ('The starry messenger') of 1616, to the *Dialogue Concerning the Two Chief Systems of the World* of 1632. This, as Brian Vickers has recently pointed out, constitutes a veritable show-case of epideictic rhetoric, where the demonstration of the truth of the Copernican system and the falsity of the Ptolemaic is interwoven with their respective praise and denigration.[29] In fact such was Galileo's faith in his rhetorical powers that he attempted a dangerous sleight of hand to disarm clerical censure: disguising his dialogue of instruction as a Ciceronian debate in which no explicit conclusions were drawn. Alas, he had forgotten to his cost that Urban VIII (a latin poet of some distinction) and the Jesuit brothers of the Collegio Romano were equally skilled in rhetoric, and not so easily deceived. His rhetorical conceit only added fuel to the fire, confirming the suspicion in the minds of Church authorities that they had been tricked into allowing the work to be published.

There is, however, no simple dividing line between rhetoric as a system of persuasion and rhetoric as a comprehensive system of thought. To adopt rhetorical forms of argument so as to gain public credence invariably involves a rhetoricization of scientific logic itself. Lisa Jardine and Maurice Finocchiaro have shown the extent to which two key figures of the new philosophy, Bacon and Galileo, relied on dialectical techniques typical of a rhetorical education. The same could no doubt be said of other natural philosophers.[30] In the final part of this essay, however, I wish to focus on an aspect of rhetoric which attracted great interest after the mid-sixteenth century, whose links with natural philosophy remain largely unexplored: the demonstrative power of metaphor and its hermeneutic status.[31]

One of the earliest and most obvious examples of the role of rhetorical figures in natural philosophy is the use of metaphor in anatomy to name the newly discovered parts of the inner body. The practice was not new to the Renaissance, and in persevering with it anatomists may have unquestioningly continued a long established tradition, or followed the precepts of rhetorical mnemonics, to make their discoveries more visually memorable. In the case of the eye, the terms 'iris' (Greek for 'rainbow') and 'cornea' (from the latin for 'horn-like') were long established, and to them would be added new ones such as 'humor vitreus' (i.e. 'glass-like fluid') and 'tunica aranea' ('spider-like tunic'). But this use of metaphor to describe the shape, colour or consistency of bones and tissues, equating the inner space of the body with the outer space of the universe beyond, or the artificial space of man-made objects, reinforced the encyclopaedic view of a cosmos ordered by correspondences and analogies. The human body became a *wunderkammer*, testifying to the wondrous perfection of the

Divine plan, but lost sight of the natural philosopher's mission to investigate 'the causes, and secret motions of things'. The identification of discrete anatomical parts could evade the question of function and focus on supposed similarities which by a perfectly circular argument were then held up as 'evidence' of the underlying epistemology. A way of describing actually implied a way of thinking.[32]

Such ways of thinking were not restricted to predominantly descriptive disciplines like anatomy. Della Porta provides numerous examples of hypotheses dependent on, or structured as rhetorical figures, to be found even in his work on magnetism and optics where, as some would have it, he is closest to the modern 'scientific mentality'. When he sets out to consider 'the natural reason of the Loadstones attraction', he rejects the view of Epicurus that the atoms of magnet and iron 'easily embrace each other' and puts forward instead his own, that 'the loadstone, is a mixture of stone and iron... and whilst one labours to get the victory of the other, the attraction is made by the combat between them'. But both explanations are founded on the rhetorical figure of hypostasis, or personification of inanimate objects and abstract concepts (as will be the different 'animistic' explanation put forward a few years later by William Gilbert).[33] Similarly, the workings of the telescope are 'explained' by the exquisitely rhetorical inference that since concave lenses produce sharp, reduced images, and convex lenses hazy enlarged ones, the two together will produce images both sharp and enlarged, a typically rhetorical assertion of the plausible whose effectiveness as persuasion is wholly due to a sort of verbal algebra founded on two rhetorical figures we have already encountered: chiasmus and antithesis, whose function was to suggest completeness and balance. It is precisely the completeness and balance which give the argument, however lacking in 'hard' explanation, its air of plausibility. Galileo will find it plausible (and therefore credible and probable) enough to reproduce it word-for-word in *Sidereus Nuncius*.[34]

Elsewhere Della Porta describes how a concave mirror will focus light and heat on a precise spot, and speculates on its potential usefulness. Mindful of classical anecdotes about such a device having been constructed by Archimedes, he immediately suggests applications that might appeal to a prince: melting gold or setting a fort on fire. Given limited contemporary understanding of reflection, the hypothesis was plausible enough, though untestable, since constructing a sufficiently large mirror was also beyond contemporary resources. Proposing applications in excess of current technology continues to be an important aspect of the rhetoric of scientific research. But as we now know, the useful range of concave mirrors is restricted by the inverse proportionality between focal length and the energy concentrated at the focal point. Della Porta lacked the mathematics or physics to arrive at that proportionality, as well as the means to demonstrate it empirically. He based his argument instead on what is once again a rhetorical figure, a hyperbole, which as the theorists explained is not simply a form of exaggeration, but an implicit argument *ad maiorem* (from the lesser to the greater, the linguistic equivalent of mathematical extrapolation), further corroborated in its plausibility by an *auctoritas*. The result is a process of reasoning perfectly encapsulating the rhetoricization of natural philosophy, since it is founded on generally accepted opinion; possible, plausible and therefore probable; directed to a practical end; intended to instruct the prince, for the good of the state.[35]

An even clearer example of how fine the distinction could be between rhetorical and scientific arguments is provided by the debate on 'atomism'.[36] This shows how effortlessly natural philosophers slid from one to the other, and how even when they were aware of the difference between rhetorical analogy and explanation they could end up confusing the two. Faced with a problem beyond the demonstrative powers of either empirical observation or deduction, supporters of atomism resorted to analogy to defend the plausibility and probability at least of the existence of atoms, and most of the analogies in question had a long-established literary pedigree. Anti-atomists argued the implausibility of the existence of discrete units of matter smaller than the eye could see, yet possessing all the qualities and accidents proper to a particular substance. This was countered by pointing out how the smallest of living organisms, scarcely visible even under a microscope, must by definition be made up of even smaller organs, whose invisibility to the eye in no way removed the necessity of their existence.

A second demonstrative analogy was that of a candle wick, whose 'atoms', diffused through the air, could fill a room with smoke, yet result in no perceivable diminution in the wick's volume. A third, which again had literary precedents, but was capable of impressive (and therefore convincing) experimental demonstration, was the passage of steam or other gaseous vapour through a very close filter (for example several layers of fine paper). Yet more popular was the argument based on the motes of dust revealed by a sunbeam, finer than the finest hair or grain of sand, and only visible by the light they reflected. But here the dividing-line between analogy and identity was extremely precarious. While some suggested that this was merely another example of how the fallibility of our senses did not preclude the existence of the invisibly small (which might be revealed in the right circumstances or with more powerful instruments), others thought these might be the atoms themselves, like Claude Bérigard, who around 1620 set up an experiment to determine whether they were dust particles or something yet finer (and decided to his satisfaction that they were indeed atoms).

Nor should the (to us) patent ingenuousness of the practical execution be allowed to detract from the general methodological value of such experimental procedures. Bacon held that 'There is no proceeding in invention of knowledge but by similitude'. The problem, as for Camillo, was how to systematize analogy, leading the imagination to identify true rather than illusory relations, and defining rules by which they might be tested. With rather more restraint Galileo argued that analogies might suggest both hypotheses and the means by which these might be verified. His view is seemingly more in line with modern opinion, separating knowledge from mere association of ideas, and attributing to the imagination a mediatory role between the senses and reason. But in fact his epistemology is more complex than this suggests, and his 'modernity' as dependent on rhetoric as on experimental observation or the construction of mathematical models.

In the introduction to his *Letters on Sun Spots* Galileo confronts the delicate question of whether it is possible to use terrestrial physics to explain the super-lunary world. His argument is different from those of the Platonists, but no less rhetorical. Rather than argue that Creation is a uniform whole, so that knowledge of things earthly must be capable of extrapolation to provide knowledge of the celestial, Galileo begins by challenging the whole concept of knowledge, denying that the earthly is any more capable of being known than the celestial. By standing earlier arguments on their heads, a typically dialectical method which, effectively, founds a syllogism upon a litotes (the rhetorical figure which consists in asserting something by denying its opposite) Galileo is attempting to resolve two problems at a stroke. He side-steps both the difficulty of demonstrating the axiom on which his interpretation of telescopic observations is based,[37] and possible theological objections to denying the Aristotelian distinction between earthly and celestial, which would have been seen in conservative Church circles as implying that the mysteries of God's Creation are accessible to man (a claim supposedly opening the way to dangerous heresies).

But if the only unity Creation can claim lies in its uniform inaccessibility, where does that leave the natural philosopher? '"Until we reach the state of blessedness" we must be content with a system of analogies, whose function however will not be to reveal the hidden causes of the universe (which must remain unknown) but to describe what is recorded by our senses'.[38] One is tempted to dismiss this as one of Galileo's many sophistries (precisely, rhetorical tricks) to sidestep church censure, culminating in the ambiguities and ironies of the ill-fated *Two Chief World Systems.* According to this view, on the one hand there is a rhetoric cunningly exploited to persuade princes and reassure popes, on the other an emerging method and a philosophy of science characterized by the refinement of experiment and the development of mathematical models to describe and predict natural phenomena. But 'description' is the operative word, and we confuse it with knowledge (in the sense of understanding hidden causes) at our peril.

The Galileo who studied the acceleration of falling bodies was well aware that 'gravity' (from the Latin for 'heavyness') was 'just a word', or to be more precise a synecdoche, the metaphorical figure by which an attribute stands in place of the thing itself. The outward manifestation, the 'weight' of an object, describes the unknown force which drew it. He went on to attempt to describe its operation, the acceleration it impresses on falling bodies, mathematically, and there is no doubt that for him that was a form of knowledge. It must also be clear, however, that this knowledge was qualitatively different from that sought by Aristotelian natural philosophy.

The difference is neatly summed up in the well-known story – it matters little whether truth or fiction – concerning Galileo's discovery of the moons of Jupiter (which he would cite as inductive evidence of heliocentricity), and the most famous of late-Renaissance Aristotelians, the Paduan professor Cesare Cremonini.[39] Faced with the 'incontrovertible' evidence of Galileo's telescope, Cremonini is supposed simply to have refused to look through it. But this should not be dismissed as an unwillingness, on the part of Aristotelians, to accept the evidence of the senses. For one thing, Aristotelianism was not automatically conservative (Cremonini was a radical who denied the immortality of the soul). For another, the truth of the evidence produced by such a 'machine' as the telescope could by no means be taken for granted in an age which was very conscious of how easily the eye might be deceived.

There was a more fundamental epistemological issue at stake, however. Aristotle had suggested that the most perfect, universally valid knowledge (as opposed to knowledge of particulars) was arrived at by deduction, logically correct reasoning based on irrefutable axioms. Truth arrived at by simple induction (extrapolation from particular instances) was profoundly problematic and could be seen as implying nothing more than a high degree of probability. It was the truth of rhetoric rather than science. Whether Aristotle himself would have gone so far as to dismiss inductive proofs is questionable, but many rigorous Aristotelians took this view. For them, looking through a telescope was irrelevant, a distraction from the serious business of deductive thought. They sought truth, where induction offered mere plausibility. There is evidence to suggest that for a mixture of philosophical, theological and political reasons this was the position assumed by Galileo's clerical censors.[40]

Galileo does not deny that the truth of inductive reasoning is merely probable. What he does seem to deny is that any truer or more useful kind of knowledge is attainable than description founded on inductive analogies. If this is so, then his position is very closely akin to that of rhetoric generally and contemporary theorists of metaphor in particular. But is it?

As historians of science have long recognized, a key difference between Galileo and earlier exponents of the primacy of sensory evidence, from Sextus Empiricus to Bernardino Telesio, lay in his use of experimental method. But what is the exact status of a Galilean experiment? Not only is it not 'nature', making its extrapolation to all the naturally occurring instances of the phenomenon it represents problematic (particularly since so often in Galileo the experiment in question is 'imaginary' rather than 'real'). Its relationship to the phenomenon under study is itself questionable, since what is reproduced is not the phenomenon itself but something which stands in its place. And that 'standing in place' of nature is even more difficult to define, since in the majority of cases what is at stake is either a demonstration by exclusion (to suggest what cannot be rather than what can), or an attempt (Galileo's other key methodological innovation) to arrive at a general mathematical formulation of a phenomenon).

Consider again the matter of gravity. It might appear that what Galileo is concerned with here is cause. But when he shows that the weight of a body is irrelevant to the speed of its fall, or experiments with inclined planes to arrive at a formula for the acceleration of falling bodies, what is really at stake is a description, answering the question, 'how does gravity manifest itself?' rather than 'what is gravity?'. And that description has a general, that is to say scientific, validity precisely because its components exist outside nature ('perfect' spheres rolling down an inclined plane; arbitrary units of time and distance). What we have here is not so much analogy or simile as pure metaphor: the construction of a model (be it mechanical, linguistic or numerical) of something which can only be expressed figuratively: not the concrete individual manifestations of a phenomenon, but that relation or set of relations which makes all its different occurrences the same.

The same is true of the other plank of Galilean science, the quantitative representation of phenomena. In so far as mathesis rather than laboratory reconstruction is Galileo's ultimate aim, we could be said to be dealing with a metaphor of a metaphor, as nature is first represented as experimental apparatus, and that apparatus in turn is replaced by mathematical symbols. Its truth, like that of the maxims extracted by the rhetorician from (carefully selected and reconstructed) examples of history or daily life, resides not in any claim to represent the absolute meaning of things, but in the capacity to predict future events and be usefully employed in the service of the human community.

Notes

1 Quoted in H. C. Lyons, *The Royal Society, 1660-1940*, London 1944, p. 81.
2 For these, and the rest of the Aristotelian *corpus* see *The Works of Aristotle Translated into English*, edited by W. D. Ross, 11 vols., Oxford, 1908-1931. References below are to the division of the text adopted in this.
3 *Rhetoric*, 1355^a-1357^a, 1393^a.
4 This at least is the inference to be drawn from his works as a whole, though it must be remembered that the Aristotelian *corpus* is both incomplete and made up of what are probably lecture notes written up by his students.
5 *De Caelo* 293^a ff. rejects such Pythagorean arguments for heliocentricity as that 'the most precious place befits the most precious thing' and 'fire... is more precious than earth'; but the proof of the earth's centrality (296^b) rests on the no less culturally loaded, undemonstrable assertion that 'the natural movement of the earth... is to the centre of the whole'.

6 See for example *Rhetoric* 1394^a, 1408a, 1412^b-13^a.

7 For the concept of rhetorical 'memory' I am indebted to the work of Lina Bolzoni, and in particular the essays now in *Il teatro della memoria: studi su Giulio Camillo*, Padua, 1984, and *L'universo dei poemi possibili: studi su Francesco Patrizi da Cherso*, Rome, 1980.

8 The question of the *Rhetoric*'s transmission and tradition is a curious one which deserves attention. The first modern edition dates from 1508, after which commentaries and vernacular translations followed relatively quickly, but though the text had been known in the Middle Ages it seems to have had virtually no influence on rhetoric.

9 *De Oratore, Brutus* and *Orator* were first published in 1465, the *Institutio* of Quintilian followed in 1470. For the chronology of humanist 'discoveries' see A. Grafton, 'The availability of ancient works', in *The Cambridge History of Renaissance Philosophy*, edited by Charles B. Schmitt and Quentin Skinner, Cambridge, 1988, pp. 767-91.

10 Sperone Speroni, *Dialogo della Retorica*, (c. 1530), and *Apologia dei Dialoghi* (c. 1574), now in *Trattatisti del cinquecento*, edited by M. Pozzi, Milano – Napoli, 1978, pp. 637-724.

11 See for example Girolamo Bargagli, *Dialogo de' Giuochi che nelle vegghie Sanesi si usano di fare* ('A dialogue concerning games played in the evening-entertainments of Siena'), Siena, 1572, and Pier Antonio Ferreri, *Il cavallo frenato* ('The horse tamed'), Naples, 1602.

12 Lauro Martines, *The Social World of the Florentine Humanists*, London, 1963, suggests this is hardly surprising, since fifteenthth-century humanists were drawn from the ruling oligarchy, a difference from their sixteenth/seventeenth-century successors which goes a long way to explaining the latter's sense of displacement.

13 Luisa Muraro, *Giambattista Della Porta mago e scienziato*, Milano 1978, pp. 21-58.

14 B. S. Eastwood, 'Descartes on Refraction. Scientific Versus Rhetorical Method', *Isis*, 75, 1984, pp. 481-502.

15 *New Atlantis*, in *The Philosophical Works of Francis Bacon*, edited by John M. Robertson, London, 1905, p. 727.

16 The *Città del Sole*, written in 1602 when Campanella was a prisoner of the Neapolitan Inquisition, was revised in line with papal absolutism and given its definitive title in 1627, when Campanella enjoyed the protection of Urban VIII. It circulated in manuscript and was only published posthumously.

17 Perhaps the most spectacular example of Jesuits combining rhetoric and science in educating the ruling class to Christian values is that of the missions to China, particularly the work of Father Matteo Ricci, for which, see Jonathan D. Spence, *The Memory Palace of Matteo Ricci*, London, 1985 (Ricci was an exponent of mnemonics).

18 Bernardino Daniello, *La Poetica*, Venice, 1546, now in *Trattati di Poetica e Retorica del Cinquecento*, edited by Bernard Weinberg, 4 vols., Bari 1970-74, I, p. 238.

19 For the most emphatic claims, see the work of Frances Yates, notably *Giordano Bruno and the Hermetic Tradition*, London and Chicago, 1964; *The*

Art of Memory, London, 1966. For a critique of this see Robert S. Westman and J. F. McGuire, *Hermeticism and the Scientific Revolution*, Los Angeles, 1977.

20 For Giulio Camillo (also known as Delminio after his Dalmatian origins) and for his place in the mnemonic tradition see in addition to Yates, *op. cit.* (n. 19), the earlier, but clearer Paolo Rossi, *Clavis Universalis*, Milan – Naples, 1960, and Lina Bolzoni, *op. cit.* (n. 7).

21 'Dell' Imitatione', in Giulio Camillo, *Due trattati*, Venice, 1544, f.27ᵛ. But the most comprehensive outline of his theatre is to be found in his *Idea del Theatro*, Venice, 1550.

22 The sephiroths were 'the attributes or emanations by means of which the Infinite enters into relation with the finite' (*OED*).

23 A model was apparently built, and inspected by Erasmus' correspondent Viglius Zuichemus in 1532 (Erasmus, *Opus Epistolarum*, edited by P. S. Allen *et al.*, 11 vols., Oxford, 1906-47, IX, p. 479). In addition to the already cited *Idea* there was a 'Discorso in materia del suo theatro' published in the first, incomplete, edition of his *Opere*, edited by L. Dolce, Venezia, 1552.

24 To my knowledge, the first explicit definition of nature as a system of relations rather than an inventory of 'substances' and 'accidents' is that of Giulio Cortese ('nature is an instinct or genius pertaining to things... to come together or move apart from each other'), in the essay 'Avvertimenti nel poetare', first published in *Rime & Prose del Sig. Giulio Cortese*, Naples, 1592, and now in *Trattati, cit.* (n. 18), III, p. 183. Cortese was a pupil of Telesio and may have been influenced by Della Porta (who also wrote an *Ars reminiscendi*, Naples, 1602); he makes several references to Camillo (though critical of his 'wanting to get involved in things to do with the Cabbala', *ibid*, p. 180).

25 See Yates, *op. cit.*, particularly pp. 355-74, and Rossi, *op. cit.*, pp. 179-258 (n. 20).

26 *The Twickenham Edition of the Works of Alexander Pope*, edited by John Butt *et.al.*, 6 vols., London and New Haven, 1970, VI, p. 317. Maurice Clavelin, *La Philosophie naturelle de Galilée*, Paris, 1968, argues convincingly that the Galilean mathematization of physical space involves neither a simple description of observable, material reality, nor Platonist representation of a higher order of truth, but a third form, conventional and man-made rather than innate, but nevertheless real in its definition of the phenomenon's universality. It seems to me that something very similar could again be said of Camillo, whose *logos* does not originate in the divine order (though made possible by it), but rather is a distillation of man-made principles of eloquence.

27 Giovan Battista Della Porta, *Magia Naturalis*, Naples, 1558 (in 4 books) and then Naples, 1589 (expanded to 20 books). I quote from the English translation, *Natural Magick*, London 1658, p. 9.

28 For this aspect of Galileo's dealings with his patrons see Richard S. Westfall, 'Science and Patronage. Galileo and the telescope', *Isis*, 76, 1985, pp. 11-30.

29 B. Vickers, 'Epideictic Rhetoric in Galileo's *Dialogo*', *Annali dell' Istituto e Museo di Storia della Scienza di Firenze*, 8, 1983, No. 2, pp. 68-102.

30 Lisa Jardine, *Francis Bacon: Discovery and the Arts of Discourse*, Cambridge, 1974; Maurice A. Finocchiaro, *Galileo and the Art of Reasoning*, Dordrecht − Boston − London, 1980; more generally, see *The Politics and Rhetoric of Scientific Method*, edited by J. A. Schuster and Richard R. Yeo, and on Descartes, R. R. Yeo, 'Cartesian method as mythic speech: a diachronic and structural analysis', *ibid.*, pp. 33-95.

31 The most notable exponent of this tendency was Emanuele Tesauro, whose *Cannocchiale Aristotelico* ('The aristotelian telescope'), Venezia, 1655 and subsequent revised editions, argued that all rhetorical 'figures of thought' (i.e. bearing an argument or 'conceit') were sub-varieties of metaphor.

32 An example in the Jesuit historian and scientific popularizer Daniello Bartoli, *Le ricreazioni del Savio*, Rome, 1659, where the reader is invited to see in the mechanisms uncovered by the new philosophy nothing more challenging than a chain of microcosm − macrocosm analogies testifying to divine order.

33 Della Porta, *Natural Magick, cit.*, pp. 191-2.

34 See Muraro, *op. cit.* (n. 13), p. 106.

35 *Ibid*, p. 25-9.

36 See C. Meinel, 'Early seventeenth-century atomism. Theory, epistemology, and the insufficiency of experiment', *Isis*, 79, 1988, pp. 68-103.

37 For example, Galileo argues that since the spots on the face of the Moon appear to change as the Moon moves in relation to the Sun, they *could be* the shadows of mountains. For an Aristotelian, such an interpretation would be impossible, since he assumed that heavenly bodies were qualitatively different from earthly ones.

38 E. Reeves, 'The rhetoric of optics: perspectives on Galileo and Tesauro', *Stamford Italian Review*, 7, 1987, p. 133.

39 It has been told by Bertolt Brecht, in his *Life of Galileo*.

40 See Guido Morpurgo Tagliabue, *I processi di Galileo e l'epistemologia*, Roma, 1981.

5

Natural philosophy and its public concerns

Julian Martin

Since the history of the sciences in Renaissance Europe can be understood as an aspect of the general cultural history of Renaissance societies, historians of science are increasingly deploying the perspectives and techniques of the social historian in explaining their subject matter. Such perspectives suggest, for example, that it is of no more value to speak of '*the* natural philosophy' of the Renaissance than to speak of '*the* Renaissance man' or '*the* Renaissance state'. We note that fifteenth-, sixteenth- and seventeenth-century Europeans experienced repeated political and social upheavals, and the struggles of regimes to assert themselves against domestic and foreign competitors often failed. Similarly, Europeans witnessed a series of competing claims about the aims, the contents, the methods, and the ownership of natural knowledge, and not until the end of our period did the vigorous promotion of certain models force others to the fringes. What is broadly characteristic of the political history of societies in Renaissance Europe is broadly characteristic of the history of their natural philosophies as well: both are the histories of contested social selections.

Such a view defines certain tasks for the social historian of the sciences; no more than the political or economic historian can we appeal to any inherent 'rightness' or 'wrongness' in particular proposals to explain their success or failure. We wish instead to explain, for example, why it was that different sorts of natural philosophers believed their own claims reasonable and themselves deserving of respect, how their various contests arose, and by what social means these were sustained or dismissed. Few would suggest that natural philosophers were oblivious to their social, political or religious upbringing and environment, and few historians would try to explain their philosophical activities without reference to such matters, but it is common to forget that despite their own blandishments about the objectivity and truth of their claims, Renaissance natural philosophers, *qua* philosophers, were partisans. In short, their blandishments often as not were disingenuous strategems; claims about the aims and methods of natural philosophy had conscious socio-political significations. I propose to show how our attention to the public, or political, concerns of Renaissance natural philosophers helps reveal what we may usefully term a political economy of natural knowledge.

1 Natural knowledge and the state

We can begin by recalling the viewpoint of Renaissance ruling elites towards knowledge peddlers in general. Among the central features of early modern Europe's political history are the struggles of ambitious rulers to wrest civil authority in nearly every sphere of life from equally ambitious feudal magnates and local gentry. These struggles shape the well-known story of the 'rise of nation-states', of the princely creation of extensive bureaucratic instruments with which to achieve them, and of the subordination of the Church (in Protestant states at least) to the secular power. A huge potential existed for social instability and political rebellion, as centralizing rulers often learned to their cost, and their anxieties in this regard extended to the production and, especially, to the dissemination of knowledge. Men publicly claiming to possess 'true knowledge' were potentially danger-ous to the authority of the ruler and the precarious legitimacy of his regime; as far as possible, such persons were to be harnessed and their products policed.

Their products increasingly came in the form of printed books, and as the material technology of printing proliferated throughout European states, so too did the administrative machinery of censor-ship. Few states (if any) failed to equip themselves with an apparatus for licensing and censoring printed texts. The Roman Church, famously, went further, issuing and periodically updating a specific *Index of Prohibited Books*, and the governors of many Italian cities developed their own versions as well. This universal concern of ruling elites for intellectual policing rested on the general presumption of the need for official mediation of knowledge. For instance, in Catholic states like France and in Protestant England as well, the maintenance of an ecclesiastical hierarchy and bureaucracy not only provided

Valois, Bourbon, Tudor and Stuart monarchs with an extensive apparatus of governance over the behaviour of their people, but it enshrined their belief that the individual's access to truth be mediated principally by persons officially licensed to reveal and interpret it. Similarly, the crimes of two Italians – a peasant, Menocchio, and a mathematician, Galileo – were no more their novel cosmological opinions than their obstreperous refusal to acknowledge official experts in such matters.[1]

Little practical distinction was made by civil authorities between religious knowledge and knowledge of the natural world. Books of natural philosophy, no less than books of theology, were vetted and licensed, and often by the same officials: for was not natural philosophy a study of God's handiwork? And equally, claims about the structure of the natural world, about the methods by which one could come to know it, and about who could legitimately participate in its discovery, could be viewed not only as having implications for religious belief, but also for the structure of the valid state, how to achieve it, and who could legitimately participate in its governance. The natural philosophies of Renaissance Europeans were not viewed as discourse innocent of civil significance, either by the makers or their political masters.

This interplay of the philosophical and the political can be further illustrated by considering the general character of the philosophic endeavours supported at royal European courts, the many schemes for philosophical communities promoted in our period, and the endemic negotiations of natural philosophers for improved social credit.

At his faction-ridden court, the Catholic King Henri III enthusiastically supported a philosophical 'Académie du Palais', a central tenet of which was that 'true learning' in the nation's leaders would be the best way to recreate a civil harmony in chaotic France and, eventually, to restore Christendom to its lost (and catholic) unity. The Huguenot King Henri of Navarre maintained a philosophical 'Académie' also, with similar (but Protestant) aims. The Holy Roman Emperor, Rudolf II, patronized occult and mathematical philosophers (including Kepler) at his court in Prague; Rudolf's personal fascination with a search for the hidden harmonies and coherence of the natural world is consistent with his lonely (and unsuccessful) struggle to maintain the political and religious unity of his fractious empire.[2] In all three cases, the roles and the utterances of the moral and natural philosophers gathered around these monarchs were tailored to the political and religious ambitions of their royal patrons. Conversely, the writings of Paracelsus are soaked in the mystical language of the disfranchised outsider, and the natural philosophical work of Blaise Pascal is best understood within the political context of the battles fought by his Jansenist religious community at Port Royal with the Jesuits dominating the counsels of the French king in Paris.[3] During our period, we find a host of visionary schemes for ideal natural philosophical communities, and for all their utopian flavour, they all manage to include persons just like their authors, and to privilege their political prejudices too. For instance, the natural philosophical community envisioned by the Dominican friar Tommaso Campanella, in later life a confidant of Pope Urban VIII, is integrated within a universal theocracy quite possibly intended to allude to an ideal papal *imperium.* The natural philosophers in Johann Valentin Andreae's *Christianopolis* are part of an idealized Lutheran community,[4] while the natural philosophers of Francis Bacon's *New Atlantis* are suspiciously akin to English royal judges (like himself). There is nothing disinterested here, and such texts speak volumes about the relentless quest of natural philosophers for respectability and public status.

While natural philosophy continued to be pursued in the universities and colleges,[5] it is a striking feature of Renaissance Europe that many of those professing natural knowledge aspired to translate themselves from the cloisters and lecture halls to the beckoning courts of secular nobles, from the schoolrooms to the private rooms of the powerful, and from a station and expectations circumscribed by the norms of the guild or of academic pedagogy to the uncertainties and opportunities provided by attaching themselves to the patronage of public figures. By accepting the courtly protocols of deference, they too could grace an audience chamber, advise a prince, and glean pensions from his purse for advice and services rendered with their pen. This social move was facilitated by the adoption of humanist doctrines by noblemen, and by the flood into Europe's universities and colleges of young gentlemen aiming to colour themselves with the new learning; it was compounded by the occasional but celebrated pursuit of natural knowledge by court nobles themselves (e.g. Tycho Brahe), which gave natural philosophy new prestige as an avocation for members of the governing elites. This process was not merely the creation of a new and socially-elevating career trajectory for natural philosophers, but an intellectual and social solvent for natural philosophy, redefining disciplinary boundaries and embracing new sorts of persons who would publicly declaim about physical reality and the means of uncovering natural knowledge.[6] The pursuit of natural philosophy and the role of the philosopher in Renaissance Europe was inextricably bound up with concerns and aspirations much wider than first meet the eye.

Such expansiveness was not viewed by all as an unmitigated good; by the early seventeenth century, it was creating considerable unease in those valuing official mediation in natural knowledge above liberality in its pursuit. In a celebrated remark, John Donne attacked the morals of natural philosophers and asserted the priority of deference to political authority and social norms.[7] Marin Mersenne's agenda for natural philosophy aimed to re-establish stable criteria of what constituted proper natural philosophy and its methodological protocols.[8] So alarmed was Francis Bacon by this expansiveness and by its dissonance with his political expectations, that he felt the need to refashion natural philosophy and natural philosophers as well. It is in Bacon that we can see most clearly that intimate relationship between the philosophical and the public which we have been tracing here, for Bacon was *both* a philosophical maker and a political master, and he was explicit in his conviction that the problems of natural knowledge were inseparable from the problems of social order.

2   An Elizabethan civil servant

With a healthy scepticism toward Francis Bacon's ceaseless self-advertisements about the newness of his natural philosophy, much modern scholarly effort has been devoted to showing the degree to which Bacon can be situated within various Renaissance intellectual traditions. For example, persuasive claims have been advanced by Paolo Rossi for Bacon's debts to traditions of natural magic, occultism, and the mechanical arts, and by Lisa Jardine for Bacon's debts to traditions of rhetoric and dialectical analysis.[9] Bacon was clearly alive to these traditions, but we should be as cautious as were these scholars about assuming that such attributions constitute in themselves a sufficient framework for explaining the content of his natural philosophy. Bacon made choices, and we need to know something about his criteria for making those choices. This demands our examining his particular circumstances and interests; and in doing so, we find there is a telling difference between Bacon and almost any other Renaissance natural philosopher we might think of. Unlike Paracelsus, Fabricius,

John Dee, William Gilbert, Kepler, Galileo or Harvey, Francis Bacon was a politician and statesman by trade, and he always regarded himself as such, and not as a natural philosopher *per se*. Bacon was a member of the English governing elite; his overriding ambition was the augmentation of the powers of the Crown in the state, and he believed his refashioned natural philosophy was but one (albeit novel) instrument by which to achieve this political aim.[10]

This self-image and ambition conditioned every detail of his natural philosophical programme. Bacon's aspirations and methods for uncovering natural knowledge were novel for natural philosophers (as he claimed) yet they were not novel in any absolute sense. His aspirations were adapted from those of his father's generation of Tudor statesmen, and his procedures for natural philosophy were adapted from procedures in the English Common Law. If Bacon's political and legal career is ignored and we attempt to explain his natural philosophy apart from it, then we shall miss the central point of his labours. In other words, Francis Bacon's chosen career was crucial in the creation and content of his natural philosophy, and his natural philosophy cannot be separated from his political ambitions. It is, therefore, not merely descriptive but explanatory to say that his was, in fact, the natural philosophy of a royal servant, a Tudor statesman.

In Bacon's writings about natural philosophy we can see the transposition of late-Elizabethan political anxieties and social prejudices, of Tudor strategies of bureaucratic state management, and of a principal intellectual resource of the landed Englishman – the science of the Common Law. This transposition was made by none other than a faithful scion of the Tudor governing elite, and in an audacious attempt to institutionalize natural philosophy in accordance with the highly-centralized apparatus of monarchical government he wished to see realized in England.

How then can we explain the development of this particular example of the intimate conjunction of political and philosophical interests? It is usually taken for granted that Francis Bacon was interested in natural philosophy, but it is not at all obvious why a man born to a distinguished family of Crown servants and educated to follow their careers should have come to devote to natural philosophy any sustained attention whatsoever. Yet by the early 1590s, when Bacon was in his early thirties, this was indeed the case. The explanation for this peculiar interest lies in the political values he shared with his father and uncles (crucial figures in Elizabethan central government) and in his own circumstances during the later 1580s and 1590s.

Sir Nicholas Bacon, Sir William Cecil, Sir Henry Killegrew, Sir Thomas Hoby, and Sir Thomas Gresham all entered the service of the Tudor dynasty during the 1530s and 1540s, and they served it throughout their lives. From the beginning of Queen Elizabeth's reign in 1558, Bacon (Lord Keeper) and Cecil (Lord Treasurer) in particular strove to implement a set of reforming policies which they had learned under Thomas Cromwell and his successors. These involved what we might call long-range state planning designed to increase the economic wealth and independence of the nation and to increase the Crown's control over the apparatus of governance throughout the country. The first of these aims was tackled by encouraging technological innovation and the domestic manufacturing of goods by means of legal privileges extended to immigrant craftsmen and English industrial entrepreneurs. The second aim required reform of the judicial system in the hope of bringing the dispensation of justice firmly under the supervision of the royal government and of bringing fractious or negligent magistrates and juries to a 'due obedience' to the Crown. Both programmes were thought necessary for increasing the stability and security of the regime, which had precarious authority and finances, and which remained vulnerable to domestic insurrection, to foreign invasion and to pressure on the wool trade (the Crown's principal source of revenue).

Sir Nicholas Bacon was convinced that his youngest son, Francis, should follow in his footsteps, first becoming 'learned in the law', and then making a career as a crown servant. At the age of twelve, Francis was sent up to Cambridge and, after less than two years there, to Gray's Inn to study the law. In 1576, he went in the company of a new Ambassador to the French royal court to experience the life and machinations of diplomats and courtiers. The Lord Keeper's death in 1579 brought Francis back to London, to more study at Gray's Inn, and to his overbearing mother who, now released from her husband's control, became an increasingly vigorous Puritan activist. She encouraged Francis to seek the patronage of the Puritan earls (Leicester, Warwick and Bedford). Yet by 1586 and the beginning of the long war with Spain, Bacon (now twenty-five) had rejected his mother's wishes, had turned away from an increasingly radical and politically marginalized Puritan movement, and had accepted offers of legal employment within the central government.

Several points here merit our closer attention. At Cambridge, Bacon was the personal pupil of the Master of Trinity College, Dr John Whitgift, the 'Hammer of the puritans', who regarded Puritans as the 'enemy within', endangering the fragile official unity of Church and State. After his brief brush with the Puritan lords in the early 1580s, Bacon's desire to serve the Crown, and his deepening anxiety about the preservation of the regime, led him into the Privy Council's security service, examining Catholics and especially Jesuit plotters from 1586 onwards. When in 1589 the radical Puritan pamphleteer known as 'Martin Marprelate', scandalized the Queen and her Council by his attacks on the Church hierarchy, Bacon wrote a lengthy defence of the bishops, in which his highly conservative and political point of view was made clear: bishops were royal officers, engaged in the tutelage of the people, and to attack them was simultaneously to attack the social and political hierarchy in general. Those who did so were politically divisive, 'sectarians', wilfully (and erroneously) believing they alone possessed religious Truth. Bacon insisted that the truths of religion must be mediated by official experts, not dispensed by the self-appointed ('voluntaries', as he called them scathingly). His increasing immersion in the problems of the security of the regime (and specifically, in the interrogation of religious zealots) led him to view 'epistemology', namely the problems of attaining sure knowledge, as *political* problems and properly a part of a statesman's concerns.

In the early 1590s, Bacon's security work was extended to include radical Puritan zealots as well, and it was at this time that he made his first pronouncements about a reform of natural philosophy. These were his reaction to new developments within English society which he regarded as politically dangerous and a potential challenge to the regime.

With the conservative retrenchment of the central government during the Spanish war, and with the death of the great lords who had at Court and Privy Council sustained and protected the Puritan movement, came the rapid destruction of the Puritans' political hopes and the hounding of their 'godly' ministers by the Crown. Many of the Puritan gentry now began a gradual retreat into self-selecting 'voluntary communities' of the 'godly' alone.[11] The Queen and her Privy Counsellors viewed this as a serious breach of what they took to be the landed gentleman's paternal responsibility for the entire local community. Furthermore, the 'godly' had always believed they could find Truth in the Scriptures without the guidance of official teachers, and several intellectual developments in the later 1580s and the 1590s mirrored this: for example, an enthusiasm for Paracelsian 'chemical philosophy'. A considerable part of the 'chemical philosophy' was devoted to denigrating traditional intellectual authorities (particularly the philosophies of Aristotle, Galen, and the university 'Schoolmen') and to singing the praises of direct personal 'experience' as the paramount way of attaining sure knowledge about the natural world. It carried forward, in other words, much of the religious reformers' assault upon the ingrained assumption of medieval and Renaissance Europeans that to attain true knowledge required the intercession of learned experts.[12] Like many members of the Privy Council (including Whitgift, who had since become Archbishop of Canterbury), Bacon viewed these beliefs with alarm, interpreting them as containing dangerous political implications. Bacon's concern to reform natural philosophy was the concern of a loyal crown servant, a Royalist reaction against independent gentlemen and what he saw to be their way of knowledge.

Bacon was deeply suspicious of the social and political pretensions of those he called 'voluntaries' in natural knowledge, and around him in the 1580s and later he could see many such independent investigators. It seemed clear, for example, that several were associated with Puritan political activists. The celebrated Dr John Dee, student of the occult and advocate of Paracelsism and 'natural magic', was patronized by Puritan politicians such as Sir Philip Sidney and the Earl of Leicester, as were Giordano Bruno and the mathematical philosophers Thomas Harriot and Thomas Digges.[13] Sir Hugh Plat, the Puritan advocate of 'philosophical experiments', was deeply immersed in alchemical lore and constantly advertising for financial support from private gentlemen for his technological 'projects'.[14] Making matters worse in Bacon's eyes, men like Dee and Plat celebrated the private creation of their natural knowledge, insisted on its private ownership, and expected rewards for any practical fruits of it.[15] Quite apart from this, Elizabethan mathematical artisans, such as Thomas Hood and Robert Norman, were claiming that their work was important as *natural philosophy*.[16] It seems that Bacon's refusal to give any priority to mathematics in natural philosophy and his repeated criticisms of Paracelsus ('that illiterate Enthusiast') may well have stemmed originally from his socially elitist and partisan political assessments of those Englishmen who were espousing the 'mathematicks', the occult or chemistry as central to natural philosophy. The study and production of natural knowledge was too important to be left in the hands of political ciphers or the disaffected.

3   A new intellectual regime

In 1592, Bacon told his uncle, now Lord Burghley, that he had 'taken all knowledge to be my province'. The Elizabethan meaning of 'province' was the same as its Latin meaning: a *provincia* was an administrative unit of the central government.[17] Bacon wished Knowledge to become a department of state, and himself to be its governor. He believed that knowledge could be harnessed to the advantage of the state *if* it was centrally governed and organized. He would, he assured Burghley, first 'purge' his province 'of two sorts of rovers, whereof the one with frivolous disputations, confutations, and verbosities, the other with blind experiments and auricular traditions and impostures, hath committed many spoils'.[18]

In Court masques in 1592 and 1595, he elaborated upon these themes, while declaring that the 'conquest of the works of nature' required 'royal works and monuments': these needed the participation and organization of a great number of carefully selected people, and for a great length of time. First, Bacon believed institutional tools, and the bureaucracy and division of intellectual labour they imply, were necessary for discovering knowledge of nature. Secondly, these institutions were to be devoted to collecting and then gleaning information. Thirdly, this was to be a royal enterprise, not merely because of the wealth and human resources that would need to be mobilized, but because Bacon desired these 'monuments' to be in the legal possession of the monarch. His advocacy of a reformed natural philosophy was first voiced during a turbulent period in which he was involved deeply in matters of security and the political stability of the regime he served, and during which he feared the social and political implications of any 'voluntary' approach to knowledge. These fears never abated. For example, notes exist of a 1617 speech Bacon made to the judges before they went out on the assize circuits; Bacon warned them of the dangers of unmediated knowledge: 'New opinions spread very dangerous, the late Traske a dangerous person. Prentices learn the Hebrew tongue'.[19]

Also from the early 1590s, Bacon began to clamour for legal reform. For Bacon, legal reform was much more than the elimination of particular laws, or the speedier dispatch of justice. He wanted the complete subordination of the central judicial system to the wishes of the monarch. At stake here was power in the state: did the judges rule, or did the King himself? Bacon planned to reduce the powers of the judges, whom he increasingly regarded as 'voluntaries', hiding behind their personal interpretations of the common law, and resisting thereby the wishes of his master. He proposed, therefore, a reorganization of the science of the law. This involved a hand-picked committee of experts (Bacon and several friends) to vet the statute-book, weeding out what they considered to be duplications and redundancies, and improving 'dubious' wordings. It also involved another expert committee to vet the Law Reports, so crucial to the practice of the common law, and to reduce them to 'order'. Official recorders would hereafter take reports on important cases and only these reports would be added to the officially recognized Law Reports. On the basis of these new reports, Bacon expected to discover *regulae juris*, the true (but hidden) principles of the law, and these too would become part of the common law. Hence the cherished 'common erudition' of the common lawyers' guild would be forever altered, and judges would be much hampered in delivering idiosyncratic interpretations and judgments. It was a comprehensive scheme to smash the common lawyers' practical control over the law, and to smash as well the ability of judges like Sir Edward Coke, whom Bacon regarded as seditious, to delimit the scope of the royal prerogative. Bacon, who described himself as a 'perfect and peremptory royalist', aimed to restructure this principal apparatus of the central government in order to facilitate a more powerful monarchy. Reform of the law, then, was an important part of this political plan, and a reformed natural philosophy was another. Both reforms were designed to severely limit the opportunities for 'voluntaries' to make claims about true knowledge.

Although Bacon believed that a centrally organized natural philosophy would discover the 'causes and secret motions of things', and although he believed it would thereafter result in inventions and useful knowledge for the state and the people, this latter claim (albeit the most memorable) was an enticement to adopt his programme rather than a claim about the essential purpose of natural philosophy. Instead, as Bacon wrote (privately) to King James in 1603, his natural philosophy would allow the discovery of the 'true rules of policy':

there is a great affinity and consent between the rules of nature, and the true rules of policy: the one being nothing else but an order in the government of the world, and the other in the government of an estate ... The Persian magic, which was the secret literature of their kings, was an observation of the contemplations of nature and an application thereof to a sense politic; taking the fundamental laws of nature, with the branches and passages of them, as an original and first model, whence to take and describe a copy and imitation for government ... I conclude there is (as was said) a congruity between the principles of Nature and Policy ... This knowledge then, of making the government of the world a mirror for the government of a state ... I have thought good to make some proof ...[20]

Because so very much was at stake here, Bacon's programme for a reformed natural philosophy emphasized a division of labour and a hierarchy of responsibility. Although he was convinced that the success of the work required the co-ordination of many helpers, these helpers were of two unequal sorts, a huge number of under-labourers and a handful of 'Interpreters of Nature'. Bacon was unshakably convinced of the worth of this socially stratified arrangement; it featured as prominently in his unfinished and posthumously-printed utopian tract *New Atlantis* (1627) as it had in his court masque of 1592. *New Atlantis* is famous for its portrayal of an institutionalized natural philosophical community, called 'Solomon's House', and the organization of labour in 'Solomon's House' clearly reflects Bacon's confidence in bureaucratic organization. Here, the work of the many assistants was directed by the elite 'Brethren' who alone devised the 'experiments', pondered their results, uncovered and possessed knowledge of the principles of Nature, and then devised useful technologies for the state (in the first instance) or for the people. As Bacon wrote in a personal memorandum of 1603, 'some fit and selected minds' alone were permitted to discover natural knowledge, and these persons would be selected by the state (represented, no doubt, by Bacon himself). The programme demanded their utter loyalty (including the eradication of their intellectual presuppositions), and their maintenance of a complete secrecy over the gradually emerging knowledge of the structure of the natural world. Because, as he often claimed,'Knowledge is power', it was potentially politically dangerous, since it provided the key not only to material prosperity, but also to the irresistible (since God's) and hidden principles of civil government. Only those officially licensed to do so could engage in a task so

111

politically sensitive.

The details of Bacon's reformed natural philosophy were modelled on his reforms of the science of the law. Although he never provided a detailed discussion of how a particular 'artificial experiment' was to be conducted, it is clear that he envisioned an 'experiment' to be closely similar to a lawsuit and its trial in a courtroom. Both in lawsuits and in Bacon's natural philosophy, the first and crucial stage was the 'establishment of facts'. A 'fact', moreover, is an artefact of the courtroom; it is merely 'alleged' in the first instance and not taken for granted: its legitimation as fact is acquired only after the application of official legal procedures. In the common law, a principal means of gathering together information relevant to the establishment of facts was the use of 'schedules of interrogatories' for witnesses, and Bacon meant to do the same in natural philosophy: 'I mean (adopting the example in civil cases) ... to examine Nature herself ... upon interrogatories'. In natural philosophy, 'interrogatories' were to be used in the preliminary stage of investigations – the gathering of 'experiments' for the 'Natural Histories' – and just as in the legal process, 'this primary history [is] to be drawn up most religiously, as if every particular were stated upon oath'. Whenever judicial torture (an exclusive responsibility of the security service of the Privy Council) was employed, its purpose was to force responses to interrogatories from unwilling prisoners, and we can recall the insistence of Bacon (an old hand at sedition investigations) that Nature best revealed herself when 'vexed' or 'tortured'. The 'Process of Exclusions' which Bacon described in *Novum Organum*, Book Two, is the same process that barristers employed in arguing before a judge. The judge (and Bacon's 'Interpreter of Nature') was to adjudicate which legal principle (or 'natural species') it was which was manifest in the case (or 'given nature') under review. Barristers engaged in a process of argument by analogy with older cases and with the principles of law former judges had announced as the basis for their decisions in those cases (hence the crucial importance of Law Reports). The process of legal reasoning was described by ancient lawyers (e.g. Cicero and Quintilian) as 'Induction'. This use of the term is different from the way it was employed by logicians and philosophers from Aristotle onwards. When Bacon declared that 'the Interpretation of Nature' was dependent upon a 'true induction', this legal usage is what he had in mind; as he well knew, 'the induction of which the logicians speak, which proceeds by simple enumeration, is a puerile thing'. Bacon's 'method' for natural philosophy, so to speak, was not a *philosopher's* method but a bureaucratic and legal process.

Both Bacon's reformed science of the law (the uncovering of the hidden principles of the law) and his reformed natural philosophy (the uncovering of the hidden principles of Nature) were to be pursued by officially-licensed investigators alone. Their work would be based on the labour of others, which had resulted in the compilations known as Law Reports and 'Natural Histories', and Bacon regarded these as similar. Law Reports, he said, were 'a kind of history ... of the laws'. Both the royal judge and the 'Interpreter of Nature' were to scrutinize such 'digests' to uncover *legum leges* and 'higher axioms of Nature'. They would sort the individual 'maxims' or 'axioms' into a hierarchy of species and genera; this is how they would gradually discover God's hidden structure of the common law or of nature.

Significantly, in both the classical Latin and in the Elizabethan English, an *Interpres* or 'Interpreter' was an official expositor of the laws. The master of Bacon's 'new instrument' (*novum organum*) was to be a new sort of royal judge, and the *Instauratio magna* (1620) justifies and describes his work. It is no coincidence that *Instauratio magna* was written by the greatest judge in the land, the King's Lord Chancellor.

4   Politics and science after Bacon

It seems clear that when Bacon fell from power in the early 1620s, his political programme fell with him. Yet it is equally clear that during the next generation, men whose social allegiances and political interests were very different from Bacon's own did explicitly appeal to Bacon the natural philosopher.

Several factors were involved in this remarkable process of appropriation. First, the political ambitions which underpinned Bacon's vision of a reformed natural philosophy were largely unknown to later readers; his programme for political reform, and the relation of natural philosophy to this grand scheme, had been expressed in private letters and memoranda which neither Bacon nor his immediate executors published. Secondly, Bacon's exalted social status – Viscount St Albans, Privy Counsellor and Lord Chancellor of England – had the effect of helping to ensure that the pursuit of natural knowledge would no longer be socially appropriate solely for ambitious 'mechanics' and cloistered or court scholars, but a respectable vocation of English gentlemen as well. Thirdly, readers have always been at liberty to read

113

E

selectively, and since Bacon's published writings on the reform of learning and natural philosophy were deliberately constructed to attract as many persons as possible to labour in his grand plan, it is unsurprising that his stirring proposals for reform would be taken in a variety of ways.

Charles Webster's splendid studies of Puritan programmes of social reform have shown just how extensive such 'Baconian' appropriations were among men whose thirst for a reformed natural philosophy was motivated by their passionately millenarian religiosity.[21] Their Francis Bacon was a Puritan prophet of social regeneration by means of a natural philosophy which required active participation by all godly men as equals and which privileged utilitarian public ends. This was a natural philosophy and a Francis Bacon appropriate to their own religious and political interests. Bacon's imperious 'Interpreter of Nature' was gone, and what for him had been the preliminary stage of his philosophical procedure – collections and 'experiments' by many hands towards 'natural histories' – was converted into its central activity. Yet the seventeenth-century Puritans, in their anxious preparations for the Second Coming, expected that the reform of society would be led and managed by the (benevolent) state. To be useful, natural philosophy required to be institutionalized as a department of this 'godly' power. John Dury's and Samuel Hartlib's ambitious 'Office of Address' is a case in point. This was a state-licensed and funded body with official and salaried 'Agents' to gather up information, and in so far as such an 'Agent' was concerned with 'Ingenuities' uncovered by natural knowledge, he was 'to offer the most profitable Inventions which he should gaine, unto the benefit of the state, that they might be publickly made use of, *as the State should think most expedient*'.[22] A historical understanding of the seventeenth-century Puritans' pursuit of natural knowledge 'necessitates attention to the character of religious motivations, of adherence to philosophical tenets which have subsequently been discarded, and of involvement in contemporary political affairs'.[23] The natural philosophical activities and the public ambitions of English Puritans were intimately and explicitly intermeshed.

A similar attentiveness to the civil significance of natural knowledge can be seen in Thomas Sprat's *History of the Royal Society of London* (1667). Sprat was at pains to justify the new, gentlemanly, institution and its philosophical work in terms of its perfect congruence with his view of the aims of the Restoration social order: it was moderate and undogmatic; political, religious, and metaphysical debate was forbidden; 'hypotheses' were rejected in favour of 'matters of fact'; *gentlemen* laboured together in a public forum and in a spirit of harmony. Yet recent historians have rightly been wary of taking Sprat's *apologia* at face value,[24] and just what was at stake for natural philosophers in the 1660s can be seen in the contests between Thomas Hobbes and Robert Boyle. Both men were profoundly anxious to repair the fragile social order which the civil wars and Republic had created and which the Restoration failed to resolve, and both were aware that natural philosophy was not isolated from this political process. Yet, their explanations and solutions for the problems of social and political order were utterly different, and consequently so too were their views on what constituted proper natural philosophy, who could legitimately participate in its production, and which methods would result in credible natural knowledge.[25] Hobbes looked to natural philosophy for certainty by means of the discovery of hidden principles demonstratively proved; Boyle proposed to investigate 'matters of fact' by means of experimental apparatus which unproblematically displayed the truth to undogmatic gentlemen. Hobbes disputed the virtues of Boyle's apparatus, and condemned both the Society and Boyle's appeal to its Fellows as sufficient witnesses: the entire programme was exclusive, partisan, vulnerable to dangerous errors, and bordering upon sectarian conspiracy. It might simply be concluded that Boyle 'won', and the Royal Society accepted his prescriptions for the production of natural knowledge because he was 'right', but for the social historian it would be much more fruitful to begin in supposing that he was better able than Hobbes to mobilize interested allies.

During the Renaissance, the social terrain in which natural philosophy was pursued expanded from the academic setting and into the wider, and volatile, public arena. The struggles of natural philosophers were not simply intellectual struggles with recalcitrant Nature and stubbornly conservative scholars. The secular process of winning both heightened status for their multifarious labours and social respect for themselves was a process of local negotiations with those who dispensed social credit. Yet securing careers and social respect was no easy business amid the political and religious instability of Renaissance societies. Our present survey suggests the history of natural philosophy in Renaissance Europe reflects the concerns and contests among the political classes about the proper nature and maintenance of the social order. Natural philosophers were in no position to stand aloof from such pressing matters. Their explicit and repeated contests about the ownership, organization, generation and applications of natural knowledge and about who could participate in its discovery are shot through with public concerns. For example, was the pursuit of natural knowledge to be conducted by (and for) the state, and within official academies, or by (and for) private gentlemen, scholars and 'ingenious' artisans, and by means of their independent inquiries? Were natural philosophers to be official 'Interpreters of Nature' (Bacon) or independent gentlemen – 'priests of Nature' (Boyle)? To what (or whose) authority should natural philosophers appeal in order to justify their assertions of truth about the natural world: to God (everyone's best claim), to the monarch, to Reason, to utility, to the support of one's colleagues? Closely related to this was the question of what (or whose) authority should justify their methods for attaining such truths, given the pervasive habit of scrutinizing epistemological claims for implicit political significations.

Although this essay has been focused upon European, and particularly English, society in the later Renaissance, and examined Francis Bacon as a particular illustration of our general themes, it should not be taken to imply that the interplay of concerns about the political and the natural order was an eccentricity of this period; it is a much more typical feature of the history of the sciences than is commonly recognized. We have already overspilled the conventional boundaries of the Renaissance by alluding to Robert Boyle and the Royal Society of London. To go further: eighteenth-century Britain saw protracted anxiety over the possible political allegiances of natural philosophy. For example, both Whigs and Tories claimed Newton's philosophy as support for their views of the social order, and Leibniz's natural philosophy was suspected by them of giving aid and comfort to religious 'free-thinkers'. Two generations later, the natural philosophical work of Joseph Priestley would be intimately connected with his beliefs as a religious dissenter and with his radical politics.[26]

In sum, natural philosophy was never a socially-disengaged, purely intellectual, activity and natural philosophical pronouncements were believed to entail assertions about the political order. If we ask how some of these pronouncements became compelling in the presence of alternatives, then we shall begin to uncover the public concerns and interests that Renaissance natural philosophers grappled with, and our histories of their sciences and their society will be enriched.

Notes

1 On Menocchio (1532-1599) and his examinations by the Inquisition, see Carlo Ginzburg, *The Cheese and the Worms: The Cosmos of a Sixteenth-Century Miller*, trans. John and Anne Tedeschi, Baltimore, 1980.

2 See F. A. Yates, *The French Academies of the Sixteenth Century*, London, 1947, and R. J. W. Evans, *Rudolph II and his World: A Study in Intellectual History, 1576-1612*, Oxford, 1973.

3 See Lucien Goldmann's (uninvitingly titled) *The Hidden God: A Study of tragic vision in the 'Pensées' of Pascal and the Tragedies of Racine*, trans. Paul Thody, London, 1964, and Peter Dear, 'Pascal, miracles and experimental method', *Isis*, forthcoming.

4 Of Campanella's many works, his *City of the Sun* and *Discourse Touching the Spanish Monarchy* (written in 1602 and 1624 respectively, and only published after many years' circulation among patrons and friends) are particularly relevant here; Campanella and Andreae are surveyed by Frank E. and Fritzie P. Manuel, *Utopian Thought in the Western World*, Oxford, 1979.

5 See, for example, L. W. B. Brockliss, *French Higher Education in the Seventeenth and Eighteenth Centuries: A Cultural History*, Oxford, 1987.

6 These points are gleaned from Anthony Grafton and Lisa Jardine, *From Humanism to the Humanities*, Cambridge, Mass., 1986; R. S. Westman, 'The astronomer's role in the sixteenth century: a preliminary study', *History of Science* 18, 1980, pp. 105-47; J. A. Bennett, 'The mechanics' philosophy and the mechanical philosophy', *History of Science* 24, 1989, pp. 1-28; Mario Biagioli, 'The social status of Italian mathematicians, 1450-1600', *History of Science* 27, 1989, pp. 41-95; and from Nicholas Jardine, *The Birth of History and Philosophy of Science*, Cambridge, 1988.

7 ''Tis all in pieces, all cohaerence gone; / All just supply and all Relation; / Prince, Subject, Father, Sonne, are things forgot, / For every man thinkes he hath got / To be a Phoenix, and that there can bee / None of that kinde, of which he is, but hee …', *The First Anniversarie*, ll. 213-18. Marlowe's Doctor Faustus seeks the attributes of secular rulers through philosophy: 'O, what a world of profit and delight, / Of power, of honour, of omnipotence, / Is promis'd to the studious artisan!', *Doctor Faustus*, I, i, ll. 53-5.

8 See Peter Dear, *Mersenne and the Learning of the Schools*, Ithaca, 1988.

9 Paolo Rossi, *Francis Bacon: From Magic to Science*, trans. S. Rabinovitch, Chicago, 1968, and *Philosophy, Technology and the Arts in the Early Modern Era*, trans. S. Attanasio, London, 1970; Lisa Jardine, *Francis Bacon and the Art of Discourse*, Cambridge, 1974.

10 This account follows my *Francis Bacon: The State, the Common Law, and the Reform of Natural Philosophy*, Cambridge, forthcoming.

11 'Voluntary communities' is Patrick Collinson's apt phrase, in *The Religion of Protestants: The Church in English Society, 1559-1625*, Oxford, 1982; on Elizabethan Puritan politics in general, see his *The Elizabethan Puritan Movement*, London, 1967.

12 Owen Hannaway, *The Chemists and the Word: The Didactic Origins of*

Chemistry, London, 1975, evokes this splendidly.

13 On the philosophical circles around Leicester and Sidney, see E. Rosenburg, *Leicester: Patron of Letters*, New York, 1955, and Frances Yates, *Giordano Bruno and the Hermetic Tradition*, London, 1974.

14 See for example, his *A briefe Apology of certain new inventions*, London, 1593. The Sloane collection of manuscripts in the British Library contains many of Plat's alchemical papers.

15 See N. H. Clulee, *John Dee's Natural Philosophy*, London, 1988.

16 See J. A. Bennett, *op. cit.*, note 6.

17 s.v. 'province', *Oxford English Dictionary*; Sir Thomas Elyot, in his *Dictionary* (1538) remarked that 'Provincia is sometyme taken for a rule or autoritye of an officer: also an office, alsoo for a countraye or royaulme.'

18 *The Letters and Life of Francis Bacon*, edited by James Spedding, 7 vols, London, 1861-74, I, pp. 108-9. A 'rover' is a pirate or marauder.

19 *Ibid.*, III, p. 315. Such fears help explain Bacon's approval of *Jesuit* colleges for lay learning; they too, he thought, were opposed to 'modern looseness' in learning and 'moral matters', and he declared 'they are so good that I wish they were on our side', *The Advancement of Learning* (1605), in *The Works of Francis Bacon*, edited by J. Spedding, R. L. Ellis and D. D. Heath, 7 vols., London, 1857-61, III, pp. 276-7.

20 *A Brief Discourse Touching the Happy Union of the Kingdoms of England and Scotland, Letters*, III, pp. 90-2. Allusions to traditions of ancient wisdom were a commonplace in the Renaissance; see D. P. Walker, *The Ancient Theology*, London, 1972. It is worth noting that Bacon is *not* drawing an analogy here between political law and 'natural law' (a much championed argument, particularly in the formulation of Thomas Aquinas); he is alluding to hidden structures ('laws') in nature.

21 For example, see Charles Webster *The Great Instauration: Science, Medicine and Reform, 1626-1660*, London, 1975.

22 John Dury, *Considerations tending to the Happy Accomplishment of England's Reformation in Church and State*, London, 1647, p. 47. My emphasis.

23 Webster, *op. cit.* (n. 21), p. xv.

24 For example, Paul Wood, 'Methodology and apologetics: Thomas Sprat's History of the Royal Society', *British Journal for the History of Science* 13, 1980, pp. 1-26; Michael Hunter, *Science and Society in Restoration England*, Cambridge, 1981.

25 See Steven Shapin and Simon Schaffer, *Leviathan and the Air Pump; Hobbes, Boyle and the Experimental Life*, Princeton, 1985.

26 Consider Margaret C. Jacob, *The Newtonians and the English Revolution 1689-1720*, Ithaca, 1976; J. Martin, 'Explaining John Freind's *History of Physick*', *Studies in the History and Philosophy of Science* 19, 1988, pp. 399-418; Steven Shapin, 'Of gods and kings: natural philosophy and politics in the Leibniz–Clarke disputes', *Isis* 72, 1981, pp. 187-215; and *Science, Medicine and Dissent; Joseph Priestley (1733-1804)*, edited by R. G. W. Anderson and Christopher Lawrence, London, 1987.

6

The Church and the new philosophy

Peter Dear

The pre-eminent social institution in the Europe of the sixteenth and seventeenth centuries was the Christian Church. A full understanding of the social place and meanings of the natural philosophical innovations that occurred during the period therefore demands consideration of the Church's role: the ways in which it incorporated those innovations, resisted them, or, more generally, constituted the context in which they were evaluated and used.

In fact, the events of the Reformation and Counter-Reformation mean that simply speaking of 'the Church' is inadequate in characterizing the cultural realities of the period; rather, the context for the ecclesiastical shaping of natural knowledge is set by the continual interactions of 'the Churches', Catholic and Protestant. That broad confessional divide is itself complicated further by the differences between Protestant denominations as well as among Catholics, especially those of different national traditions; some of the fine-structure of the issues to be examined in this chapter turn on precisely such disagreements. Two main themes will be considered. One is doctrinal or ideological, concerning ideas about God and the created world together with the policing of those ideas by the ecclesiastical apparatus. The other is institutional, concerning the formal social settings in which knowledge about nature was produced and taught: universities, colleges and schools run or heavily influenced by religious orders or denominations.

1 Natural philosophy and theology in the early sixteenth century

Contrary to a longstanding myth owing much to anti-ecclesiastical

Enlightenment thinkers of the eighteenth century, the high Middle Ages did not witness the imposition of a monolithic world-view by the Catholic Church that stifled independent creative thought about nature. In the thirteenth century the adoption of the works of Aristotle by the newly founded universities had provoked a crisis in traditions of academic theology that required accommodations to be made between views of the physical world as the product of God's free creation and criteria of rational inference about nature found in Aristotelian philosophy. A variety of approaches resulted. The two greatest synthesisers in the scholastic tradition were the Dominican Thomas Aquinas and the Franciscan John Duns Scotus. Aquinas, the greatest of the Christianizers of Aristotle, held that philosophy, in the sense of the independent use of human reason, was a proper handmaiden to theology, in the sense of a science based on divine revelation. There could be no conflict between truths of philosophy and truths of theology; the two were complementary and harmonious. The truths of philosophy, furthermore, were those indicated by Aristotle, including ideas about the natural world. Scotus differed from Aquinas in certain aspects of his philosophy, being in particular less strictly Aristotelian and more sympathetic to the older Platonic currents in Christian theology stemming from St Augustine.

The adoption of their differing philosophical perspectives in the fourteenth century did not entail the exclusive sanctioning by the Church of one at the expense of the other; different religious orders and different individual thinkers selected for themselves what they found useful or convincing. The early fourteenth-century Franciscan philosopher and theologian William of Ockham was considerably more controversial, denying basic elements of Aristotelian thought in favour of a philosophy that stressed God's absolute freedom to make the world in any way He chose – in contrast to Aquinas's belief in the power of human reason to discover much of God's work unaided by divine Revelation. Despite the political and theological difficulties in which he became embroiled, however, Ockham's philosophical ideas (known as 'nominalism') became widely influential over the next two centuries, in Italy as well as in northern centres such as Paris. These facts indicate both the lack of enforced philosophical uniformity by the Church throughout the later Middle Ages and early Renaissance and the lack of a single set of all-encompassing doctrines to which all elements of the Church subscribed.

By the beginning of the sixteenth century, therefore, ecclesiastical control of the philosophical ideas promulgated in the universities was, as earlier, restricted in practice to securing the boundaries of admissible doctrine rather than determining its content. There were certain positions that it was not permissible for philosophers to uphold, because of direct conflict with central tenets of theology – such as the immortality of the soul or the creation of the world. However, both of these, and others, were denied in Aristotle's writings, which formed the basis for philosophical instruction. Scholastic philosophers had developed a number of ways to deal with the problem. Aquinas had concentrated on 'correcting' Aristotle without falsifying him, so as to maintain a basic conformity between Aristotelian philosophy and theology; the Church leadership was already tending to favour Aquinas's position, but this was not in any sense an absolute orthodoxy.[1]

None the less, practically all educated Europeans in 1500 subscribed to one basic cosmological picture. Derived from Aristotle's account of the universe, but sharing its broadest features in common with that of Plato, it placed a spherical earth at the centre of a finite universe bounded by the sphere sustaining the fixed stars. The sun, moon and planets revolved on additional nested spheres of their own serving to carry them around the stationary earth. On the most common (but by no means canonical) interpretation, the celestial spheres, made of immutable 'aether', rotated by virtue of disembodied intelligences, usually identified with angels, which moved them in the same way as the soul moves the body. Aquinas had adapted an argument of Aristotle's to demonstrate on that basis the existence of God. Aquinas held that the ceaseless and unchanging rotation of the spheres cannot be explained (as Aristotle could explain the self-moved progress of animals) by the striving of their associated intelligences for their intended destinations, because the motion is circular; it does not go anywhere. Furthermore, there is no activity of motion beyond the spheres to provide any external motive force. The cause of the constant, uniform rotation must therefore be the striving of the intelligences to imitate the ceaseless, unchanging perfection of some transcendent Prime Mover, identifiable with God.

Although its static, hierarchical and theologically-integrated character, mirroring the authority and social structure of the Church itself, guaranteed this picture its high degree of established acceptance, it did not constitute official Church dogma. Its undogmatic status is rendered especially evident by the case of Nicholas of Cusa in the fifteenth century. Nicholas argued for a non-Aristotelian cosmology, rooted in theological and metaphysical considerations concerning God's omnipotence, which involved an infinite universe and an infinite multiplicity of other worlds. These unorthodox ideas, appearing in 1440, did not, however, prevent Nicholas from being made a cardinal in 1446 and from holding a number of other important ecclesiastical positions. Clearly, philosophically heterodox views concerning the natural world were not in themselves seen as a serious threat to theological orthodoxy.[2]

Deviance in natural philosophy began to appear more dangerous with the advent of the Protestant Reformation. While not necessarily conducive to heresy and social disorder, departure from tradition in matters concerning any aspect of learned culture could plausibly be seen by Catholic authorities as potentially disruptive of established authority. Nicolaus Copernicus's remarks in the dedicatory letter prefixed to his *De revolutionibus orbium coelestium* ('On the Revolutions of the Celestial Orbs') in 1543 express sentiments appropriate to a period of acute doctrinal anxiety. Addressing Pope Paul III, Copernicus, a Church canon, acknowledged that his doctrine of the earth's daily motion on its axis and annual motion around the sun was liable to meet with stern opposition. Indeed, the dedication to the Pope himself rather than to a local patron may perhaps be seen as a pre-emptive strike. The theological dangers were, Copernicus thought, real; he warned of 'idle babblers' who, though ignorant of mathematics, might condemn his work on the basis of 'twisting' scriptural passages.

Two years later the Council of Trent was convened to address those scandals which had been partly instrumental in provoking the ecclesiastical rebellion of the Protestants, and to respond to the Protestant challenge itself. The crisis impelled the leaders of the Catholic Church to impose strict disciplinary controls on moral errancy within its ranks; at the same time, the newly-flexed hierarchical muscle was applied to doctrinal matters. Just as concubinage was now firmly ruled to be an unacceptable interpretation of celibacy, so theological theses and devotional practices were subjected, after lengthy debate, to many restrictions and clarifications. Above all, the Counter-Reformation for which Trent came to stand signalled the determination of the Catholic Church to control dissent. Henceforth, many things would matter that once had been tolerated or ignored.

2 Doctrinal and institutional developments of the second half of the sixteenth century

The Council of Trent met intermittently from 1545 to 1563. Among its decisions relating to ideas about the natural world and how to know it were the following. First, because the large number of Catholic miracles served in polemical attacks on miracle-bereft Protestants as evidence of divine approval (continuing to do so throughout the sixteenth and seventeenth centuries), Catholic stress on this argument rendered careful management of miracles of great importance. Accreditation of miracles was accordingly brought under formal ecclesiastical control, whereby a judicial procedure was established to review claimed instances of miraculous events. Since miracles were defined as events caused directly by God that suspended or violated natural processes, nature itself had on this view to be seen as governed by firm and unalterable regularities: miracles were only possible if there were natural regularities to violate. Catholic orthodoxy therefore encouraged a knowledge of nature that stressed order and intelligibility rather than disorder and caprice. At the same time, divination and magic, including astrology, were firmly repudiated.[3]

Secondly, the Council approved rules for the proper interpretation of Scripture. These were designed to combat the Protestant view that the individual believer, properly illuminated by the Holy Spirit, could determine the true meaning of the text unaided by ecclesiastically-sanctioned authoritative readings. The Council made the authority of the established interpretation of any particular biblical passage paramount, with especial weight given to the consensus of the early Church Fathers.[4]

These decisions were to have considerable consequences for dealing with apparent conflicts between biblical statements about the natural world and assertions of natural science, conflicts of exactly the sort foreseen by Copernicus.

The effort to control theological doctrine as a means of demarcating admissible Catholic teaching from heresy also involved the elevation in 1567 of St Thomas Aquinas to the status of an official Doctor of the Church.[5] His Aristotelian natural philosophical worldview together with its epistemological foundations therefore became implicated in the new Catholic regime. Although variance from Thomistic-Aristotelian teachings in natural philosophy did not usually itself constitute religious deviation, adherence to those teachings became for the most important of the teaching orders, the Dominicans and the Jesuits, a matter of prudence and propriety. One central theological doctrine, however, raised the possibility of direct conflict with natural philosophy. In determining the proper Catholic response to various Protestant interpretations of the Eucharist, the Council of Trent had elevated the characterization first developed by Aquinas into a matter of orthodoxy: the doctrine of transubstantiation was interpreted in Aristotelian philosophical terms that established precisely how it should be regarded as miraculous. The bread and wine, contrary to a number of alternative Protestant views, truly became the body and blood of Christ.[6] The miracle consisted in the retention of the 'accidents', the sensible properties, of the bread and wine in the absence of their actual 'substance', the stuff itself, to act as the possessor of those properties (the accidents of the bread and wine were not allowed to subsist in the body and blood of Christ itself). This central element of Catholic doctrine thus became closely associated with Aristotelian teaching on the metaphysics of substance, which necessarily involved questions of the nature of matter.

Finally, the *Index of Prohibited Books* was established in this period as a policing mechanism to control publication, possession and reading of heretical literature – the chief medium by which the rulings of Trent might be undermined.[7]

Theological and ecclesiastical features of the newly emerging Protestant world also had implications for the study of nature, although the apparatus of enforcement was less effective than that of the Catholic Church. Most famously, in 1539 Martin Luther had casually condemned the opinions of Copernicus as those of a fool who 'wants to turn the whole art of astronomy upside down'.[8] However, this was an isolated instance of such an attitude, although the anonymous preface to *De Revolutionibus* inserted by the Lutheran theologian Andreas Osiander (without Copernicus's knowledge) stressed that the motion of the earth was just a hypothesis designed to yield accurate predictions; it should not be taken as physically true. Osiander's apparent misgivings, similar to Copernicus's – though handled in a less bold manner – were borne out in the early reaction to Copernican astronomy by the intellectual pilot of Lutheranism, a man responsible for the establishment or reform of a number of Lutheran universities in Germany, Philipp Melanchthon. Melanchthon at first dismissed it as gratuitous novelty, but soon softened his attitude in the face of fruitful application of Copernicus's mathematical

– although not cosmological – ideas by astronomers at his own University of Wittenberg. Thus, while rejecting the motion of the earth on scriptural grounds, he accepted Copernicus's book as a useful tool in a true reform of astronomy.

Natural philosophy at the Lutheran universities, as with other fields, followed Melanchthon's prescriptions and suggestions closely. His first impulse when embarking on this task of educational reform was to replace Aristotelian natural philosophy in the curriculum entirely, because of its close association with Catholic scholastic wrangling over its theological implications. He therefore tried using Pliny's *Natural History* as the central text for an entirely new approach. Before mid-century, however, this had proved impractical, being both too restrictive and too disruptive of established patterns of university pedagogy. In the event, Melanchthon himself wrote school commentaries on a number of Aristotle's natural philosophical works.[9]

The direct relationship between the Churches and the study of nature in the second half of the sixteenth century, therefore, was one of potential rather than actual ecclesiastical shaping of natural philosophical orthodoxy. When, in 1553, Jean Calvin ordered the execution by burning of Michael Servetus, nowadays best known for his novel ideas on the movement of the blood through the lungs, he did so because of Servetus's heretical views on the Trinity, not for physiological unorthodoxy. Much the same can be said of Giordano Bruno, burnt at the stake in Rome in 1600 not for his advocacy of Copernicus's moving earth, as used often to be claimed, but for heresy it would have been difficult for the Roman authorities to overlook.

3   A case-study of intellectual and political tensions: the Galileo affair

The potential shaping of natural philosophy by the Churches became actual conflict in the celebrated case of Galileo's condemnation. The central issue concerned the rules of biblical exegesis. Most theological dogma, such as the divinity of Christ or the immortality of the soul, could be acknowledged and ignored by students of the natural world: little of what they had to say could be deemed as challenging that dogma, and throughout this period, with one or two notable exceptions, little was. The arena of relevant intersection became much enlarged, however, if the entire text of the Bible was made the touchstone for doctrine even on non-theological matters.

The Council of Trent explicitly invoked the interpretations of the Church Fathers as supreme authorities in biblical exegesis so as to avoid the freedom of individual interpretation claimed by the Protestants. Unlike Catholic orthodoxy, which placed Church tradition at least on a par with holy writ as a source of religious authority, Protestant theology of whatever stripe treated the Bible, together with the divinely illuminated believer's act of reading it, as the only true basis of Christianity. In practice, however, no one maintained that every word should be read literally, since that would create absurdities such as the attribution to God of human characteristics – unstable emotions like anger, or even the possession of hands or a face – or the untenability of a literal reading of the Song of Solomon. Guidance on appropriate exegetical techniques had therefore always formed an important part of Christian tradition. For Protestants, who insisted on the corruption of the Church since the time of the Church Fathers in the early centuries of Christianity and whose reforms aimed at restoration, St Augustine was an especially favoured authority in this as in all matters.

Augustine's views, expressed in his commentaries on Genesis, required as the first, preferred option in interpreting a biblical passage a literal reading. But other options were also available. In those cases where a literal reading conflicted with the informed judgement of an educated man, as with the attribution of certain inappropriate human characteristics to God or, indeed, the description of the heavens as a tent sheltering the earth (in clear conflict with the demonstrations of Greek astronomical science), it became permissible to take account of the historically-situated expectations of the audience. God, speaking through His divinely inspired servants and prophets, was thus portrayed as a classical rhetorician. The common people to whom, for example, Moses spoke would only have been distracted from the central message if ordinary language and ordinary perceptions of things indifferent to the theological or religious message at hand were replaced by strictly accurate speech. Describing God as angry served its purpose even though it was only a metaphorical characterization; describing the heavens as a tent fitted the simple ideas of the Israelites where speaking of a geocentric spherical universe would have confused them.

This line of argument, subsequently used by Aquinas and well-established in the sixteenth century, offered a powerful tactic to any natural philosopher who might be challenged on grounds of conflict with some statement in the Bible. The (admittedly somewhat heterodox) Lutheran Johannes Kepler invoked it in the introduction to his *Astronomia nova* of 1609, the book in which he set out his first two laws of planetary motion. He maintained that 'the Sacred Scriptures, speaking to men of vulgar matters (in which they were not intended to instruct men) after the manner of men, so that they might be understood by men, do use such expressions as are granted by all, thereby to insinuate other things more mysterious and divine'.[10] Kepler's arguments met with little ecclesiastical opposition, and his astronomical work never led him into difficulties; Protestant churches in general ignored the matter, along with other natural–philosophical novelties. But the clarity, forcefulness, and authority of Augustine's statements also appealed to the Catholic Galileo. His deployment of them under the auspices of a Church that guarded jealously its privileges of authorising theological discussion, and that had the power, especially in most of Italy,[11] to enforce its will, led to a confrontation that Kepler never had to face. The Catholic Church had a greater ability to impose its stamp on its territories than the majority of Protestant denominations could achieve; the differences were not essentially theological.

The story of Galileo's condemnation in 1633 and his earlier encounters with Church authority has been told often, and there are now a number of reliable treatments to replace the sectarian polemics of much of the last century. Galileo had a talent for making enemies as well as friends, and in 1614 a Dominican priest called Tommaso Caccini issued an unauthorized denunciation from the pulpit of his well-known Copernican views. Caccini used the weapon of scriptural quotation, deploying such standard passages as Joshua's command to the sun to stand still (Joshua 10: 12-13), which apparently implied that the sun genuinely has a proper motion through the sky rather than merely reflecting the diurnal rotation of the earth. The issue of the role of Scripture in judging Copernicanism had cropped up in dinner conversation during the previous year among a group that included the Grand Duchess Christina, the mother of Duke Cosimo II de' Medici of Florence (Galileo's patron), and Benedetto Castelli, a protégé of Galileo. Castelli reported the exchange to Galileo, who determined to compose a response to the scriptural challenge. The result, a much-expanded version of his response to Castelli, was Galileo's famous *Letter to the Grand Duchess Christina*, finished in 1615 (not printed, although widely circulated in manuscript, until 1636).

The Letter addresses both the general issue and, more specifically, points made in 1615 by Cardinal Robert Bellarmine, in a criticism of another champion of Copernicanism, the Carmelite priest Paolo Antonio Foscarini. Bellarmine's importance in this matter stemmed from his enormous power in the Catholic Church: he was papal theologian, Father Commissary to the Holy Office (the Inquisition), and wielded immense influence in Vatican affairs. Consequently, Galileo was obliged to confront his views if Copernicanism were to be inserted comfortably into the overall cultural complex of the Catholic world. Bellarmine adhered to a fairly hard line on the issue of scriptural interpretation: of course a literal reading should always be the first resort, to be modified only when there were adequate reasons to do so; but those reasons had to be unusually good, such that even in questions of natural philosophy the words of Scripture needed solid demonstrative arguments, not just likely ones, to justify metaphorical or figurative readings. Thus he told Foscarini that Copernicanism might indeed require reinterpretation of some apparently conflicting biblical passages, but only if it were to be proved with demonstrative certainty – something he doubted could ever be done.

This left Galileo in a bind. With characteristic self-confidence, he accepted Bellarmine's criteria (which also squared with certain Augustinian passages), and then insinuated that Copernicanism could indeed be demonstrated. He made a visit to Rome from Florence in late 1615 and early 1616 especially to forestall any official Church action against Copernicanism in the face of Bellarmine's opposition and the agitation caused by Caccini and others. However, despite his touting of a purported proof of the motion of the earth from the tides, Copernicus's opinion was declared 'formally heretical', Foscarini's tract condemned, and Copernicus's book placed on the Index 'pending correction'. The corrections, finally issued in 1620, concerned those statements in *De revolutionibus* clearly showing that Copernicus regarded the doctrine of the motion of the earth as a fact rather than as a hypothesis convenient for making calculations.[12] Galileo's name appeared nowhere in the 1616 condemnation, although he was certainly at the centre of the business and had private meetings with Bellarmine. He asked for, and received, a certificate from Bellarmine confirming that Galileo had not himself been condemned for supporting Copernicanism. However, there also exists an unsigned and unwitnessed document in the relevant Vatican file to the effect that Galileo had been enjoined by Bellarmine (and had, naturally, acquiesced) never again to maintain or even discuss Copernicanism. This document was later brought forth as the central legal plank on which Galileo's condemnation was based; it has variously been accounted a forgery, or, most recently and persuasively (by Richard S. Westfall), a quiet enjoinder typical of Bellarmine's political style.

Galileo met his fate through over-confidence. Chastened by the 1616 ruling, he kept quiet about Copernicanism for several years. In 1623, however, a fellow-Florentine, Maffeo Barberini, was elected Pope, and he granted Galileo a number of private audiences during the following year which left Galileo believing that he had been permitted once again to discuss the question. He gained this impression from Urban VIII's observation that the motion of the earth, although unlikely to be demonstrable, had never been condemned outright as heretical ('formally heretical' clearly meant something importantly different for Urban). The final result of Galileo's new confidence, the vernacular *Dialogue Concerning the Two Chief World Systems*, appeared in Florence in 1632. Perhaps partly due to suggestions that he was being personally ridiculed, Urban took strongly against Galileo: in a section at the close of the book an argument that Urban had upheld against Galileo is presented by the dialogue's frequently ridiculed straw-man character. It is probable also that Urban felt pressured into responding sternly to Galileo's presumption because of his political position: powerful Spanish interests in Rome disliked Urban's friendliness towards the French and criticized his perceived liberal attitudes; the Galileo affair obliged him to demonstrate his willingness to impose the Church's authority. Thus, in 1633, on the legal basis of the aforementioned document by the long-deceased Bellarmine, Galileo was forced to renounce his views and condemned to house arrest for life. An affair inflamed by issues of proper biblical exegesis ended with an exercise of power by those who guarded the privilege of discussing them.

The fallout from the 'Galileo affair' is difficult to judge. Perhaps the greater power of the Vatican's arm in Italy restrained Copernican speculation there; certainly, it was not a prominent item on the philosophical agenda subsequently in the century (although see below on the Jesuits). In France, however – perhaps the most philosophically vital territory in the Catholic world outside Italy – response to the Church's action was fairly restrained. The independent Gallican traditions of the French Church, always suspicious of Roman interference, seem to have left the matter of official suppression of

Copernicanism a dead letter; the condemnation was never promulgated there. After the initial consternation of French natural philosophers such as Marin Mersenne and Pierre Gassendi – both priests – the ruling was quietly overlooked; sympathetic discussion of the motion of the earth, albeit with due caution, took place throughout the 1630s and 40s without interference. However, Mersenne's frequent correspondent René Descartes, by that time living in the Netherlands, took sufficient alarm at the news of Galileo's fate that he suppressed a treatise detailing the structure of the universe because it required the motion of the earth. Only eleven years later, in 1644, did he publish a greatly elaborated version of that world-system wherein he took care to stress his principle of the relativity of motion so as to avoid implying the reality of absolute terrestrial motion. For Protestants, Galileo's condemnation had a different consequence: it served as a useful piece of anti-Papist propaganda.[13]

4   New natural philosophies and new theories of matter: the problem of transubstantiation

We saw in Section 2 how the Tridentine reforms of the Catholic Church tightened up doctrines on transubstantiation in the miracle of the Eucharist. In particular, the philosophical basis of the doctrine developed along Aristotelian lines by Thomas Aquinas, although not made part of dogma, took on an air of orthodoxy that made speculation on alternative views of matter a delicate undertaking. From the beginning of the seventeenth century onwards, classical Greek atomism began to attract attention from repudiators of Aristotle's natural philosophy. Atomism involved the denial of real qualities in bodies, viewing them as simply artefacts of the sensory process. For Aristotle, properties such as redness or hotness were objective qualities of bodies, residing in them whether or not they were observed. Atomism, as developed by Leucippus, Democritus and, especially, Epicurus, held on the contrary that the appearance of such qualities arose from the effect on the senses of atomic particles of different shapes, sizes and motions – these being the only true properties of material things. Thus, for example, bodies were not in themselves hot; the sensation of heat arose from the agitation against the flesh of appropriately shaped sharp particles, so that heat reduced to something else. However, denial that most human categories of perception corresponded to distinct realities in the world in effect challenged the Thomistic account of the Eucharist.

Aquinas had maintained that the appearances, or 'accidents', of the bread and wine remained even though the 'substances', the underlying things themselves, transformed into the body and blood of Christ. This conceptualization may be compared to the transformation of a Georgian private house into a medical clinic: the appearance of the building may remain basically unchanged, even inside, but its nature has changed from one kind of thing, a house, into another, a clinic. In the miracle of the Eucharist, however, the relevant accidents were almost all qualities of the kind that atomists asserted to be illusory – taste, colour and so forth. On this view, the Aristotelian claim of a real distinction between a thing's accidents, or properties, that our senses detect, and its substance, in which the accidents subsist and which constitutes the thing in itself, disappears. Accidents become merely subjective correlates of objectively identifiable atomic characteristics; if their atomistic substrate – in effect, the 'substance' – changes, then so do they. Thus the philosophical assumptions on which Aquinas's account of transubstantiation was based would no longer hold. Although the Thomistic interpretation was not an article of faith, so that in principle an alternative could always be developed, any implicit challenge to it without the presentation of a theologically acceptable substitute risked the appearance of heresy, since it could be held to deny the reality of transubstantiation altogether.

The seriousness of the issue is revealed in criticisms levelled, once again, at Galileo. In the course of a dispute spanning several years, a leading Jesuit mathematician at the Collegio Romano, the flagship of the leading Catholic teaching order, drew attention to the problems for understanding orthodox doctrine on the Eucharist that were implied by Galileo's remarks, in his book *Il Saggiatore* ('The Assayer'), published in Florence in 1623, about the physical basis of qualities such as taste, odour and colour. Orazio Grassi, in a treatise of 1626, draws attention to Galileo's atomistic account of qualities as sensations within us caused by the motion and shape of particles. He then expresses misgivings derived from 'what we have regarded as incontestable on the basis of the precepts of the Fathers, the Councils, and the entire Church' concerning the miraculous maintenance of the qualities of the bread and wine in the absence of their substance. Since Galileo denied that these kinds of properties were objectively real, being nothing but names attached to subjective impressions, Grassi asks 'would a perpetual miracle then be necessary to preserve some simple names?'[14]

An anonymously-authored document recently discovered among papers at the Vatican relating to the 'Galileo affair' and presumably also dating from the 1620s contains similar observations about Galileo's atomistic speculations. Although it seems unlikely that this problem was the main cause of Galileo's troubles, its potential for undermining precisely the sorts of ideas that were becoming characteristic of the new views in natural philosophy in the early seventeenth century is clear. The question of the Eucharist became a significant weapon in the public arena of debate on the nature of matter and qualities soon afterwards, with reaction to Descartes's *Meditations* of 1641.

Antoine Arnauld, who in 1641 became a Doctor of Theology at the Sorbonne and subsequently a leader of the heterodox Jansenist movement in French Catholicism, wrote the fourth of a series of six 'Objections' appended, together with the author's replies, to the first edition of Descartes's work. Arnauld concludes a number of queries about Descartes's metaphysical arguments and the religious delicacy of the issues they confront with remarks very similar to those raised by Grassi against Galileo, worrying that 'according to the author's doctrines it seems that the Church's teaching concerning the sacred mysteries of the Eucharist cannot remain completely intact'. The problem is that Descartes 'thinks there are no sensible qualities, but merely various motions in the bodies that surround us which enable us to perceive the various impressions which we subsequently call "colour", "taste" and "smell".' Sensible qualities are unintelligible without their underlying substance; after transubstantiation there would be no qualities of bread and wine for God to sustain miraculously.[15]

Descartes's attempts at solution of the difficulty both in his 'Reply' to Arnauld's objections and in letters of 1645 and 1646 to a Jesuit correspondent, Denis Mesland, centred on two principal ideas. One involved stressing that, on his view of the origins of sense-perceptions, only the surfaces of the particles of bodies create sensory impressions. Surfaces, or superficies, are really only interfaces, not substances (in Descartes's terminology, they are 'modes'). Hence replacement of one substance by another, as in transubstantiation, could leave the appearances unchanged as long as the superficies of the particles were unaltered. This still left the problem of what it would mean for the substance to change if the figures and motions of the particles remained the same, in so far as on Descartes's view of matter only those latter properties served to give a substance its nature and characteristics – matter itself was homogeneous and identical with spatial extension. However, Descartes came up with an alternative account of transubstantiation whereby Christ's soul joined with the bread in just the same way as human souls are usually joined with human bodies. The standard scholastic view of human beings was that they consisted of matter informed by soul to yield substance: the soul is what makes humans human. On this point, if on few others, Descartes agreed with the scholastics. He exploited the position by arguing in effect that if Christ's soul informed the bread, then that bread became, by definition, Christ's body. Descartes also suggested, most comprehensively of all, that both of these views of transubstantiation could be combined, so that Christ's soul combines with bread that has itself retained only its immaterial, modal superficies after a physical substantial change.

Descartes did not himself suffer for his questionable reinterpretations, even though they came dangerously close to Protestant views of the Eucharist. His correspondent Mesland was not so lucky: his superiors rewarded his sympathetic contact with Descartes by sending him in 1646 to serve in the Jesuit mission in Canada, where he remained until his death in 1672. In fact, dispute about Cartesian transubstantiation did not become serious until after Descartes's death in 1650, when Cartesianism took strong hold in France as a fashionable new philosophy. Followers of Descartes found that the Eucharistic issue was their most vulnerable point, since it involved the most fundamental aspects of Cartesian physics and metaphysics.

5   The Catholic Church and Jesuit science

The rapid rise of the Jesuit order to pre-eminence in Catholic education from its foundation in 1543 to the end of the sixteenth century is witnessed by both the founding of dozens of Jesuit colleges all over Catholic Europe (the number continuing to grow throughout the seventeenth century) and the reputation for excellence the Jesuits had acquired as educators even, grudgingly, among Protestants.[16]

The full curriculum offered by the Jesuits (although not necessarily covered in its entirety at all colleges) called for nine years of intensive, disciplined study by the mostly teenage boys who formed their clientele. The first six years, the 'course of letters', provided training in Greek and Latin grammar and rhetoric; the final three years, the 'course of philosophy', logic, ethics, mathematics, Aristotelian-style physics, and metaphysics.[17] The priestly teachers of these subjects were themselves typically of high scholarly attainment, and many of them, especially professors of mathematics, played a central part in the innovations of the Scientific Revolution.

The label 'mathematics' covered more than just the so-called 'pure' branches of arithmetic and geometry. It also included mathematical sciences of nature, pre-eminently astronomy (in its instrumental guise as a calculatory and computational science associated with, but not restricted to, practical matters concerning calendars and navigation), and optics. The establishment of mathematics so construed as an important part of Jesuit pedagogy owed much to the efforts of Christopher Clavius, until his death in 1612 professor of mathematics at the Collegio Romano (the same chair later occupied by Grassi). Clavius was the respected chief architect of the Gregorian calendar, which replaced the ancient Roman Julian calendar and put the Catholic world back in tune with the sun in 1582 (and left Protestant countries to catch up; Britain, for example, resisted until the eighteenth century). Clavius pushed hard for mathematics in the 1580s, during debates over a standardized curriculum for the colleges, chiefly on the grounds of its utility, both practical and interpretative (for understanding better the mathematical references in classical authors).

Consequently, from the early seventeenth century onwards there emerged Jesuit professors, themselves products of this educational system, highly adept in astronomy and optics and with corresponding attitudes towards the value of mathematical and instrumental approaches to nature. The official Thomistic Aristotelianism of the order tended to be interpreted quite flexibly: innovation appeared in all areas, including the work of the metaphysician Francisco Suàrez at the turn of the century and the slightly earlier theological ideas of Luis de Molina on free will and predestination. Not surprisingly, therefore, Jesuit astronomers were prepared to compromise in the face of Galileo's telescopic discoveries published from 1610 to 1613, which were presented as destructive of the incorruptible Aristotelian heavens and supportive of Copernicus. When Galileo came to Rome in 1611 to demonstrate his new discoveries of the satellites of Jupiter and the mountainous appearance of the moon, the astronomers of the Collegio Romano fully endorsed his claims. It was clear to them that the

existing orthodoxy was liable to change, and after the 1616 condemnation of heliocentrism the choice of Jesuit astronomers tended to
be for some version of a Tychonic rather than Copernican system,
wherein the earth would remain central and motionless but the planets
would orbit the moving sun. Lectures in the 1570s by Robert
Bellarmine, himself a Jesuit, had already allowed that the heavens were
corruptible (an opinion he later retracted); his discussion of the
heavens as composed primarily of elemental fire rather than aether is
matched in a major work of 1630, *Rosa ursina*, by the leading Jesuit
astronomer Christophorus Scheiner, where sunspots are the chief
object of discussion.[18] Jesuit mathematical scientists generally welcomed innovations such as Galileo provided, although within certain
prudential constraints: works of 1615 and 1620 by Clavius's former
pupil Josephus Blancanus (Biancani) were censured by his superiors
because of their open sympathy for Galileo's ideas.[19]

The new mathematical model soon found its way into Jesuit
natural philosophy proper. Galileo's mathematical studies of motion,
for example, appealing as they did to the model of Archimedean statics,
were examined by a number of Jesuit natural philosophers. These
included Roderigo de Arriaga at the college in Prague and Honoré
Fabri in Lyons as well as the astronomer Giambattista Riccioli.[20]
Much Jesuit work in natural philosophy from the 1620s onwards,
including studies of the phenomena of magnetism and electricity by
Niccolò Cabeo (who also investigated free fall), Athanasius Kircher
and others shows the influence of experimentalism and the use of
mathematically-structured arguments (expressed as 'theorems',
'propositions' and so forth) found in the classical mathematical
sciences, especially optics. By mid-century, Kircher and his pupil
Gaspar Schott were producing works that also investigated themes of
natural magic, especially mechanical and hydraulic phenomena, very
close to the interests characterizing the new mechanically-oriented
natural philosophy of such men as Galileo, Descartes, Mersenne,
Hobbes and Pascal.[21]

It should not be surprising, therefore, that both Descartes and
Mersenne had been educated by the Jesuits (at the Collège de La
Flèche), or that Galileo had from early on in his career close relations
with Jesuits of the Collegio Romano, especially Clavius. The Jesuits,
the intellectual vanguard of the Catholic Church, participated fully in
the new philosophical developments of the seventeenth century even
while having more sensitive regard to such issues as matter theory and

Copernicanism, and to the importance (pedagogical as well as doctrinal) of at least appearing conservative in their attitude towards Aristotelianism. It may even be that the stress on miracles as witnesses of the truth of Catholicism against the Protestants encouraged, as the backdrop against which genuine miracles could occur, a view of nature as a system of knowable regularities – the proper subject of scientific investigation. Priestly participation in philosophical novelty was, in fact, widespread even among non-Jesuits – Fathers Gassendi and Mersenne were among the first to test Galileo's claims about the behaviour of falling bodies.

6 Science and the churches: conflict, mutual accommodation or disinterest?

Tommaso Campanella, a renegade Dominican of the early seventeenth century, recommended natural philosophy as a Catholic tool by which to reconvert the Protestants through distraction and subversion.[22] Although Campanella got himself into considerable trouble for doctrinal deviance (on other grounds), the institutional role of the Catholic Church in the natural philosophical developments of the period up to 1650, and even beyond, was certainly in keeping with his recommendations: it fostered educational norms that embodied a view of the importance of the scientific study of nature, and allowed clerics to pursue that study both as teachers and independent investigators. As long as the new 'mechanical philosophy' associated especially with the work of Descartes remained no more than a contentious speculation, the vanguard of natural philosophical innovation could be maintained even while remaining within the boundaries of post-Tridentine Catholic orthodoxy. By the second half of the seventeenth century, however, a new European scientific community was coalescing around much stronger commitments to a world-view clearly transgressing the limits of scholastic Aristotelianism. The resultant officially-sanctioned reactions against it in Catholic countries thus pushed the Scientific Revolution beyond ecclesiastical succour.

There is no similar picture for the Protestant churches. Protestant church authorities seem to have lacked a certain sureness of touch, or effectiveness, compared to their Catholic counterparts in France, Italy or Spain. The frequent uncertainty attending the design of official university curricula after the break from Rome partly resulted from a suspicion of Aristotelian philosophy because of its association with traditional Catholic theology – as is indicated by Melanchthon's flirtation with Pliny. If any generalization were to be made, perhaps, it might take the negative form of a stress on the weaker shaping of science, and hence a greater latitude for change, in Protestant territories: the introduction of Cartesianism into the universities of the Calvinist Netherlands in the 1640s or into the Academy of Geneva somewhat later, for example, contrasts with the resistance to it in Catholic France. It has also been claimed that the weakening of controls and censorship associated with the activities of Puritan reformers of the Church during the English Civil War and Interregnum of the 1640s and 50s was a factor in the success of English natural philosophy.[23]

Had it not been for the schism of the Reformation and the resultant need for the Catholic Church to define orthodoxy, and hence heresy, more strictly, the study of the natural world might have proceeded more freely than it actually did – at the least, there would have been no 'Galileo affair'. The newly-tightened authoritarian line on biblical exegesis and the greatly heightened sensitivity to potential challenges to the doctrine of transubstantiation formed significant dimensions of the Catholic parameters of legitimate natural science. But on the Protestant as well as Catholic side, knowledge of nature during this period was created in societies powerfully structured by ecclesiastical forces: the Church was a part of the life and thought of everyone.

Notes

1 However, the Fifth Lateran Council in 1513 ruled that the doctrine of the soul's immortality could be positively proved philosophically, thereby putting philosophers on notice: Charles H. Lohr, 'Metaphysics', in *The Cambridge History of Renaissance Philosophy*, edited by Charles B. Schmitt and Quentin Skinner, Cambridge, 1988, pp. 537-638, p. 584.

2 *Ibid.*, pp. 548-56.

3 *Canons and Decrees of the Council of Trent: Original Text with English Translation*, edited by H. J. Schroeder, St. Louis and London, 1941, pp. 217, 276.

4 Robert S. Westman, 'The Copernicans and the churches', in *God and Nature: Historical Essays on the Encounter Between Christianity and Science*, edited by David C. Lindberg and Ronald L. Numbers, Berkeley, Los Angeles, London, 1986, pp. 75-113, on p. 89; *Canons, cit.* (n. 3), pp. 18-20.

5 William R. Shea, 'Galileo and the church', in *God and Nature, cit.* (n. 4), pp. 114-135, p. 115.

6 For Trent see *Canons, cit.* (n. 3), pp. 73-80.

7 Preliminary version 1559; so-called Tridentine Index 1564: Paul F. Grendler, 'Printing and censorship', in *Cambridge History, cit.* (n. 1), pp. 25-53, on p. 46.

8 Westman, *art. cit.* (n. 4), p. 82.

9 Charles G. Nauert, Jr., 'Humanists, scientists, and Pliny: changing approaches to a classical author', *American Historical Review*, 84, 1979, pp. 80-1.

10 Marie Boas Hall, *Nature and Nature's Laws: Documents of the Scientific Revolution*, New York, 1970, p. 74 (from Thomas Salusbury's translation of 1661).

11 With the most notable exception being fiercely independent Venice; Galileo was offered asylum there when called to trial in Rome in 1632: Giorgio de Santillana, *The Crime of Galileo*, Chicago, 1955, p. 208.

12 Owen Gingerich, 'The censorship of Copernicus' *De revolutionibus,' Annali dell' Istituto e Museo di Storia della Scienza di Firenze*, 7, 1981, pp. 45-61.

13 Shea, *art. cit.* (n. 5), p. 132.

14 Grassi translated in Pietro Redondi, *Galileo: Heretic*, Princeton, 1987, pp. 335-6.

15 *The Philosophical Writings of Descartes*, trans. John Cottingham, Robert Stoothoff and Dugald Murdoch, 2 vols., Cambridge, 1984, II, 152-3.

16 John L. Heilbron, *Electricity in the 17th and 18th Centuries: A Study of Early Modern Physics*, Berkeley, Los Angeles, London, 1979, p. 102.

17 There were variations in the curricular structure between colleges and over time; this is the curriculum at the college of La Flèche when Descartes and Mersenne were there: Peter Dear, *Mersenne and the Learning of the Schools*, Ithaca, 1988, pp. 12-13; Heilbron, *op. cit.*, pp. 101-2.

18 William A. Wallace, *Galileo and His Sources: The Heritage of the Collegio Romano in Galileo's Science*, Princeton, 1984, pp. 282-4; Westman, *art. cit.* (n. 4), p. 101; Christophorus Scheiner, *Rosa ursina sive sol*, Bracciano, 1630, p. 690 and *passim*. A letter of 1634 by the Jesuit Athanasius Kircher suggested that Scheiner, among others, would have been a Copernican if not for the condemnation: Santillana, *op. cit.* (n. 11), pp. 290-91.

19 Wallace, *op. cit.*, p. 147 n. 156 and p. 269. Scheiner, on the other hand, was, like Grassi, a Jesuit who had little personal affection for Galileo.

20 Peter Dear, 'Jesuit mathematical science and the reconstitution of experience in the early seventeenth century', *Studies in History and Philosophy of Science*, 18, 1987, p. 174; Stillman Drake, 'Impetus theory and quanta of speed before and after Galileo', *Physis*, 16, 1974, pp. 47-65.

21 Heilbron, *op. cit.* (n. 16), pp. 180n., pp. 183-5.

22 James R. Jacob, *Henry Stubbe, Radical Protestantism and the Early Enlightenment*, Cambridge, 1983, p. 86: the argument was used in 1670 as the pretext for an attack on the Royal Society.

23 Michael Heyd, *Between Orthodoxy and the Enlightenment: Jean-Robert Chouet and the Introduction of Cartesian Science in the Academy of Geneva*, International Archives of the History of Ideas, 96, The Hague/Boston/London, 1982, 'Conclusion'; L. W. B. Brockliss, *French Higher Education in the*

Seventeenth Century and Eighteenth Centuries: A Cultural History, Oxford, 1987, pp. 345-50; Charles Webster, 'Puritanism, separatism, and science', in *God and Nature, cit.* (n. 4), pp. 192-217, esp. pp. 209-10.

PART II

Ture learning, useful arts, foolish superstitions

7

Society, culture and the dissemination of learning

Paolo L. Rossi

In early modern Europe there was no simple relationship between the various social groups, culture (in the widest sense), and the means by which information and knowledge were disseminated. This essay is intended as an introduction to the way technology, politics, religion, philosophy and learning interacted. Aspects of the Reformation debate demonstrate how the popular belief in a variety of prophetic mechanisms could be taken up and manipulated to further a particular cause. Magic and astrology were refined and transformed by the elite into symbols and talismans of power. The new interest in the natural world, fed by the influx of exotica from the voyages of exploration, gave rise to the setting up of collections and the printing of massive compendia which combined the real with the fictitious. Empirical methods, originating in the practices of artisans, were taken up by the elite and used to investigate the laws of nature. Nothing was impossible and everything was worthy of study in this desperate and doomed attempt to order the ever increasing corpus of knowledge so as to reflect, as a microcosm, God's plan for the universe.

1   Eschatology, prophecy and propaganda

In the early Renaissance there was a popular[1] culture common to all levels of society. One strand was the belief in the supernatural; it included such things as astrology, magic, the healing powers of particular stones, herbs, incantations, the symbolic importance invested in all kinds of unusual natural phenomena, and the traditions of mystical prophecies.

Different classes, of course, formulated and understood these beliefs in different ways. In the case of Renaissance astrology, the elite made use of complicated mathematical techniques with an illustrious pedigree. Astrology, as part of astronomy, was enshrined in the *Quadrivium*, the superior course of university study which also comprised arithmetic, music and geometry. It was an essential part of the practice of learned medicine, again an elite preoccupation. For the lower classes, astrology was often restricted to the phases of the moon. The popular calendars and almanacs left out the mathematical and philosophical arguments and gave a brief, digestible summary of those points relevant to agriculture, the weather and medicine. There was however a common denominator, and an exchange of information between all levels of society. This open line of communication allowed the manipulation of public opinion in the first decades of the sixteenth century. The convergence of the new technology of printing, mathematical predictions of potentially cataclysmic planetary conjunctions, and the politico-religious struggles of the period provide an insight into the relationship between power, technology and popular culture.

Political unrest following the French invasions of Italy, the call for a council to reform the Church, the Lutheran Reformation and the great flood predicted for 1524, caused an unprecedented explosion in the production of broadsheets and pamphlets. There had already been a rash of prophecies in the fifteenth century. The year 1484 had been regarded with trepidation by both learned astrologers and popular preachers. Their apocalyptic visions, fired by astrology, pseudo-Joachimite mysticism and millenarian expectations, foretold a radical transformation of the existing social order and a reformation of the Christian world. Every major event was hailed by a host of new prophecies. An anonymous Florentine apocalypse of 1502-3 condemned the foreign invasions of Italy, and spoke out against tyrannical rule and the current state of the Church, prophesying a new republican regime.[2] Others took the opposite view and used their prognostications to welcome the French invasions.

What was new about the wave of prophecies of the early sixteenth century was the use of the printed word to disseminate and highlight the debates, and to combine them with other political, religious and social issues. Between 1500 and 1530 at least 10,000 editions of pamphlets were produced in the Holy Roman Empire; given an average print run of 1,000, then ten million pamphlets were distributed to a population where, at least in Germany, the average literacy rate was five per cent (higher of course in the towns).[3]

The prophecies concerning the flood which began to circulate at the end of the fifteenth century were instigated by the professional men of the universities and courts. It was the ephemerides of Johannes Stoffler and Jacob Pflaum, published in 1496, that announced twenty planetary conjunctions, sixteen of them in the sign of Pisces, for February 1524. Because they were said to mirror the position of the planets in Aquarius during the Great Flood of Noah in the Old Testament, the sluice gates opened, and more than sixty authors produced more than 160 treatises on the imminent disaster.[4] The world of learning was ransacked for supporting evidence and scholars, mathematicians, philosophers and theologians all contributed their expertise.

These learned treatises were unsuited for popular consumption. The printers, always ready to exploit a chance for profit (indeed their venality was blamed for the popular unrest which followed), began to issue broadsheets and pamphlets which, instead of transmitting information, now became vehicles of propaganda. Fears already latent in popular beliefs were used to promote and defend political, religious and social issues. The broadsheets and pamphlets, though written by learned men, were issued in the vernacular (sometimes alongside Latin versions for elite consumption) and the detailed learned apparatus was excluded. They were aimed primarily at literate townsfolk but in the country too the itinerant chapman sold small, inexpensive works. In Germany by 1520 vernacular pamphlets began to outnumber those in Latin.[5]

In town and countryside alike oral transmission (readings by the literate members of society, especially craftsmen, preachers and priests) played its part in the dissemination of information. The high social standing of literate groups helped to give credence to the material divulged. As well as readings, discussion and conversation were an important aspect of this information exchange, particularly regarding such inflammatory material as prophecies of disasters, wars, monstrous appearances and the last days. The scope for lugubrious and inventive embroidery must have been endless but, alas, has passed unrecorded.

A further means of reaching the illiterate majority was the increasing number of woodcut illustrations on the title page of the pamphlets, which gave the message in visual terms. The pamphlets became a sophisticated multi-media vehicle for the rapid dissemination of ideas and propaganda, particularly in Germany. In the Reformation war of words the propaganda potential of this print medium was initially ignored by the Catholics. Where the Protestants employed a wide range of forms to present their message: songs, satires, verse and woodcut illustrations, all common aspects of popular culture, the Catholic pamphlets were poorly presented, written in a ponderous, verbose style and had little appeal.[6] This intellectual arrogance was indicative of the confidence of the Church in its traditional authority. If the sermon, sacred image and ritual had worked in the past, they would surely do so again. What the Catholics, and to a certain extent the Protestants, had left out of their political equation was the power of the printed word to stimulate public debate.

The 'great flood' debate elicited two quite different kinds of prophecies; one was the consolatory prediction, mitigating the effect to a localized and not a universal flood, and offering salvation in return for repentance. This line was taken largely by the pro-papal astrologers. Pope Leo X would hardly have promoted himself as presiding over the catastrophic end of the world, especially when his own astrological destiny, in which he obviously had great faith, foretold peace, harmony, and a new Golden Age.

The more apocalyptic prognostications came from the Protestant camp. The Pope was seen as the Antichrist, who would be swept away, and the faith of the reformers would ultimately be triumphant. Where the flood was prophesied as a localized event, the astrologer located it in the territory of the 'enemy'. The Lutherans placed it in Italy, whereas the Italians saw it affecting Germany. Each side accused the opposing regime of causing the coming catastrophe. After the crucial day in February had passed, reports circulated in Italy that the flood had actually struck Germany, a case of unfulfilled wishful thinking.

Both the religious and secular powers were anxious to turn the tide of popular unrest to their advantage, and though the astrologers can be seen as belonging to separate camps, a clear division between them is not possible. There were factions within each group and many of the treatises have more to do with rivalry between astrologers than the presentation of a particular point of view. Some Catholics promoted reform, and some Protestants were traditionalists. Nor was there a different kind of prophecy for the lower and upper classes, as consolatory and alarmist works were aimed at both.

The response to the flood warnings was widespread apprehension and distress. There is some evidence that the upper classes were disturbed, and tried to make provision for the coming disaster. A report from Vienna in 1523 speaks of the wealthy neglecting their affairs and selling property because they could transport money more easily to the mountains, out of reach of the water. From Rome we learn of noblemen going to the hills on the pretext of hunting, though this report reveals a certain embarrassment about giving credence to the flood warnings.[7]

As 1524 approached the increased number of prophecies reflected the tensions in Europe: the Valois–Hapsburg wars, the unrest in Germany and the spread of the Reformation. Though the eschatological prophecies were centred in Germany and Italy there were echoes in France, and Spain.

The flood literature soon became entangled with the eschatological ferment concerning the coming of the Antichrist, which was stirred by wandering preachers. They drew attention to the evidence of the wrath of God in the natural world and introduced another prophetic strand, in the significance of omens, which revolved around prodigious births, unnatural monsters and natural disasters. Encyclopaedias of prodigies began to appear at the beginning of the sixteenth century, and were used as further signs of divine anger and imminent retribution. Like astrology, prodigies and monsters fueled the politico-religious debate, and were probably more effective at capturing the public imagination. The chronicles of the period abound with reports of monstrous humans and animals, like the mule which in 1514 was reported to have given birth to an elephant in St Peter's square. As God was infinitely powerful and nature infinitely varied, these reports, and the illustrations that appeared in the broadsheets and pamphlets, were accepted as factually accurate by all social groups.[8]

The interpretations of these signs quickly achieved great popularity and both Luther and Melanchthon added their voices to the ensuing battle of images. The discussion of unnatural and natural phenomena was linked to an already existing tradition concerning prodigies and led to vastly different interpretations of particular phenomena, depending on the stance taken.

The case of a deformed calf born in Saxony in 1522 is a good example of this flexibility. Luther, in the pamphlet which he wrote with Melanchthon entitled *On the Papal ass of Rome and the Monk Calf of Freyberg* (fig. 2), cites the prodigious birth of a calf with a bald patch resembling a monk, and uses it in a biting satire criticizing the Catholic clergy. Yet in the *Cronaca* of Tommasino Lancellotti (entry for 8 December 1522) we read

Deuttung der grewlichen figur des Munchkalbs tzu Freyberg in Weyssen gfunden.

D. Martin. Luther.

2   The Monk Calf (*Deuttung der zwo grewlichen figuren Bapstesels zu Rom und Munchkalbs zu Freyburg*, Wittenberg, 1523).

there was brought to Modena a painting of a monster born in Saxony, which has an almost human head, and it has a tonsure and a cloak of hair like that of a friar's habit, and its front arms and its legs and feet are like those of a pig; they say it is a friar called Martin Utero who is dead, who some years ago preached heresy in Germany.[9]

Flood prophecies were even regarded by some as a veiled reference to Luther.

Luther himself did not ascribe to belief in astrology but was quite willing to acknowledge that the devil could intervene in human affairs, and that exceptional natural phenomena and prodigies could be portents. Melanchthon accepted all of these, including astrology; for him it was theologically sound for man to investigate the world of nature in that he saw 'natural and historical events as being a hidden symbolism which related to God's providential design for man's salvation'.[10]

It was among the lower classes, especially in Germany, that the pamphlet war had its most remarkable, if tragic, influence. Luther's appeal to the common man had taken root and flourished. The pamphleteers had helped to spread political consciousness and engender a feeling of involvement in events which had been missing in the past. They clarified issues, and made them the subject of debate which helped to focus a dissatisfaction with the *status quo*. This finally exploded in the violence of the peasants' revolt in 1525. Luther was horrified. His adherence to the existing social order demanded condemnation of the very people he had so greatly influenced and espoused. The representation of the common man as a peasant, which had been a feature of the early Reformation writers and illustrators, came to an end with the peasants' war and the image of the peasant disappeared from popular literature.[11]

The popular culture shared by both the lower classes and the elite allowed political and religious propaganda to be grafted on to prophetic and eschatological beliefs which were already a part of the common mentality. There was a strong link between the high levels of pamphlet production and social tension, and though there was no causal effect between the new technology and the Reformation, it is undeniable that the volume of pamphlet production and the use made of text and illustrations represents a 'typographical revolution coinciding with the early dynamic years of the Reformation'.[12]

2   Printing, authority and censorship

A weapon as powerful as the printed word could not be left uncontrolled. *The Index Librorum Prohibitorum* (list of prohibited books) issued in 1559 is often quoted as the first move in the Counter-Reformation offensive against the spreading of heretical and morally dangerous books. This is not strictly true. Since the invention of printing in the previous century, both the secular and religious authorities had been aware of the dangers posed by this new medium, which by 1500 had already put some 20 million volumes into circulation.[13]

The first laws were enacted in Germany, at Cologne (1478), and Mainz (1485). Successive popes also sought to control the circulation and reading of this rapidly growing mass of books; decrees were issued by Sixtus IV (1479), Innocent VIII (1487), Alexander VI (1501) and Leo X (1515, 1520). After Luther, the secular authorities feared heresy, not only as a corrupter of souls but for the civil unrest it might provoke. A number of towns issued their own laws independent of Rome: Milan (1523, 1538, 1543), Venice (1527, 1543), Mantua (1544), Lucca (1545), Siena (1545), and Naples (1544). Outside Italy, laws and lists were published in Louvain (1542), Cologne (1549, 1550) and the Sorbonne (1542, 1544). In Paris, early censorship was bound up in the relationship between the King, the Sorbonne and the *Parlement*, and issues concerning the French Church predominated. Britain did not escape the net; the bull of Innocent VIII included *Angliae et Scotiae* as places where printing should be controlled. Henry VIII published the first English list of prohibited books and there followed a series of laws by Henry, Edward VI and Mary.

In the face of such an array of bulls, laws and proclamations, we might tend to assume that a solid, impregnable bulwark had been erected by the Church and the secular authorities to keep at bay the new heretical ideas. Nothing could be further from the truth. The humanist respect for knowledge, the lack of administrative follow-up, the struggle for jurisdiction and the financial repercussions on an important sector of the community all contributed to a war of attrition between Church, State and text.

Some measures were aimed at single books, for contingent reasons: the Cologne action of 1478, for example, concerned a work questioning the city council's right to reduce the economic privileges accorded to the clergy. Papal edicts were more general in character, but in the context of an expanding book trade they proved vague and unsatisfactory, giving little in the way of detailed advice on what kinds of books were to be censored, or how this should be effected. They were not even widely circulated. The bull of 1487 was published once during Innocent VIII's lifetime. Alexander VI's bull of 1501 was apparently never published at all, though it did circulate in manuscript form. That of 1515 promulgated by Leo X was printed once individually, and once with the collected decrees of the Lateran Council.[14]

These early pronouncements appear more as guide-lines, and lack any true sense of a response to an immediate threat. The Reformation brought greater urgency, as heretical books were brought into Italy in large numbers and, what is worse, began to be translated into the vernacular. Even then, however, the new laws do not appear to have had any lasting effect. One factor was that the Church still had hopes of a reconciliation with the Protestants. Rome had initially been unwilling to appear dogmatic and intransigent. As positions hardened at the end of the 1530s, the attitude changed. Paul III established the Roman Inquisition (1542), and an attempt was made to control all Inquisitorial activity from the centre. But once again, as Paul Grendler has shown in the case of Venice, the major centre of Italian publishing, the Church's attempts to regulate the printing industry came to nothing.[15] The opposition of the patriciate, and the protests of the bookmen, hindered the enforcement of all pre-1559 attempts at censorship.

In 1548, the Venetian government ordered all works containing material against the Catholic faith to be handed over; a successful appeal was made for the tolerance of non-Christian authors and lamented the effect censorship would have on scholarship. In 1549 the papal nuncio to Venice, Bishop Giovanni Della Casa, was instrumental in producing a list which was the most accurate, to date, in Italy. Della Casa wrote to Rome asking for support, but his list was attacked by the Venetian patriciate and was suppressed. Della Casa himself was a man stranded between two worlds – that of the early Renaissance humanist and that of the Counter-Reformation prelate. He had taken holy orders to further a career, while at the same time continuing his literary and academic studies. His literary production included pornographic verse, hardly a fitting activity for a member of the new, rigid, orthodox Church, and he was ultimately unsuccessful in achieving his ambition to become a Cardinal. Ironically his own poetry was banned in the Index of 1559. Another Index was drawn up in

1552. This time the bookmen questioned why books of a non-religious nature should be banned merely because printed by a non-Catholic, and pointed out how imprecise and vague some of the entries were, and how grave a burden financially this would be for the printers. They also pointed out, quite rightly, that, as this Index was not enforced in Rome, why should Venetians suffer, and not the Romans too? This new Index was also suspended.

The fifth decade of the sixteenth century saw the ascent of zealous reformers to positions of power within the Church hierarchy. Men like Michele Ghislieri, Commissioner General of the Holy Office, and Giampietro Carafa, who was to be elected Pope in 1555 as Paul IV, wanted the Church to have complete control over the faithful and the authority to exercise that control. The Index of Paul IV, issued in 1559, was the first serious attempt to enforce censorship on the whole of Italy. It was not issued first in draft, like the others, inviting comments, but set out the books, publishers and authors to be banned, the rules to be followed, and demanded obedience.

Venice was not alone in its reluctance to carry out the Church's demands. Duke Cosimo I in Florence was careful to do as little as possible when the 1559 Index was issued. He made a show of burning books on 18 March 1559, in front of the Cathedral and in Piazza Santa Croce, but did little else. His main concern was to find out what was being done elsewhere, particularly in Venice, Milan and Naples, and only then to take a decision. This policy was followed by his son Francesco who, when faced with a request to implement the new 1564 Index, wrote his reply at the bottom of the letter: 'Let us wait and see what they are doing elsewhere, so that we shall not be the first'.[16]

The Pauline Index showed just how much church policy depended on individual initiative. On Paul IV's death in 1559, it, too, was suppressed. With Pius IV moderation replaced severity, and the bookmen and scholars were reprieved. Nevertheless, with all these edicts, it is not surprising that individual cases of intellectual 'starvation' should arise, particularly as a result of the prohibition to hold books written by Protestants, or printed by Protestant printers. In his discussion of natural history in the Renaissance, F. David Hoeniger suggests that 'no pope or inquisitor, no Calvin or Puritan, as far as I know, wished any student of plants or animals to be silenced, or any of his books to be destroyed or censored. If some ... experienced censure, they did not do so on account of their botany but because of their religious views.'[17] This assessment needs to be examined carefully.

After the publication of the 1559 Index,the protest against such harsh measures relative to non-religious works brought some concessions from Rome. Cardinal Ghislieri permitted Marco Corner, Bishop of Spoleto, to keep, for a full six months, Leonard Fuchs's *De Historia Stirpium Commentarii* (Commentaries on the History of Species)(1542); as a work printed in Germany, it must have been banned and necessitated a special permit. When the Venetian bookmen explained to the Holy Office the effect which the 1555 list would have on their business, they singled out nine authors whose banning would be unjust and cause financial hardship. Their memorandum lists works by Conrad Gesner and Otto Brunfels in natural history. Brunfels's *Herbarium Vivae Icones* (A Herbal of living images)(1530-31) which had accurate illustrations by Hans Wieditz, a student of Dürer, displayed the new interest in the visual description of plants, and Gesner's *Historia Animalium* (1551) was an encyclopaedic work on animals.[18]

The eminent Bolognese naturalist, Ulisse Aldrovandi, friend and correspondent of dukes and princes, was forced to submit his library to the Inquisition for inspection and had many books confiscated. Aldrovandi felt his studies were being compromised by the Church's pronouncements, and sought exemption via his confessor Francesco Palmio who wrote in February 1559:

> Ulisse Aldrovandi public professor of philosophy and a good and learned person, and who with his lessons gives great service to the University ... is seeking permission to own books set out in the enclosed list ... I pray ... that you will do this because ... with his studies and theories he gives and will continue to give great service to those in his profession.[19]

Aldrovandi had cause to be careful: in 1549 he had appeared before the Roman Inquisition on a charge of heresy. This was eventually dropped, but it was best not to invite further suspicions. Mindful of its university's reputation the city of Bologna had already sought permission, in 1559, for everyone to use non-religious books; this had been refused, and it was up to the individual to ask for permission to consult specific works. Aldrovandi's difficulties with the Inquisition continued after his death in 1605, when he willed his writings to the University. They were sequestered by the Holy Office, which prevented publication until 1615, when one of his former Dutch students, Johannes Cornelius Uterverius was charged with checking the many manuscripts for 'errors' before a permit for printing was allowed – a task that took three years.[20]

The problem was also felt in medicine. The name of Janus Cornarius, an author of medical works, including commentaries on Hippocrates and Galen, had appeared in the Venetian memorandum. In Florence the doctors were quick to make representation to Duke Cosimo concerning the pointless banning of the works of Hippocrates and Galen because they had been edited or printed in Germany. They asked that 'scientia should be left alone, it is so useful to man for his well-being'. Montaigne tells us how his books were taken and examined by the Master of the Sacred Palace, who confiscated a work on Swiss history because it had been translated by a heretic. In 1601 the Congregation of the Index even banned the printed catalogue of the libraries of Cambridge and Oxford Universities.[21]

The discussion of censorship and non-religious works has centred on Italy and it would be interesting to follow this line of inquiry for other countries. Heretical works may have been banned everywhere, however there appears to have been no common attitude towards non-religious works and the subject deserves further investigation. Owen Gingerich's study on the censorship of the editions of Copernicus, points out that this was vigorous, at least in seventeenth-century Italy, less so in France under the Jesuits, and that there was little or no censorship in Spain or Portugal.[22]

Any assessment of the effect of the censorship of the Church and the Indices to learning must also take account of the blanket prohibitions of authors classed under the rubric *Opera Omnia*, where both religious and non-religious works were condemned, and books edited or printed by Protestants were banned, regardless of their subject matter. It is impossible, at present, to quantify the effect of all these measures on study and scholarship. The protests by bookmen, academics and professional men such as doctors linked to the efforts made by local rulers to mitigate the laws passed by Rome, indicate a real problem with no simple solution. Publications were delayed, new ideas did not circulate freely and the book trade may have found the financial risk of carrying suspect stock intolerable. In the sixteenth century the secular authority often cast a blind eye, and relied on the inefficiency of the local Inquisitor. Books were also collected at centres of learning, and retained, awaiting correction, but no *Index Expurgatorius* was produced until 1607. The task of correcting such a number of books would in any case have been impossible. There was neither the man-power nor the information available to delete by hand every offending word, sentence or paragraph. The Inquisition sometimes resorted to desperate measures. At Bologna in 1606 all books coming from Venice were to be confiscated and burned.[23] This, of course, led to subterfuges like the printing of books with false title pages and colophons.

By the time the 1664 Index came to be compiled, the amount of printed matter was such that there could have been little hope of putting together a comprehensive list. In any case a list to be used for seeking out heretics seemed to be no longer the issue. The Index ceased to take an active part in the ideological struggle; it became an historical document. It was evidence that the Church had taken action, a condemnation had been made, and duly recorded.[24]

3   Power, magic, and ritual

The Reformation and the great flood debate brought many of the popular beliefs of the period to the fore. The intervention of popes, princes and intellectuals gave them an institutional stamp of approval. In the world of the elite, these beliefs permeated, and sometimes dictated activities in politics, religion, medicine, court festivities and ceremonies.

Charms, amulets and talismans common in the popular culture of the Middle Ages took on a deeper significance in the revival of Neo-Platonism in Florence, centred on the figure of Marsilio Ficino, in the mid-fifteenth century. A client of Medici patronage, particularly Cosimo the Elder and Lorenzo, his interest in Neo-Platonism, Hermeticism and magic laid great emphasis on the power of the stars which could be used to invest objects with special properties, and to create propitious conditions for specific undertakings. In his *De Triplici Vita* ('On the Threefold Life'), Ficino expounds at length on the role of talismans in sympathetic magic and tapping the power of the stars. There was an accepted philosophical foundation for his theory, as J. D. North writes: 'the standard methods used stones, metals, plants or anything conceived as having the same elementary qualities, or music, incantations prayer, writing, or whatever had the same rational structure, or consonance. Such magical methods using natural substances were acceptable even for Aquinas.'[25] But he drew the line at using engravings on objects, writing or incantations, as these could invoke demons, a fact Ficino himself was aware of. Nevertheless, the objects with which the Medici and other great families surrounded themselves were crowded with talismanic images.

3   Early Italian engraving.

A good example is the helmet shown in an early Italian engraving, which has astrological references to Capricorn and Mars (fig. 3).

Talismans were frequently used in medicine where, as Ptolemy had said, 'the wise man can dominate the stars' to affect the bodily humours.[26] Thus, properly made, talismans in the form of medals, jewels, engraved glass, rings, amulets, pins, and rare stones, were ubiquitous in Europe. The awe in which these magical objects were held is well attested to by the Florentine, Bartolomeo Masi. In his memoirs (entry for 5 April 1492) he tells us how six bolts of lightning hit the lantern of the cathedral, and caused great damage to the church. According to Masi, it was believed that the incident was brought about by Lorenzo de' Medici, who, being ill, had decided to let loose a spirit, which he had kept for many years trapped in a ring. Masi was not a member of the elite, nor were his *Ricordanze* (Memoirs) a work of literature. His report reflects what the people of Florence believed about the activities of Lorenzo and his circle, and the magical power of the objects with which they surrounded themselves.[27]

This was a society in which magic, in its most multifarious forms, was an integral part of the world picture. It may have been perceived and expressed in different ways, but all the readers of, and listeners to, Ariosto's *Orlando Furioso* would have seen the connections between the two worlds, of fiction, and reality.

Astrological symbols were also a means of conveying complex personal and public information. The great painting cycle in the Palazzo Schifanoia in Ferrara executed for Borso d'Este in 1467-70, consists of a complex astrological programme representing the personification of the planetary deities. Borso's brother Leonello is reputed to have changed his clothes each day of the week so that the colours would correspond to the planets ruling that day. In the sixteenth century cycles of astrological paintings became widespread. They had a number of functions, the most important being the glorification of the patron's virtues, and his destiny. In the spectacular decoration of the *loggia* in the Villa Farnesina, owned by the Pope's banker, Agostino Chigi, the precise astronomical data indicated the nativity (the positions of the signs of the Zodiac and planets at the time of birth) of the patron, and set out his characteristics as a child of Virgo; – he would be studious, learned, eloquent, shy, visionary and infertile – not an altogether inaccurate portrait of Agostino Chigi.[28]

Important ceremonies often had to be delayed until the astrologers were satisfied that the heavens would beam down rays of good will. Practically every seigneurial court had its own official astrologer. The marriage of Guidobaldo da Montefeltro and Elisabetta Gonzaga was put off until such a time as court astrologer Ottavio Ubaldini decreed the stars were in a favourable conjunction. It is some testimony to the awe in which astrologer-magicians were held that Ubaldini was later accused by courtiers of having used magic to make Guidobaldo impotent, and of wishing to extinguish the Montefeltro line.[29] After Pope Julius II was elected on 31 October 1503, the actual coronation was delayed until the astrologers were consulted and gave their approval. In Florence, the foundation stones for the Palazzo Strozzi and the Fortezza da Basso were laid at times specified by the astrologers. On the advice of Marsilio Ficino, Lorenzo de' Medici began the construction of his villa at Poggio a Caiano on 12 July 1490, at precisely 9 hours. The villa itself is astrologically orientated so that on Lorenzo's birthday the *oculus* (circular window) in the *salone* is aligned with the rising sun.[30]

Occasionally the patron's time and date of birth presented a quandary for the astrologer. Federico II Gonzaga, the lord of Mantua, was born on the 17 May 1500. As Lippencot has pointed out, this would have given his ascendant sign, crucial in Renaissance analyses, as Taurus, placing him under the influence of the Pleiades. This would have indicated 'transvestism, devotion to luxury, and a devotion to emotional display'. These were not the attributes Federico expected from his astrologer. The solution was simple. Luca Gaurico re-cast the horoscope to give the ascendant in Gemini in a beneficent conjunction with the sun.[31] Luca Gaurico was also responsible for changing Luther's date of birth, so that he could point to a connection between Luther's birth and an evil conjunction in Scorpio.

No ruling house made more consistent use of astrology in all its different forms than the Medici: as a straightforward pictorial representation of the heavens; as personifications; with human beings and animals acting out the Zodiac and planetary conjunctions; in oblique, highly learned allusions, such as the mythological paintings of Botticelli, and Signorelli. These intimate, private possessions are particularly difficult to decipher as the patron was the only person who needed to know the key. Lorenzo's son Giovanni who was elected Pope, as Leo X in 1513, apparently chose both his name and number for astrological reasons.

Leo X's great astrological cycle in the *Sala dei Pontefici* in the Vatican spells out the effects of his spiritual and temporal powers ordained by God and the stars, but should also be read as a 'monumental protective talisman'. Such an attempt at astral magic could have been termed sacrilegious, particularly at the very heart of the Christian Church. The paradox is, that although the clergy condemned such practices in public, many were deeply involved in them in private.

Pope Pius II, who feared the circulation of Hermetic ideas, as a possible danger to orthodoxy, closed the Accademia Romana in 1468, and attacked the astrological symbolism in the church of San Francesco (*Tempio Malatestiano*) in Rimini. Other popes, from Julius II right through to Urban VIII in the seventeenth century also had strong links with astrology. The astrological cycles at the end of the fifteenth century, and up to the mid-sixteenth century, were refined and sophisticated with elegant emblems, and, if propagandistic, were aimed at a very limited audience. Like exotic bookplates, they indicated ownership and learning; even though the astrological messages concerned dynastic themes, they were basically personal and intimate.

Complete Medici domination of Florence and Tuscany came about with the accession of another Cosimo (Cosimo I) to the recently established Duchy of Florence (transformed in 1569 by Pope Pius V and Philip II into the Grand Duchy of Tuscany). The informal rule of earlier Medici was put on a more solid constitutional footing and Cosimo I reorganized the administration on absolutist lines. To these ends he took over the imagery of Lorenzo and Giovanni (Leo X) de' Medici, and used it to the full in a determined propagandistic onslaught. On the title page of Paolo Giovio's *Dialogue on Military and Amorous Emblems*, (fig. 4) we see a goat, Capricorn (Cosimo I's ascendant sign), hovering over Florence. Medici emblems and insignia are to be found in medals, books, villas, churches and palaces. They all became settings for the proclamations of Cosimo's legitimacy as preordained ruler, all the more important because he was descended from a cadet branch of the Medici rather than in direct line from Cosimo the Elder, and had in fact come to power in questionable circumstances. Cosimo used astrology in public far more that any of his predecessors. He employed astrologers to draw up his natal horoscope, charts for his birthdays, and for important events. Important ceremonies connected with his political programme were organized to coincide with particular anniversaries, such as the baptism of his first son, which was arranged to coincide with the anniversary of the battle of Montemurlo.

This astrological propaganda was certainly more widespread and

4   Capricorn over Florence (Paolo Giovio, *Dialogo delle imprese militari et amorose*, Lyons, 1559).

public than that of his predecessors, yet, at the same time, Cosimo and his Court were becoming more separated from the rest of society. The final statement of Cosimo's rise is the painting of the *Apotheosis of Duke Cosimo* in the 'Salone dei Cinquecento' in the Palazzo Vecchio. He is represented as the Divine Augustus, surrounded by astrological symbolism in a quasi-religious icon. Cosimo has been raised above other men, and, like God, can only be approached through intermediaries. In the same way his three palaces, the Palazzo Vecchio (the seat of government), the Uffizi which housed the administration and the courtly Palazzo Pitti were linked by a corridor which is emblematic of the insularity and remoteness of the Italian Court at the end of the sixteenth century. The Grand Duke could move between his various duties without ever coming into contact with the people.

The astrological power which had been vested in painted images began to overflow into extravagant theatrical productions which not only incorporated astrology, but also complex allusions to

160

Hermeticism and a new epistemology. The great Medicean theatre festivals of this period take on the function of a talisman. They are not simply to be seen as a number of separate entertainments such as music, stage machinery, dancing, costumes, and decoration, but as a complex hieroglyph. It was a statement of superiority and detachment exclusively for the elite, and even there, only a few would hold the key of revelation.

Having achieved political hegemony, the court had become a stage on which the ruler articulated his existence. As the separation of Court and State increased – and it can be seen in Mantua as well as Florence – spectacles became an important part of court life, and were planned with great care. The productions were full of erudite and ingenious symbols. The visual display set out the prince's dynasty, ambitions, learning, and power. The Court was asked to contemplate itself, its structure and hierarchy, as well as assuming self-congratulatory praise for recognizing the learned references and emblems. The stage was the seat of government, and the perspective lines of the setting radiated, like the rays of the sun, from the ruler's seat at the centre. The play becomes not so much a representation of the marvellous and the magic, but rather, a magical act in itself.[32]

This magic stemmed from the Neo-Platonic/Hermetic theories concerning the harmony of the Universe; by putting together different elements, one could capture celestial harmony and retain its influence, like the demon in Lorenzo de' Medici's ring. Music was essential to this act. The 'effectiveness of music for capturing planetary or celestial spirit rests on two principles ... that both the universe and man, the macrocosm and microcosm, are constructed on the same harmonic proportions [and] that there is a music of the spheres ... of man's body, spirit and soul ... of voices and instruments'.[33] When, in the fifteenth century, the dome of Florence Cathedral was dedicated, the motet composed for the occasion by Dufay repeated in its proportions those of the architecture. This kind of harmonic relationship probably also exists in the pieces written for theatre festivals. Unfortunately there appears to be a lack of written evidence setting out the ideas behind these compositions.[34]

There is another important aspect of these productions which has not been adequately explored. Strong is correct in saying that the theatre festivals 'encapsulate current Florentine philosophical attitudes [and] scientific ones'.[35] Strong was thinking particularly of the mechanical science displayed in the stage machinery, but in the case of

Cosimo's son Francesco I, the science did not just reside in the elaborate decoration or the spectacular automata. It manifested itself in the performance as a whole, and had more to do with his particular view of *scientia* (knowledge). Francesco was an enthusiastic collector (see below) of *naturalia* (things of nature) and *artificalia* (things made by man). He corresponded with the most important naturalists of his day, and supported the study of natural history in both Pisa and Florence.

Baldinucci's description of the play *The Faithful Friend*, written by Giovanni de' Bardi for the wedding of Virginia de' Medici and Cesare d'Este, shows us how the same interests informed Medicean spectacles. He tells us that the theatre was full of exquisite marbles, varieties of fruit trees, shrubs and flowers, giving off exquisite perfumes; there were mechanical birds and animals, and at the beginning of the performance live birds were set free. The scenery was the city of Florence where perfumes came out of the false chimneys.[36]

Given Hermetic theories concerning the affinities and harmonies in the universe, the act of collecting and ordering could lead to knowledge. The court spectacles, then, with their ritualistic display of *naturalia* and *artificalia*, their music, dancing, singing, costumes and emblems acted out this epistemology. They were also talismanic evocations of harmony, which was, of course, the object of the exercise; to bless the wedding couple, and guarantee their future.

As this court culture developed, the distance between it and the people increased. The Court did not depend on the State, nor was it identified with the State. Courtiers from different centres and countries had more in common than they had with their own countrymen. The Court's spectacles were not just an expression of power, but were the embodiment of that power, brought about by a rite which had its basis in the Neo-Platonic revival of the fifteenth century, with all its Hermetic and Cabalistic ramifications. To this a new interest in the natural world was added. The aim was not to understand nature *per se*, but to explore all knowledge as an aid to revelation.

4   Collecting, *Meraviglia* and experiment

The amount of factual data entering European intellectual circles in the sixteenth and seventeenth centuries presented a serious problem. The new species of plants, minerals, and animals, flooding in from the four corners of the earth did not conform to the knowledge handed

5   The Museum of Francesco Calzolari (*Museum Francisci Calceolari*, Verona, 1667).

down from antiquity. Having collected it, the question was how to organize it. The epistemological structure which evolved to cope with this superabundance of information was rooted in the philosophical currents of the fifteenth century, but it did give rise to new kinds of social and institutional activities.

The framework was provided by the 'museum' or *wunderkammer*, an eclectic collection of natural and man made objects, from the most mundane to the exotic, from the valueless to the priceless (fig. 5). This represented 'a conceptual system through which collectors interpreted and explored the world they inhabited'.[37] Collecting was nothing new. The ruling elite had always gathered precious objects, gems, paintings, etc. which were used to display their wealth and education. Medieval churches were full of sacred relics displayed for the veneration and wonder of the faithful, but this kind of *mirabilia* was not to be the essence of the new collections. It was the encyclopaedic tradition of Pliny the Elder's *Natural History* (AD 77), Isidore of Seville's *Etymologies* (seventh century), Hrabanus Maurus's *De Universo* (ninth century) that was to inspire and lend expression to these new needs. Pliny's work encompassed all aspects of the natural world – astronomy, meteorology, geography, zoology, botany and mineralogy.

Collecting did not constitute a discipline in its own right. It was an activity which embraced all disciplines, *naturalia* and *artificialia*. It eventually became pansophism, the philosophical synthesis which, it was hoped, would lead to a new understanding of the universe. These collections became places of study, and great care was taken in ordering the objects, as this order could reflect the greater order of nature. The microcosm–macrocosm analogy is crucial to an understanding of why these collections were created.

The rationale behind these collections was generally mystical-religious; the Neo-Platonic, Hermetic ideas of correspondences, affinities and harmonies stated that all things were ultimately connected with one another. The task was to establish these connections and thereby achieve revelation. In this desire to order, catalogues became important. When these, in turn, became voluminous, they were reduced to catalogues of catalogues. It was as if these catalogues and lists themselves would produce some new insight. Calzolari's catalogue was even called a printed museum. The macrocosm was continuously distilled, as if its final alchemical essence would emerge as a highly concentrated philosopher's stone.

The collectors, unless the clients of some important patron, were from the upper middle classes, or the elite. The activity was expensive, and, for some, became obsessive. They exchanged specimens, asked for favours, and corresponded with each other in a language which, accepting the rhetorical niceties, transcended their social status. Ulisse Aldrovandi, whom we have already encountered in connection with book censorship, became an enthusiast and corresponded with Medicean Grand Dukes and Francesco Maria II, Duke of Urbino. The exchange of letters between Aldrovandi and Francesco I illuminates their relationship well, the tone of the whole correspondence being that of participating enthusiasts and not of a patron and client.[38] Aldrovandi wanted to collect everything. His 'museum' would be complete in the fullest sense. Filippo Costa, another naturalist, wrote '[that Aldrovandi] to the benefit of all, and to his own glory brings together in his great museum all the simples that can be found in the world, as many vegetable things as minerals, and creatures of sea and land even petrified ones'.[39]

This world also included monsters. These, according to Aldrovandi, were due to a superabundance of male seed, but he also saw them as portents for socio-historical events. In this category of phenomena, God's wrath, and not nature, was still strongly in command. Aldrovandi's works contain illustrations of monsters which he believed existed in foreign lands. These were now being explored for the first time, and were a source for the most exaggerated descriptions (fig. 6). What we find in the late Renaissance is a revival of late medieval interests in the study of monsters which can also be found in the fine arts such as the garden decoration and architecture at Pratolino, Bomarzo, and the Boboli gardens.

The treatises composed by the great naturalists of the period like Aldrovandi and Conrad Gesner were still within a traditional framework. Aldrovandi's, monumental *Pandechion Epistemonicon* (Compendium of Knowledge), strayed little from the Aristotelian categories, though, as a lexicon on everything natural and man-made, it vied with Pliny, whose *Natural History* boasted 20,000 facts gleaned from one hundred authors. This is equally true of Gesner's *History of Animals*, where there is little attempt to organize the animals further than listing them in alphabetical order, after dividing them into classes according to Aristotle.[40] The last great representative of this confident vein of philosophical research instigated by Ficino and Pico della Mirandola was the Jesuit Athanasius Kircher who, in the mid-seventeenth century set up a *wunderkammer* in Rome, and produced a

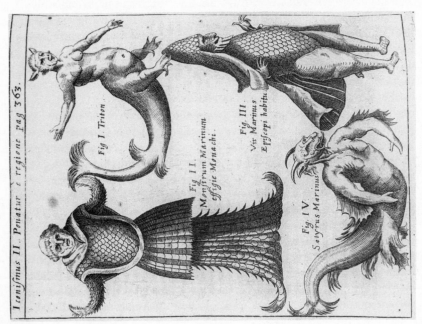

6   Monsters (Caspar Schott, *Physica curiosa sive mirabilia naturae*, Wurzburg, 1667).

series of monumental encyclopaedic works which sought to reunite post-Tridentine Catholicism, Humanism, and the cultures of the world. It included Hermeticism, alchemy, Egyptian hieroglyphics, and numerology – all aimed at a pansophic corpus which, by including everything, would reveal everything.

The sixteenth- and seventeenth-century collections, together with the encyclopaedic mentality that produced the illustrated books of plants, animals, and rocks did, however, result in some important contributions for the study of the natural world. The books contained accurate illustrations, made possible, for the first time, by the printing press. They included not only examples, from the New World of the Americas, and the Indies, but the description and recording of local species. The natural historians checked all the available written material, and in the humanist erudition which they brought to bear on these topics, gave rise, by the very nature and quantity of the material under discussion, to a debate concerning nomenclature and classification.[41]

The desire to organize knowledge,though from a totally different standpoint, was also felt by Alessandro Piccolomini (1508-1579). He came from an upper class family which could boast a Pope, Pius II, among its members, he himself was to become an archbishop. Piccolomini was determined to show that the vernacular was a proper language for discussion. According to him Italian was 'expanding and becoming noble' and capable of expressing every concept. His works did not just contain knowledge, they were concerned with how to divulge it; he gave clear instructions on the correct sequence for reading the text, and referred the reader to other works for clarification He was convinced that knowledge should be available to everyone, and tried to break the vicious circle of privilege defended by the learned elite. He complained that they had attacked his attempts to make information accessible, because 'they do not wish that knowledge should be found in the multitude'.[42]

Collecting was not just an Italian phenomenon. It was particularly popular in Northern Europe, and one of the great centres for this was at the Imperial Court of the Hapsburgs in Bohemia. The Emperor Ferdinand I, according to a Venetian ambassador, was 'a most curious investigator of nature, of foreign countries, plants and animals'.[43] His son Maximilian II, was educated in the *studia humanitatis*, and interested in natural history, particularly that of plants, and collected books and antiquities.

It is, however, their successor Rudolf II who has captured the imagination of historians. During his reign, from 1576 until 1612, he indulged in collecting, alchemy, and occult mysticism. His patronage encompassed artists, astronomers, (Johannes Kepler's astronomical tables, the *Tabulae Rudolphinae*, published in Ulm in 1627, were named after him), and astrologers. Men with a reputation for occult investigation such as the Englishmen John Dee and Edward Kelley were welcomed at his Court.

Rudolf's ideas were not separate from his public, political and religious stance. His education in the Counter-Reformation atmosphere of Spain did not stop him opening his borders to refugees from the religious wars, nor did a man's religion take precedence over his learning. Trevor-Roper talks of Rudolf's attitude as one of religious indifference, but if this was so, it was an indifference born out of a belief that harmony between men could be achieved through alchemical- Hermetic investigation of the natural world. All activities, whether gem cutting, the creation of automata, or the study of natural history, metals, plants, painting and sculpture, would ultimately lead to an insight into God's programme for mankind. Rudolf's collection incorporated those of his father and grandfather, and in 1612, was valued at 17 million gold pieces; unfortunately there is little information about the contents, which were dispersed when Prague was sacked by the Swedish army.[44]

One of the most misunderstood figures at this Court is the painter, Giuseppe Arcimboldo. Trevor-Roper dismisses him: 'No-one can really take Arcimboldo seriously'.[45] His metamorphic portraits, made up of fruits, vegetables, fish, animals and books are often regarded as virtuoso performances, the work of a court jester. This is to misunderstand the works of Arcimboldo, and to read them with our own aesthetic prejudices. They are not peripheral, but come from the very heart of the interests in the natural world, and played an important part in at least one Imperial ceremony. Arcimboldo was not just another jobbing painter. His roots were securely upper middle class. His grandfather had been Archbishop of Milan, and his father an artist in the service of Milan Cathedral. Arcimboldo worked for all three Emperors, and was intimately acquainted with their interests in collecting and natural history. His paintings of the *Seasons* (fig. 7) and the *Elements* were given to Maximilian on New Year's Day in 1569. Each painting portrays a human head made up from a number of examples from a particular category of the natural world; *Spring* uses

AVTVNNO   INVERNO

7   Etching after Arcimboldo, *Autumn, Winter*

flowers; *Autumn* shows fruit and vegetables; *Water* has a head made from different fish, and *Earth* is composed of animals.

Within these paintings are specific emblems relating to Imperial themes.[46] The golden cloak of *Winter* emblazoned with the fire iron is the emblem of the Imperial Order of the Golden Fleece. In the painting *Earth* we have a golden fleece linked to the lion skin of Hercules, a symbol of Bohemia. In *Fire*, the weapons are a reference to the Imperial campaigns against the Turks. There is also good reason to see these paintings as forming one great cycle. The seasons of the year, the elements of the earth (earth, water, air and fire), the animals of the land and sea, and the fruits of the earth represent the sublunar world encapsulated in paint. What we have here, *in parvo*, are the interests and contents of the Imperial collections. The painting of *Water* has twenty-four different species of vertebrates, and twenty-one different species of invertebrates, some of them not indigenous to Europe. The microcosm–macrocosm analogy is used to reduce the variety of creation to its hieroglyph on canvas. The paintings in their selection and assembly of objects create a new intellectual synthesis, and may even have been regarded as talismans.

169

In 1571, for the wedding of the Archduke Charles of Styria to Mary of Bavaria, the paintings of the *Seasons* and *Elements* were acted out in a great allegorical procession, designed and orchestrated by Arcimboldo, where Maximilian II appeared dressed as Winter. The procession comprised the four elements, the four seasons, the four ages of man, the four metals, the four European nations, followed by the great rivers of Europe. According to Gian Battista Fontana, who was involved in organizing the celebrations, the procession was to be read as the Emperor ruling over the microcosm of man; this microcosm, set out harmoniously with each aspect in its correct place in relation to every other, symbolized the harmonious rule of the Hapsburgs.

Although collections sprang up throughout Europe, very few were open to the public, nor would they have been very useful for the dissemination of knowledge; except to evoke wonder at the variety of nature. There was, though, in the sixteenth century, a new way of spreading technical information, which became available to all classes of society. This was the rise in popularity of the 'books of secrets'.[47] They were popular manuals written in the vernacular, offering advice and recipes on a wide range of topics. They represent an example of how authors could act as 'intermediaries between the culture of the lower classes and that of the upper middle-class'.[48]

The writing of these books was a response to a demand for practical information in a cheap and easily understandable form. They were an immediate success, and from 1530, there issued a stream of booklets on a wide variety of topics from astrology and veterinary medicine to cooking and book-keeping. The important point about these works was the absence of technical jargon, and obscure philosophical rhetoric. They contained straightforward recipes which could be easily followed towards a practical end. Printers like Christian Egenolff in Frankfurt (1502-1555), used their education and contacts to virtually create a new readership. The books on metallurgy gave precise instructions for the annealing and hardening of steel. Advice was given on different inks for dyeing, and solvents for removing stains. The impression is one of information based on empirically verified experiment.

The data were of course also available to the upper 'learned' classes, and due to their presentation as recipes, invited experimentation. They soon produced their own books. Recipes were tested, the ingredients presumably altered to see what would happen, and the results recorded. Girolamo Ruscelli listed 1,245 recipes in his 'Secrets'

(*Secreti*, Venice, 1567), each of which was said to have been subjected to experimental verification three times. Men like Ruscelli, who came from the upper classes, did not view the 'books of secrets' as a means of furthering one's career in a trade, or, picking up practical household hints. For him, and others like him, it opened up a field of activity. The mentality that gave rise to the 'museum' and '*wunderkammer*', where collection, classification and cataloguing were used as a means of exploring the secrets of nature, welcomed this added investigative approach.

Academies, based on the model of the numerous literary academies which proliferated in sixteenth-century Italy, were set up where such experiments were carried out. Ruscelli's was called the Accademia Segreta. Another was the almost identically named Accademia dei Segreti of Giovan Battista Della Porta, who gives us an idea of the upper class nature of this pursuit in the preface to his *Natural Magick*:

> nor were the labours, diligence and wealth of the most famous nobles, potentates, great and learned men waiting to assist me. I never wanted [i.e. lacked] also at my house an Academy of curious men, who, for the trying of these experiments cheerfully disbursed their moneys, and employed their utmost endeavours.[49]

Della Porta's aim was also to try out recipes and publish his findings. However, the acceptance of all secrets, curiosities and marvels, with no distinction between possible and probable, affirmation and demonstration, makes it very difficult to grasp the principles of his methodology.

The interests of Grand Dukes Cosimo and Francesco de' Medici in alchemy can be linked to the kind of experimentation found in the 'books of secrets' tradition. Evidence of Cosimo's interest can be gleaned from a manuscript compiled by his secretary Bartolomeo Concini where he mentions the tempering of metals, infusions, recipes concerning majolica and the qualities of metals. In the workshop of Francesco I there were experiments to make rock-crystal and research was conducted to produce an artificial porcelain. We know he was interested in ballistics, experimented with explosives, and had worked on a form of shrapnel. Montaigne reported : 'The same day we saw the palace of the duke, where he himself takes pleasure in working at counterfeiting oriental stones and cutting crystal: for he is a prince somewhat interested in alchemy and the mechanical arts'.[50] His son, Don Antonio, wrote two volumes of *Secrets* dealing with medicine, inks, working precious stones, and how to get rid of head lice.

By the end of the sixteenth century there were certain currents in elite culture which included both the Aristotelian concern with classification, and neo-Platonic/Hermetic philosophy. This gave a framework to the new interest in collecting and experimentation. It was an attempt to explore and discover the secrets of nature, not with the intention of laying bare its rules through a process of induction, or to understand how each separate part works, but to achieve insight into universal principles by creating an harmonic, microcosmic synthesis. There was concern for writing in the vernacular, and for clarity of exposition and expression. The popularity of the 'books of secrets' gave practical advice on how to do things, as well as recipes for further experimentation. All these are evidence of a new epistemology, and of the desire to impart knowledge to a wider popular audience. This process of education was carried out via the written word, and in the visual feast set out in the many collections which would be opened to public view in the seventeenth century.

Notes

1 For a definition and discussion of this term see P. Burke, *Popular Culture in Early Modern Europe*, Aldershot, 1988, pp. 91-115.

2 D. Weinstein, 'The apocalypse in sixteenth century Florence: the vision of Albert of Trent', in *Renaissance Studies in Honour of Hans Baron*, edited by A. Molho and J. A. Tedeschi, Florence, 1971. See also M. Reeves, *The Influence of Prophecy in the Middle Ages*, Oxford, 1969; J. D. North, *Stars, Minds and Fate*, London, 1980, pp. 79-83.

3 H-J. Kohler, 'The flugschriften and their importance in religious debate: A quantitative approach' p. 154; D. Kurze, 'Popular astrology and prophecy in the fifteenth and sixteenth centuries: Johannes Lichtenberger' p. 177, both in *Astrologi hallucinati*, edited by P. Zambelli, Berlin, New York, 1986. p. 154. Burke, *op. cit.*, p. 251, gives a useful sketch of literary distribution in early modern Europe: 'craftsmen were more literate than peasants, men more than women, Protestants more than Catholics, and western Europeans more than eastern Europeans'.

4 See P. Zambelli, 'Many ends for the world, Luca Gaurico instigator of the debate in Italy and in Germany', in *Astrologi hallucinati, cit.*, p. 239; O. Niccoli, Il diluvio del 1524 fra panico e irrisione', in *Scienze Credenze Occulte Livelli di Cultura*, Firenze, 1982, p. 385.

5 See C. Andersson, 'Popular imagery in German Reformation Broadsheets',

in *Print and Culture in the Renaissance*, edited by G. P. Tyson and S. S. Wagonheim, London, Toronto, 1986, p. 122; H.-J. Kohler, *art. cit.* (n. 3), p. 155.

6 R. G. Cole, 'The Reformation in print: German pamphlets and propaganda', *Archiv für Reformationsgeschichte*, 66, 1975, pp. 93-5.

7 For the flood debate and reactions see P. Zambelli 'Fine del mondo o inizio della propaganda' and O. Niccoli, 'Il diluvio del 1524 fra panico e irrisione' in *Scienze, Credenze Occulte, cit.* (n. 4), pp. 291-368 and 369-92.

8 R. Wittkower, 'Marvels of the East', *Journal of the Warburg and Courtauld Institutes*, 5, 1942, pp. 184-7; M. Brusatin, *Arte della Meraviglia*, Torino, 1986, p. 51.

9 From O. Niccoli, *Profeti e Popolo*, Quoted in S. Bertelli, review in *L'Indice*, 4, 1987 (n.7), p. 27.

10 S. Caroti, 'Melanchthon's astrology' in *Astrologi hallucinati, cit.* (n. 3), p. 120.

11 P. Burke, *op. cit.* (n. 1), pp. 259-60; K. P. F. Moxley, 'The Function of Peasant Imagery in German Graphics of the Sixteenth Century: Festive Peasants as Instruments of Repressive Humour' in *Print and Culture, cit.* (n. 5), p. 175.

12 R. G. Cole, *art. cit.* (n. 6), pp. 96, 102; H.-J, Kohler, *art. cit.* (n. 3), p. 157.

13 P. Burke, *op. cit.* (n. 1), p. 250.

14 R. Hirsch, 'Bulla super impressione librorum', *Gutenberg Jahrbuch*, 1973, p. 251.

15 For Venice I have followed the excellent account by P.F. Grendler, *The Roman Inquisition and the Venetian Press 1540-1605*, Princeton, 1977.

16 Archivio di Stato, Firenze, Carte Strozziane, xxii, Serie I, c. 2. For documentation concerning censorship in Florence see A. Panella, 'L' introduzione a Firenze dell'Indice di Paolo IV', *Rivista Storica degli Archivi Toscani*, 1929, i, pp. 11-25.

17 F. David Hoeniger, 'How plants and animals were studied in the mid-sixteenth century', in *Science and the Arts in the Renaissance*, edited by J.W. Shirley, F. David Hoeniger, London, 1985, p. 131.

18 P.F. Grendler , *op. cit.* (n. 15), p. 99, 119; for text of memorandum see p. 296 ff.

19 Quoted by M. Scaduto, 'Lainez e l'Indice del 1559, *Archivum Historicum Societatis Jesu*, xxiv, 1955, p. 20.

20 A. Clarke, *Giovanni Antonio Magini (1555-1617) and Late Renaissance Astrology*, unpublished PhD. Thesis, Warburg Institute, University of London, 1985, p. 93.

21 See A. Panella, *Art cit.* (n. 16), p. 23; Michel de Montaigne, *Italian Journal*, in *The Complete Works of Montaigne*, London, n.d. p. 956; J. A. Tedeschi, 'Florentine documents for the history of the Index', in *Renaissance Studies, cit.* (n. 2), p. 597.

22 O. Gingerich, 'The censorship of Copernicus' *De Revolutionibus'*, *Annali dell'Istituto e Museo di Storia della Scienza di Firenze*, 6, 1981 (n. 2.) p. 57.

23 J. A. Tedeschi, *Art cit.* (n. 21), p. 594; A. Battistella, *Il Sant'Ufficio e La Riforma Religiosa in Bologna*, Bologna, 1905, p. 156.

24 See N. Longo in *Letteratura Italiana – Le Questioni*, vol. 5. Torino, 1986, pp. 980-2.

25 J.D. North, *op. cit.* (n. 2), p. 296.

26 Quoted by M. Mugnai Carrara, 'Fra causalità astrologica e causalità naturale', *Physis*, 21, 1979, p. 40.

27 B. Masi, *Ricordanze*, edited by O. Corazzini, Firenze, 1906, p. 17.

28 M. Quinlan-McGrath, 'The astrological vault of the Villa Farnesina: Agostino Chigi's rising sign', *Journal of the Warburg and Courtauld Institutes*, 47, 1984, pp. 91-105.

29 P. Castelli, 'Gli astri e i Montefeltro', *Res Publica Litterarum*, 6, (December) 1983, p. 85.

30 See J. Cox-Rearick, *Dynasty and Destiny in Medici Art*, Princeton, 1984, pp. 160, 258; C. Rousseau, *Cosimo I de' Medici, Astrology and the Symbolism of Prophecy*, unpublished PhD. Thesis, Ann Arbour, Michigan, 1983, p. 152-77.

31 K. Lippencott, 'Astrological decoration of the Sala dei Venti in the Palazzo del Te', *Journal of the Warburg and Courtauld Institutes*, 47, 1984, p. 220.

32 T. M. Green, 'Magic and festivity at the Renaissance court', *Renaissance Quarterly*, 40, 1987, p. 648.

33 D. P. Walker, *Spiritual and Demonic Magic from Ficino to Campanella*, London, 1975, p. 14.

34 C. W. Warren, 'Brunelleschi's dome and Dufay's motet', *Musical Quarterly*, 59, 1973. I am indebted to Professor R. W. Bray, currently researching this topic, for his comments.

35 R. Strong, *Art and Power*, Woodbridge, 1984, p. 136.

36 Text in L. Zorzi, 'Il teatro e il Principe', p. 167, in *Idee Istituzioni Scienza ed Arti nella Firenze dei Medici*, edited by C. Vasoli, Firenze, 1980.

37 See P. Findlen, 'The museum: its classical etymology and Renaissance genealogy', *Journal of the History of Collections*, 1, 1989, no. 1, for terminology and function of the museum.

38 On 10 September 1577, Aldrovandi wrote:

> I would like to ask a favour of your Excellency; [could you send me] a painting of those two live snakes that you gave me, since I do not have my artist on hand [he was outside Bologna, and ill at the time]. I have not been able to have them painted, and one has died ... and I will let it exsiccate so that I can put it in my Museum.

The Grand Duke's reply is dated 7 October:

> My dearest friend ... I am very pleased with the natural objects you sent me along with the four drawings of the Indian plants which are really beautiful ... and with regard to your request, I have commissioned the painting of the two snakes and will send it to you along with some small examples of my porcelain and crystal. P.S. I shall not be able to send you the above until I return to Florence.

(Reproduced in *Ulisse Aldrovandi e la Toscana*, edited by A. Tosi, Florence, 1989, p. 224, 247.)

39 *La Scienza a Corte*, Rome, 1979, p. 42.

40 For the *Pandechion* see P. Findlen, *art. cit.* (n. 37), p. 66. For *Historia* see F. David Hoeniger, *art. cit.* (n. 17), p. 133.

41 See F. David Hoeniger, art. cit., pp. 133-45.

42 A. Piccolomini, *La Prima Parte delle Filosofia Naturale*, Roma, 1551, see dedication. See also the *proemio* to his *La Prima Parte della Teoriche o vero Speculationi dei Pianeti*, Venice, 1558.

43 H. Trevor-Roper, *Princes and Artists*, London, 1976, p. 88.

44 R. J. Evans, *Rudolf II and his World*, Oxford, 1973, p. 178.

45 H. Trevor-Roper, *op. cit.*, p. 90.

46 For the iconography see T. Da Costa Kaufman, 'The allegories and their meaning' in *The Arcimboldo Effect*, edited by J. Chapman, C. Rathman, J. Roberts, London, 1987,pp. 89–107; F. Porzio, *L'Universo Illusorio di Arcimboldi*, Milano, 1979; R. J. Evans, *op.cit.*, p. 174.

47 I have followed accounts in W. Eamon, 'Arcana disclosed: The advent of printing, the books of secrets tradition and the development of experimental science in the sixteenth century', *History of Science* 22, 1984, pp. 111-50; W. Eamon and F. Paheau, 'The *Accademia segreta of Girolamo Ruscelli*', *Isis*, 75, 1984, pp. 327-42.

48 C. Ginzburg, 'The Dovecote has opened its eyes: popular conspiracy in seventeenth century Italy' in *The Inquisition in Early Modern Europe*, ed. G. Henningen and J. Tedeschi, Dekalb, 1986, p. 192.

49 I quote from the English translation, *Natural Magick*, London, 1658, p. 3.

50 A. Allegretti, *De La Trasmutatione de Metalli*, ed. M. Gabriele, Roma, 1981, p. 16 and p. 17; Montaigne, *Italian Journal, ed. cit.*, (n. 21) p. 930.

8

The challenge of practical mathematics

J. A. Bennett

A new view of the cosmos and a new investigative approach to the natural world are recognized among the most significant intellectual changes of the sixteenth and seventeenth centuries. There is less agreement about the sources of intellectual change. The thesis presented here relies on a contemporary connection and a contemporary disjunction, both of which are characteristic of Renaissance intellectual life and neither of which are familiar today. The connection is between mathematics and certain practical aspects of life, which were becoming increasingly important in Renaissance society. The disjunction is between mathematics and natural philosophy, which dealt with the nature and operation of the natural world. While mathematics had useful applications, it had scarcely any role in natural philosophy, since the world was neither constructed nor organized according to mathematical categories. We must leave behind us the modern images of serious mathematics impossibly distant from the real world and of mathematics supplying the logical and inferential structure for scientific theory.

Renaissance developments in practical mathematics predated the intellectual shifts in natural philosophy. Further, they advanced alongside social, economic and political changes, such as the growth of trade, of exploration and of colonization. Mathematicians came to see their subject as characterized by progress, one whose techniques could be beneficially applied in a number of related practical disciplines known as the mathematical sciences. On the strength of their successes in navigation, cartography and surveying, they asserted its importance and widespread relevance. In time these claims impinged on natural philosophy, and it is significant that the reformed natural philosophy adopted practical mathematical techniques in its new methodology.

Historians have been reluctant to allow crudely constructed links between social and intellectual developments in the period, despite the initial plausibility of a relationship of some kind. Practical mathematics offers a possible medium.

We shall look first at the development of the related mathematical sciences of astronomy, navigation, surveying and cartography, and will then be better placed to see what relevance these may have had to reform in natural philosophy.

1   Astronomy, navigation, surveying

Intellectual historians are naturally reluctant to identify beginnings, but in the case of Renaissance astronomy most would be reasonably content to point to the mid-fifteenth century project of Georg Peurbach and Johannes Regiomontanus – a programme of not only retrieval, but also renewal and reform. No longer was the recovery of *Almagest*, Ptolemy's classic astronomical text of *c.* AD 150, an end in itself; the goal was now the revival of the ancient practice of astronomy, of the Ptolemaic tradition in theory and in observation. This would be realized through the preparation and publication of editions and commentaries, and also through the establishment of an observatory by Regiomontanus, aimed at providing a new observational base for astronomy.

The practical aspect of the mathematical sciences formed part of this Renaissance movement from its inception. Both Peurbach and Regiomontanus were involved with the design of instruments (such as the former's geometrical quadrant for measurement in astronomy and surveying, and the latter's rectilinear dial for time telling); indeed Regiomontanus was a maker and established, along with his printing press, a workshop for the manufacture of instruments. He was naturally attracted to Nuremberg as the most important centre in Europe for metalwork and other crafts, and he cited the availability of instruments as a reason for his settling there in 1471. The observatory he established at Nuremberg was equipped with such traditional instruments as Ptolemy's rulers and the *radius astronomicus*, and his published works included tracts on his versions of the armillary sphere and the torquetum.[1]

The practical interests of Regiomontanus were shared by a number of mathematician-craftsmen in the late fifteenth and sixteenth centuries, and a similar career pattern was followed by such important

177

figures as Bernard Walther, Johannes Werner, Georg Hartmann, Johannes Schoener, Peter Apian, Gemma Frisius and Gerard Mercator. Walther and Werner maintained the tradition established in Nuremberg by Regiomontanus. Both were active in the development of instrumentation, and Walther established his own observatory with instruments similar to those of Regiomontanus. Werner was a designer and maker of instruments, as well as a writer and publisher of books on mathematical science. Schoener was an astronomer, printer, globe maker and cartographer, who settled in Nuremberg in 1526. One of the earliest Nuremberg makers with a successful commercial business was Hartmann, who we know made astrolabes, quadrants, globes, armillaries and clocks. Talented and innovative makers were soon established in other centres, such as nearby Augsburg with Erasmus Habermel and Christopher Schissler. Apian became professor of mathematics in Ingolstadt. He too was involved with the design and manufacture of instruments, and like Regiomontanus had a printing press. Apian was not alone, as we shall see, in combining astronomical and cartographic interests.

The astronomy–cartography connection reappears in a second centre for practical mathematics, established in Louvain by Gemma Frisius, who edited the popular edition of Apian's *Cosmographia*.[2] Gemma's interests fit the now familiar pattern comprising astronomy, instrument design, workshop practice and publication. In charge of the Louvain workshop for more than twenty years was the instrument maker and cartographer Gerard Mercator, whose output included armillary spheres and astrolabes, and he was succeeded by Gemma's nephew, the celebrated instrument maker Walter Arsenius.

One aspect of the practical astronomy of the sixteenth century which is dealt with in the standard textbooks is the work of Tycho Brahe, who built an outstanding observatory on the Danish island of Hven and carried through the kind of observational programme formerly envisaged at Nuremberg. His enterprise, which often seems almost miraculous in its originality, is placed in perspective by the activities of earlier practical astronomers. In particular, Wilhelm IV, Landgrave of Hesse, founded a well-equipped observatory at Kassel with the aim of providing a new star catalogue, and here he employed the instrument maker and designer Joost Bürgi.

Tycho spent the period between 1562 and 1575 in study and travel, deliberately familiarizing himself with instruments and techniques currently in use, and it is clear from his *Mechanica*,[3] where he described his observatory and its instruments, that he knew the work of many of the figures mentioned here. We have already encountered a number of the features usually associated with Tycho's programme: the goal of a new observational basis for astronomy, the design of new instruments together with the use of Ptolemaic patterns, the production of instruments in a workshop associated with the observatory, the existence of a printing press as part of the enterprise so as to disseminate its intentions and results, the use of mechanical clocks for timing observations (previously tried by Walther and by Wilhelm).

Thus the practical aspect was an integral part of Renaissance astronomy, and the achievement of Tycho can legitimately be viewed as its culmination. The vitality of this practical tradition was not confined to astronomy. Indeed many of the same figures appear in the contemporary histories of navigation and surveying.

The infiltration of mathematical techniques into practical navigation began in the early fifteenth century, when exploratory voyages outside the Mediterranean and away from familiar coastlines raised problems for European seamen. Finding latitude was essential to exploring the African coast. Astronomers had long been familiar with relationships between geographical latitude and the appearance of the heavens, and such relationships were built into portable instruments like the armillary sphere, the astrolabe, the torquetum and the 'old' quadrant (*quadrans vetus*).[4] The simplest technique would be to measure the altitude of Polaris (its angle above the horizon, greater in more northerly latitudes) and correct for the star's daily circuit of the pole according to a rule known to the seamen as the 'Regiment of the North Star'. Alternatively a measurement of the meridian or noonday altitude of the sun would require adjustment from solar declination tables – the 'Regiment of the Sun' – which corrected for the variation in the sun's height at different times of the year.

The astronomers provided the necessary rules and tables and their instruments were simplified for navigational use, to produce a portable quadrant for seamen, the mariner's astrolabe and the crossstaff (a portable adaptation of the *radius astronomicus*). The nocturnal was a further marine instrument derived from astronomy, used to indicate the position of the celestial sphere in its daily rotation, and thus, with an instrumental correction for the date, to give the time at night. At the end of the sixteenth century the backstaff, for measuring the noonday altitude of the sun, was added to the range of instruments for latitude.

Longitude presented a more difficult problem, but the same general approach was tried: the astronomers proposed techniques based on relationships familiar to them. In this case they suggested exploiting the corrections that had to be made for local times at different meridians. According to Johannes Werner, the lunar position on the celestial sphere could be measured by the seaman, who would be provided with tables calculated for time at a standard meridian. The difference between standard time and local time (also found astronomically) would be a measure of longitude. Gemma Frisius suggested the simpler expedient of carrying a portable timepiece keeping standard time.[5]

Both these longitude methods were theoretically sound, though not practicable given the astronomical and instrumental resources of the sixteenth century. But even if the solution was incomplete, some celestial techniques were adopted and navigation became a mathematical science, with the seaman dependent on the astronomer and the instrument maker for his tables and his tools. Sixteenth-century mathematicians then tried to extend their sphere of influence into surveying, and they adopted a very similar approach. The textbooks of the period illustrate the ambitions of the mathematicians, who were laying claim to an ever-broader domain of subjects to which their techniques were relevant. Surveying was one of the more obvious candidates.

Surveying, however, was a reluctant convert to geometry, and the history of the subject in the sixteenth century shows how the ambitions of the practical mathematicians were met by the resistance of the ordinary surveyors. The surveyor's traditional emblem of office was a pole; the mathematicians sought to change this to a theodolite, and to introduce the surveyor to the mysteries of angle measuring and trigonometry. This, of course, would enhance the mathematician's influence, but as often as not the surveyor preferred to be left alone with his traditional methods.

The approach that had proved partially successful in navigation was tried in surveying: astronomical instruments were adapted to surveying needs. Indeed the same instruments – the quadrant, the astrolabe and the cross-staff – were modified and recommended to the surveyor by mathematicians such as Apian, Gemma Frisius and the Englishman Leonard Digges. Gemma explained that through the use of triangulation, traditional linear measurement would be minimal and efficiency spectacularly increased.[6] The most appropriate instrument would be the basic components from the back of an astrolabe – a circle in degrees and a centrally pivoted alidade or sighting rule – to form a simple theodolite, i.e. one for measuring only horizontal angles.

A sufficient number of surveyors was reasonably content to go this far, but some mathematicians took their enthusiasm for mathematizing surveying further still and tried to promote impossibly exotic instruments. Among these were the recipiangle and trigonometer types of 'universal' surveying instruments,[7] introduced by a number of maker-designers, among them Joost Bürgi, Erasmus Habermel and Philippe Danfrie. Then there were the altazimuth theodolites,[8] designed by the cartographer Martin Waldseemüller and by the English mathematician Thomas Digges. These designs came at a time when practical mathematicians were making ambitious claims regarding the importance and scope of their discipline, based on a confidence that mathematical science, properly applied, would effect enormous improvements having important social and economic consequences. At this point, however, the mathematicians' enthusiasm met the stubborn resistance of the surveyors, and the more exotic instruments were not adopted.

None the less surveying practice was reformed, and a working compromise established, where the surveyors accepted the simple theodolite along with a few other practical instruments. Among these were the circumferentor (a form of surveying compass)[9] and the plane table. The plane table comprised a flat surface to which was secured a sheet of paper, and a detached alidade, and quickly became popular with the surveyors, not least because its use did not need to involve angle measurement and subsequent 'protraction'. Bearings were marked directly onto the map. The mathematicians were naturally concerned that such an instrument would compromise their programme; its use required no knowledge of angle measurement and the map could be constructed in the field by easily-learned techniques. Surveying of such a kind was scarcely a mathematical science, but in the end the mathematicians had reluctantly to accept the plane table's popularity, though as 'an Instrument onely for the ignorante and unlearned, that haue no knowledge of Noumbers'.[10]

These outline accounts of practical astronomy, navigation and surveying illustrate the contemporary vigour of the mathematical sciences. By the late sixteenth century their practitioners felt themselves to be part of a thriving enterprise that had proved itself in a number of areas and was poised to extend into others. A great many textbooks had appeared, and running through them notions of improvement and progress are unmistakable. Some specifically set out to explain the potential extent of the mathematical programme, and many are strident in their claims for its importance. The most celebrated English example of this kind of promotion is John Dee's extensive and ambitious 'Mathematicall Preface' to the first English translation of Euclid's *Elements* in 1570.

One of the most striking achievements of the practical mathematicians had been the development of instrumentation. Ancient and medieval instruments had been few in number, and mostly confined to astronomy and time telling. The fifteenth and sixteenth centuries saw the addition of new sundials, new types of universal astrolabe,[11] the astronomer's rings of Gemma Frisius,[12] new designs of observatory instruments, the nocturnal, the cross-staff in forms appropriate to navigation and surveying, the back-staff, the mariner's astrolabe, the simple and altazimuth theodolites, a number of universal designs of surveying instrument, the sector as an instrument for surveying and calculating, the circumferentor, the plane table, the graphometer,[13] and so on. The instrument had become centrally identified with practical mathematics, and its development was evidence of the vigour and confidence of the mathematical programme.

2   Cartography and Ptolemy's *Geographia*

We can reinforce this perception of vigour from a different point of view. Contemporary cartography involved many of the figures we have already encountered, such as Werner, Apian, Waldseemüller and Mercator. The French mathematician Oronce Fine is another significant example where cartography combines with astronomy, surveying and instrumentation. Modern disciplinary divisions have distanced cartography from the histories of astronomy, navigation and mathematical instruments, but it is clear from the careers of many leading practical mathematicians that in the period it was part of the same domain. Further, it brought to this domain an important literary resource in Ptolemy's *Geographia*, whose history is crucial to our understanding of characteristics of the practical mathematical programme.

While the *Almagest* provided a fundamental resource for medieval and Renaissance astronomy, the *Geographia* played a similar role for cartography, but we shall see that there were important differences. Medieval cartography in the Latin West was, by Ptolemaic standards, very unsophisticated. The typical medieval world map was a symbolic or allegorical representation, typically circular, surrounded by water, and organized by a T'-shaped division into Europe to the left, Africa to the right and Asia above. Such maps are generally found in religious manuscripts such as psalters, and they scarcely relate to practical geometry at all.

There is a very significant exception to this unrealistic approach to map making, namely the sea charts or portolan charts, which are vellum manuscript charts of the Mediterranean, or centred on the Mediterranean, and whose production dates from the beginning of the fourteenth century. We know of individual cartographers, for example, in Genoa early in the century.

Portolan charts were not based on any projection of the sphere, and in this sense were plane charts, but they were real navigational tools; they were carried on board ship and used to plot courses. It was therefore important that, unlike the symbolic land maps which had other lessons to teach, the charts should conform as closely as possible to the seaman's experience of the world. Further, the activities of the seamen would reveal their inadequacies, so that, again unlike the land map, there was an inbuilt process of improvement and refinement. Lives and livelihoods, ships and their cargoes, safety and efficiency at sea all depended on the reliability of the portolan charts. Here is an example of empirical regulation in the practical mathematical sciences, and it has a significance extending beyond cartography.

While the *Almagest* was rediscovered in Europe from Islamic sources in the twelfth century, the recovery of the *Geographia* came in the fifteenth century from Byzantium. Their influences respectively on astronomy and cartography were equally dramatic. The *Geographia* was a sophisticated cartographic work, with explanations of latitude and longitude and accounts of various projections of the sphere, and so represented a considerable technical jump. Quite apart from the geometry, the maps that accompanied the various manuscripts were strikingly different from existing land maps and their detail and content immediately signalled that this work demanded attention.

In spite of the gap between the first Latin translations of the *Almagest* and the *Geographia*, their widespread impacts were not similarly spaced. The first printed edition of *Almagest* appeared in 1515, whereas the first printed *Geographia*, without maps, was Vicenza in 1475, followed by a Bologna edition with maps in 1477. At least ten subsequent editions of the *Geographia*[14] predated the first printed *Almagest.*

This was not a static tradition, rather it was developing rapidly. The different surviving manuscripts of the *Geographia* were accompanied by different maps from the hands of different cartographers. This diversity was taken over immediately into the printed editions, and the fact that there was no definitive *Geographia* in the first place seems to have encouraged innovation. From the beginning cartographers tried to improve on the originals by including new and better maps and projections, and these new maps were often very fine.

The Strasburg edition of 1513, for example, had no fewer than twenty new maps by Waldseemüller, one of the practical mathematicians caught up in the cartographic enterprise. Another was Werner, who prepared the translation for the Nuremberg edition of 1514. Cartographic activities are common among our practical mathematicians and the links with their other interests are obvious. Not only did map projection depend on a knowledge of geometry, but the actual production of the maps relied on the same engraving skills as the production of mathematical instruments. The enterprise was also closely linked with navigation and surveying. Cartography was part of the practical mathematical sciences, and was one of the most exciting branches in the early sixteenth century.

From the beginning, then, the mathematical science based on the *Geographia* was changing and developing, while that of the *Almagest* was, relatively speaking, much more static. It was recognized from the start that Ptolemy could be improved on, and the improvements that were made encouraged the notion of progress through practical mathematics, and enhanced the confidence of its practitioners. All of this was naturally reinforced by the voyages of discovery – emphatic demonstration that contemporaries had outstripped the *Geographia* and would have to rely on their own resources and initiatives.

Cartography reinforces the picture of a remarkable development in practical mathematical science, especially in the sixteenth century. By the mid-century a confident and assertive tradition was established, supported by an extensive range of textbooks, and had reached England. John Dee complained that in the 1540s, finding no appropriate expertise in England, he had had to learn his mathematics on the Continent, spending time particularly with Gemma Frisius and Gerard Mercator.[15] It was partly due to his enthusiastic promotion that by the end of the century English mathematical science was thriving, with many books published in the vernacular and a new instrument making centre established in London.

3 Longitude, magnetism and natural philosophy

It is important to note that in describing the vigorous development of the mathematical sciences, we have made no reference to natural philosophy. The two domains of activity were largely separate in the Renaissance, but areas of tension and interaction between them were to develop, and in time the ambitions of the mathematicians would extend, not only to practical activities amenable to mathematical techniques and instrumentation, but into natural philosophy itself. The seventeenth century was, of course, characterized by profound change in natural philosophy, and the eventual dominance of theories and methods that privileged mathematics, mechanism and experimentation. If this suggests at least one source in the earlier expansion of the practical mathematical sciences, the emergence of that particular novelty, the instrument of natural philosophy, as a symbol of the conjunction of mechanism and experiment, reinforces this suggestion. The instrument had previously been the exclusive symbol of mathematical science.

There were at least two important areas of intersection between mathematical science and natural philosophy. One has a large secondary literature, the other hardly any. The former is the problematic relationship between the mathematical science of astronomy and the natural philosophy of the heavens. If the detailed, predictive, geometrical account given by the mathematician dealt with the same subject as the qualitative cosmology of the natural philosopher, how did the two relate to each other, and what happened when they were contradictory? By the Renaissance period, the problem already had a long history, and also an established working solution achieved by a separation of responsibilities. The demarcation between questions appropriate to physical cosmology and to mathematical astronomy was well understood, while the subjects co-existed on the basis of a few very basic common premises, such as the central and stationary earth. When mathematical astronomy cast doubt on even these assumptions, the whole question of the relationship was raised again, since a new natural philosophy of the heavens might be required to parallel the new mathematical science.

Historians have for long seen this intersection between mathematical science and natural philosophy, and in particular Kepler's affirmation that the two depend intimately on each other, as the site where new models of scientific practice were born. Mathematical modelling would become linked with physical explanation. Less attention has been paid to a second intersection, which arose within the mathematical science of navigation. Astronomy encroached on natural philosophy when its postulates had implications for the nature of the earth and the planets; navigation did the same when it became concerned with investigating the earth's magnetic characteristics.

The realization that magnetic variation might be different in different locations led to the introduction of a specialized navigational instrument, the variation compass, early in the sixteenth century. This was used to measure variation directly, by comparing solar azimuths with magnetic azimuths, and the local value could then be applied as a correction to the steering compass. Navigational theorists became more particularly interested in global patterns of variation with the suggestion that its measurement might be used as an indication of longitude, since longitude ignorance was the greatest problem of contemporary navigation. Martin Cortés proposed that variation was due to a fixed displacement between the celestial and magnetic poles, so that variation altered in a geometrically predictable way according to latitude and longitude. Since local variation and latitude could both be measured, longitude could thus be found.[16]

Though by no means universally approved, this proposed method for longitude became known in England, following the translation of Cortés's navigational manual in 1561, and it had a number of advocates, who included John Dee. It gave rise to a considerable amount of speculation and investigation, which had natural philosophical implications, but was carried on within the community of mathematical practitioners. It therefore benefited from their techniques, such as the application of mathematics and the use of instruments, while the very problem itself derived from practical experience. From the beginning, then, practical and empirical considerations, as they had influenced the form of the portolan chart, established the ground rules for geomagnetic investigation.

The best known early work in this genre is Robert Norman's book *The newe attractive* of 1581. Norman was particularly concerned with magnetic inclination or 'dip',[17] which for him, as a maker of magnetic compasses, was a practical nuisance, since it was difficult to make compass needles rest level. His book describes an experimental investigation of magnetic dip, aimed at a general, synthetic account. It is self-consciously experimental, and genuinely so judged by the more celebrated examples of the seventeenth century. Norman insists that experience is the guide to reason, and that his conclusions rest on 'exacte triall, and perfect experimentes';[18] one example of an experiment he performs is to weigh a piece of steel before and after magnetization in 'a fine gold balance',[19] and this was repeated by Robert Hooke in 1668 in the unequivocally experimental theatre of the Royal Society.

What is particularly striking about Norman's book is his strong affirmation in the preface that, although an 'unlearned Mechanician', he and his peers have an important role to play in the programme of the mathematical sciences. Indeed, the mechanicians are more effectively placed, having 'the vse of those artes at their fingers endes', than are 'the learned in those sciences beeyng in their studies amongst their bookes'.[20] It is easy to see from this antithesis, that the situation and concerns of the practical mathematicians would commend to them the notion of acquiring sound natural knowledge through experiment. Further, Norman applies the characteristic tool of the mathematical practitioner in his experimental natural philosophy. Instruments were previously confined to practical mathematics, but Norman designs and makes a dip circle, by combining the degree scale on the back of an astrolabe with a horizontally-pivoted magnetic needle. Instruments of natural philosophy have rightly attracted the attentions of historians of the experimental philosophy of the seventeenth century; they are especially potent symbols of a new attitude to the natural world. Here is one considerably earlier, in the context of the mathematical science of navigation.

A similar approach is found in the complementary work by William Borough, *A discovrs of the variation of the cumpas, or magneticall needle*, published together with *The newe attractive*, and here the variation compass is used as an investigative, natural philosophical instrument, rather than as a device for correcting the steering compass. It extends to other such books, such as Thomas Blundeville's *Exercises* of 1594 and William Barlow's *Navigator's supply* of 1597. Works such as these established an experimental and instrumental approach to geomagnetic questions, derived from the nature and methods of the practical mathematical sciences, and since the questions derived from the problems of navigating ships, the study could scarcely have been other than empirically grounded. It was this approach that was taken on by William Gilbert in his *De magnete* of 1600, which became such an influential model for the practice of experimental natural philosophy in England.

It is interesting to note that Gilbert's theories fitted perfectly with changing attitudes to the longitude by variation method in the navigational community. Here the sceptics were prevailing, and the grand geomagnetic theories of Dee's time no longer fashionable. Actual reports of variation measurements were showing that the distribution was unpredictable, and the Dutch mathematician Simon Stevin suggested that position finding by variation would have to be based on a contingent mapping of the variation pattern, derived from extensive measurement rather than theory. Edward Wright translated Stevin's *Haven-finding art* in 1599, and wrote a preface to *De magnete*, pointing to its importance for navigation.[21] Gilbert's theory reinforced this scepticism with a natural philosophical account that attributed variation to local irregularities in the shape of the earth – as contingent as the seamen were finding variation to be in practice.

Navigation, longitude and geomagnetism thus formed a meeting point between mathematical science and natural philosophy, an intersection understood to be relevant to pressing practical problems of the period, and across which individuals engaged in such problems could transfer attitudes and techniques. This intersection continued to attract the interests of the practical mathematicians associated with Gresham College, the centre for the mathematical sciences in England in the seventeenth century. From their starting point in practical mathematics, the interests of successive groups connected with the college moved progressively into natural philosophy as the century advanced, until 'the promotion of natural knowledge' was the dominant concern of that formally constituted Gresham group, the Royal Society.

At one point – as early as 1633 – the Gresham circle explicitly advocated that the demonstrated progress of practical mathematics should be a model for reform in natural philosophy. This came in the context of an expedition by Thomas James to find a north-west passage, in a ship impressively well equipped with instruments for astronomy, navigation and surveying. Many of these instruments were of designs produced or promoted at Gresham, and many more related to geomagnetism. A published account included observations of variation and a longitude determination on land by the simultaneous observation of a lunar eclipse at Hudson's Bay by James and at

Gresham by Henry Gellibrand. An appended address to the University of Cambridge argued the natural philosophical value of such enterprises, and advocated the overthrow of Aristotle, 'that the same improuement may by this meanes accrew vnto our Physicks, that hath aduanced our Geography, our Mathematicks, and our Mechanicks'.[22]

When a lasting reform was established in natural philosophy, its prominent characteristics were indeed closely related to mathematical science. It was based on mathematics and mechanical causality. It was practised experimentally, and claimed to admit no results unfounded in experience. The new experimental and mechanical approach to nature was most emphatically evoked by the use of instruments, formerly exclusively associated with practical mathematics.

Historians of the early modern reform of natural philosophy have failed to appreciate the significance of the prior success of the practical mathematical programme, itself a response to the social, political and economic changes of the Renaissance. Outside astronomy, the mathematical programme has seemed dubiously beyond the academic Pale, or, as John Wallis famously put it, 'the business of Traders, Merchants, Seamen, Carpenters, Surveyors of Lands, or the like'.[23] In fact, the propagandists for the coherence and competence of the mathematical sciences became confident and assured, strident in their claims to importance and relevance in a wide spectrum of human activity. By the end of the sixteenth century the programme was sufficiently established as to become a resource for change in an increasingly insecure and discredited natural philosophy, and must figure in an explanation of why the new dogma of the seventeenth century embraced mathematics, mechanism, experiment and instrumentation.

Notes

1 Ptolemy's rulers formed a fixed instrument, comprising three straight rules pivoted together, used for measuring zenith distance. The *radius astronomicus* was a versatile instrument, used to measure any angle in the sky, where a cross-piece, sliding on a graduated staff, was arranged to subtend the required angle at the eye. The armillary sphere was an instrument used for demonstration, occasionally for measurement, comprising a series of nested rings representing the celestial circles. The torquetum was a portable instrument, ostensibly used for measurement in the different celestial co-ordinate systems, but probably more useful for teaching mathematical astronomy.

2 P. Apianus, *Cosmographia, per Gemmam Phrysum ... restituta*, a number of editions from 1529 (Antwerp) onwards.

3 T. Brahe, *Astronomiae instauratae mechanica*, Wandesburg, 1598.

4 The astrolabe was an instrument for calculation and demonstration, based on a planispheric projection of the heavens; alternative plates were usually provided, engraved with the local co-ordinate system projected for different latitudes. The old quadrant was an horary or time-telling quadrant with built-in adjustments for latitude and solar declination (the annual variation in the height of the sun).

5 For the longitude problem, see D. Howse, *Greenwich time and the discovery of longitude*, Oxford, 1980.

6 In a tract appended to his 1533 edition of Apian's *Cosmographia* (see n. 2).

7 A universal instrument is one claimed to suffice for all surveying needs; both these types depended on forming, with the instrument, a triangle similar (in a geometrical sense) to the one required on the ground.

8 Instruments where, as with a modern theodolite, vertical and horizontal angles (altitudes and azimuths) are measured by a single sighting.

9 The magnetic compass was flanked by a pair of fixed sights; as the sights were rotated the compass card (later needle) indicated the bearing.

10 L. and T. Digges, *A geometrical practise, named Pantometria*, London, 1591, p. 55.

11 One that could be used in any latitude, without the need of different plates.

12 A portable astronomical instrument, also used as a sundial.

13 A surveying instrument popular in France, similar in principle to the simple theodolite, but based on a semicircle rather than a full circle.

14 Rome 1478, Florence 1482, Ulm 1482 (reprinted 1486), Rome 1490, Rome 1507 (reprinted 1508), Venice 1511, Strasburg 1513, Nuremberg 1514.

15 *Philobiblon Society: bibliographical and historical miscellanies*, 1, London, 1854, pp. 6-7; *Johannis, confratis & monachi, Glastoniensis, chronica sive historia de rebus Glastoniensibus*, edited by T. Hearne, 2 vols., Oxford, 1726, ii, 'The compendius rehersal of John Dee his dutiful declaration, and proofe of the course and race of his studious life ...', p. 501.

16 M. Cortés, *The art of navigation*, trans. R. Eden, London, 1579, ff. 64^v-66^v.

17 The angle a magnet, supported at its centre of gravity, makes with the horizontal.

18 R. Norman, *The newe attractive, containyng a short discourse of the magnes or loadstone*, London, 1581, sig. Aiii^v.

19 *Ibid.*, 11.

20 *Ibid.*, Bi^v.

21 W. Gilbert, *De magnete, magneticisque corporibus, et de magno magnete tellure*, London, 1600, address to the author by E. Wright; for an English translation, see W. Gilbert, *De magnete*, trans. P. F. Mottelay, New York, 1958.

22 T. James, *et al.*, *The strange and dangerous voyage of Captaine Thomas James ...*, London, 1633, sig. S4.

23 *Peter Langtoft's chronicle*, edited by T. Hearne, Oxford, 1725, i, pp. 147-8.

9

Doctors and healers:
popular culture and the medical profession

John Henry

Throughout the Middle Ages medicine was in many respects a domestic art. A working knowledge of medical theory and a rudimentary skill in treating the more common accidents and ailments of life were regarded to be essential accomplishments for ordinary life. And, as with other domestic requirements, such as the supply of food, clothing, heat, there was an elaborate network of services available in society at large to support and augment the medicine of home and hearth. During the Renaissance, however, the first concerted efforts were made to remove medicine from the realm of popular culture and establish it as the preserve of a restricted profession. A diverse group of healers, no longer content to take their chances in competition with all other sorts of healers in what has aptly been called the 'medical market-place', began to emphasize the esoteric nature of 'sound' medical knowledge and the dangers of trusting one's health, much less one's illnesses, to practitioners of medicine who were, it was increasingly vigorously proclaimed, ignorant and unskilled. The full story of the professionalization of medicine would have to be pursued into the eighteenth and nineteenth centuries but moves towards it can be clearly discerned in the Renaissance. Medicine had already been established, during the late Middle Ages, as one of the higher faculties in the university alongside theology and law, but the Renaissance saw the widespread introduction of licensing procedures in an attempt to ensure minimum standards among medical practitioners. The aim of licensing, like that of a university degree, was to show that the holder had undertaken a particular kind of vocational training for a certain (often lengthy) time and had satisfied stipulated tests. It also had the effect of enhancing the practitioners' status and their market value.

1   The elite rhetoric of hierarchy

According to the licensed practitioners' own idealized perceptions, or their own rhetoric, there was a hierarchy of different classes of medical practitioner. The university trained physician presented himself as the supreme medical expert, the one with the most complete knowledge of medical matters. Below him were surgeons and apothecaries, both of whom were supposed to have only limited expertise and, accordingly, were expected to defer to the rulings and advice of physicians. The physician was an elite consultant who would diagnose a patient's trouble and draw up a 'regimen' to restore the patient to health. The regimen would often include a course of drug treatment and some-times even a surgical intervention. In either case the patient, or his proxies if he was bedridden, would then have to go to or call in an apothecary or surgeon to carry out the required treatment. In the case of surgery, it was often performed under the scrutiny of the physician to ensure that the surgeon performed his duty properly. An incidental – but by no means accidental – result of this system, of course, was that the patient would have to pay two or even three bills for health care.

Now, there was undoubtedly some truth in the perception and projection of medical institutions as a tripartite hierarchy, but much of the evidence tends to suggest that this was a system honoured far more in the breach than in the observance. In London, for example, the most prestigious licensing body was the Royal College of Physicians, founded in 1518 by order of Henry VIII to promote medical learning and to aid the Crown on matters of public health. The college became the licensing authority for medical practice in London and for seven miles around London. This licensing power was later (1523) extended to the whole country although the college was never able to put this into effect. When the Barber-Surgeons' Company was established in 1540 it was held to be under the protection and control of the Royal College. The Worshipful Society of Apothecaries was formed in 1617 when the Society, separating itself from the Grocers' Guild, was granted a monopoly for dispensing drugs. Here again, however, it was subject to the College of Physicians in so far as its members and their shops were liable to College inspection. On the face of it, therefore, the institutions of London medicine did reflect a three-part division of medical practice. But this was by no means universally true. In Norwich, for example, the Company of Barber-Surgeons became known, after 1550, as the Company of Barber-Surgeons and Physicians. Similarly, in Glasgow the medical licensing authority, established in 1599, was the Faculty of Physicians and Surgeons, while in Edinburgh the licensing authority, one of the oldest in Europe (founded in 1505), was simply the Incorporation of Surgeons and Barbers. Although a College of Physicians was not established in Edinburgh until 1681, this does not mean there were no 'physicians' there in the sixteenth and seventeenth centuries.

Even so, the rhetoric of a tripartite medicine was widespread. For the most part the image was promoted by physicians. John Securis, for example, could take it for granted in the title of his *A Detection and Querimonie of the daily enormities and abuses committed in Physick, Concerning the three parts thereof: that is, The Physicians part, The part of the Surgeons, and the arte of the Poticaries* (London, 1566). Furthermore, it was all too easy to justify the separation of physician from surgeon by reference to the so-called Hippocratic Oath. This oath, which is still included in the collection of writings known as the Hippocratic corpus, is now known to have been written not by a Hippocratic doctor but by a medical follower of the ancient religious sect of Pythagoreans. The Pythagoreans were opposed to all killing and were strict vegetarians, and as the oath makes clear, they were opposed to surgery or inflicting any pain. As Securis quoted it: 'I will not cut those that have the stone, but I will commit that thyng onely to the Surgions... And I wil refraine willingly from doyng any hurt or wronge'. The Pythagorean origin of this oath was completely unknown at this time and most university trained physicians saw it as an ideal laid down by their supreme authority, Hippocrates.[1] There were exceptions, however. Andreas Vesalius, arguably the greatest of the Renaissance anatomists, held the three-fold division of medical science to be the result of intellectual degeneration after the fall of the Roman Empire:

> After the devastation of the Goths particularly, when all the sciences, which had previously been so flourishing and had been properly practised, went to the dogs, the more elegant doctors at first in Italy in imitation of the ancient Romans began to be ashamed of working with their hands, and began to prescribe to their servants what operations they should perform upon the sick, and they merely stood alongside after the fashion of architects ... And so in the course of time the technique of curing was so wretchedly torn apart that the doctors, prostituting themselves under the names of 'Physicians', appropriated to themselves simply the prescription of

drugs and diets for unusual affections; but the rest of medicine they relegated to those whom they call 'Chirurgians' and deem as if they were servants.[2]

It is worth pointing out, perhaps, that the designation 'physician', which Vesalius sneers at here, derived from the Latin word *physicus*, meaning a natural philosopher. That is to say, the university trained medical practitioner saw himself not merely as a healer but as one knowledgeable in all natural phenomena. The Latin word for a healer was simply *medicus*.

Vesalius no doubt exaggerated the case in the preface of his great work but the university education of physicians was extremely bookish in its orientation. When, from the twelfth century, medical faculties began to appear in universities, they modelled themselves on the existing faculties of theology and law. Just as students in those faculties studied ancient texts, the Scriptures or the collections of Roman Law, *The Digest* and *The Institutes*, so medical students were expected to study ancient medical texts. In the same way that theologians and lawyers were trained in rhetoric, for the pulpit and the lawcourt, so the physician was trained to debate before his peers the merits of a particular proposition from the ancient texts. Once this is grasped it is easier to understand how it was that medical knowledge, as taught in the universities, failed, until the sixteenth century or even later, to progress in any significant way beyond the teachings of the foremost ancient Greek medical writings. The idea of a theologian or a lawyer wishing to discard the ancient writings on which they spent so much intellectual energy and try to reach eminence in their calling by some other way is simply absurd. Theology is defined by the sacred writings and full mastery of the law means a perfect familiarity with all the recorded precedents. If medicine was to be a legitimate study within the university system it too must be embodied in ancient texts. Moreover, there was a very widely held conviction among scholars from the twelfth to the seventeenth century that the older a text was the closer it was to true wisdom. Adam may have known all things but man's knowledge had been diminishing since the fall. The ancient medical texts which had been best preserved through the so-called Dark Ages (the period from the collapse of the Roman Empire to the beginning of the high Middle Ages, roughly the fifth to the tenth century AD), and which enjoyed the highest reputations, were the collection attributed to Hippocrates and the writings of Galen, perhaps the leading Hellenistic Greek doctor of the first century (he was personal physician to the Roman Emperor Marcus Aurelius). Perfect understanding of the ancient texts would lead to a rebirth, a renaissance, of man's original wisdom.[3]

There has been a regrettable tendency among historians of science and medicine to regard the text-centredness of pre-modern universities as proof of their stagnant and moribund nature. This is simply another case of Whiggism – judging the past by our own standards – arising from a failure to recognize what university studies were intended to be about. Concentration on textual exegesis did not mean that students in the medical faculties believed that was all there was to medicine. They were not stupid. Fully recognizing that there was a practical side to medicine, they simply did not feel, for the most part anyway, that it was an appropriate focus for university study. As William Salmon wrote in the Preface to his translation of the *Pharmacopoeia Londinensis*, 'A man may read all his life long, and the very choice of all Authors; but, if he has not the Practick part, be at a great Loss when he comes to a Patient'.[4] In order to acquire the 'practick' part of medicine, university medical students might (and frequently did) set themselves to practise while still engaged on their studies, and therefore unqualified. A pharmacological knowledge, of *'materia medicamentorum'*, could be gleaned, as Sir Thomas Browne wrote to Henry Power in 1646, 'in gardens, fields, Apothecaries' and Druggists' shops', while other practical aspects of medicine could be learned by watching 'what Apothecaries do' and 'what Chymistators do in their officines'.[5] The more prestigious medical schools actually provided for practical training. The major Italian universities were admired for the opportunities they offered for clinical experience, and in Paris every baccalaureate student was attached to a 'doctor-regent' for two years to observe his practice.

The more historians uncover of the actual details of medical practice in the Renaissance the more the three-fold division of medical labour is revealed as a rhetorical notion, dearly held by the top-most echelons of university physicians, of how things ought to be. For the most part licensed medical men seem to have operated as general practitioners, combining the skills of physician, surgeon and apothecary. A number of factors contributed to this state of affairs. First, university-trained physicians were not so numerous and so powerful in every region of Europe that they could dominate the larger numbers of apprenticeship-trained surgeons and other medical practitioners in any given vicinity. Secondly, even the most prestigious and powerful licensing bodies, like the Royal College in London or the Medical Faculty in Paris, were unable to exercise a total monopoly on medical practice in their areas of jurisdiction, much less in the vast hinterlands between one such licensing body and the next. Thirdly, there were nearly always alternative forms of medical licence which could undermine the monopoly and the arbitrary standards of any local licensing institution. And fourthly there were large numbers of unlicensed medical practitioners of many different kinds any one of whom would always be willing to act as a 'general practitioner' and thereby ensure that the patient only had one bill to pay, not two or three.

For example, it has been estimated that in London between 1580 and 1600 there were approximately 50 physicians holding licenses from the Royal College, 100 licensed surgeons, 100 apothecaries in the Grocers' or other Companies, and 250 other practitioners, some with alternative forms of license and others unlicensed.[6] It should be noted that these figures imply a high proportion of unlicensed practitioners. If unlicensed practitioners were able to flourish in London, under the very noses of the College Fellows, there can be little doubt that they formed the majority in rural areas. Alternative forms of licence for physicians and surgeons could be obtained in England from diocesan bishops, or in London from the Dean of St Paul's. In France the university medical faculties of Paris and Montpellier both claimed exclusive rights for graduates of their medical schools to practice medicine *'ubique terrarum'* ('everywhere in the world' – the 'world' being France, of course), so even the powerful Paris faculty had difficulty exercising its monopoly over Montpellier graduates. Finally, it was even possible to be granted a royal privilege to practise. The efforts of institutions like the Royal College of Physicians to exercise their licensing monopoly were undermined not just by swarms of elusive 'quacks' and 'empirics' but by significant numbers of university-trained practitioners who simply refused to apply for licences. Qualified doctors from foreign universities were not automatically licensed to practice medicine, as graduates of Oxbridge were, even though their medical training might be far superior to that offered at either Oxford or Cambridge (neither of which had high reputations for medical education at this time). Some physicians educated abroad chose to 'incorporate' their degrees at one or the other of the English universities and so acquire a licence, but many simply did not bother.

If these were the institutional difficulties hampering the establishment of a tripartite medical profession, they were perhaps not so significant as the pressures brought to bear from the collective individual demands of patients. There were potential patients, and plenty of them, at all levels of society and there was evidently no shortage of would-be healers to whom they could turn in time of need. Sometimes helping out of a sense of Christian charity, 'irregular' healers would try to help where they could at no charge. Others would at least offer their services cheaply and thus provide the only possible help for the lower orders of society. Needless to say, such 'irregulars' were despised and disparaged by qualified practitioners but it is worth reminding ourselves that, from the point of view of therapeutic success, there may not have been anything to separate regulars from irregulars. William Clowes, a learned surgeon, complained that it was easy to see 'tinkers, toothdrawers, pedlars, ostlers, carters, porters, horse-gelders, and horse-leeches, idiots, apple-squires, broomsmen, bawds, witches, conjurers, soothsayers and sow-gelders, rogues, rat-catchers, runagates, and proctors of Spittlehouses, with such other like rotten and stinking weeds ... in town and country ... daily abuse both physic and surgery.'[7] This sounds like a sorry state of affairs but it might not have been so. After all, the first Caesarean section performed on a living mother (who went on, incidentally, to have further children and lived to the age of 77 years) was carried out by Jakob Nufer, a Swiss sow-gelder.[8]

2 Popular knowledge of medicine

Striking and exceptional though the example of Nufer is, it is only one of many which could be used to indicate that certain members of the general population possessed a high level of medical knowledge. Perhaps the strongest evidence of this can be seen in the statute passed during Henry VIII's reign to protect those who might well be seen as the English equivalents of Jakob Nufer (of whom there must have been significant numbers) from interference by members of the Incorporation of Surgeons (this statute is sometimes called 'The Quack's Charter'):

> the company and fellowship of the surgeons of London mindful only of their own lucres, and nothing of the profit of the diseased or patient, have sued, troubled, and vexed divers honest persons, as

well men as women, whom God hath endued with knowledge of the
nature, kind, and operation of certain herbs, roots, and waters, and
the using and administration of them to such as being pained with
customable diseases ... and yet the said persons have not taken
anything for their pains or cunning, but have ministered the same to
the poor people only for neighbourhood and God's sake, and of pity
and charity. And it is now well known that the surgeons admitted
will do no cure to any person but where they shall know to be
rewarded with a greater sum than the cure extendeth unto:-
Be it enacted, that it shall be lawful to persons having knowledge
and experience of the nature of herbs, roots, and waters, to practise,
use, and minister to any outward sore, uncome, wound,
apostemations, outward swelling, or disease; any herbs, baths,
pultes, and emplaisters, according to their cunning, &c; or drinks for
the stone, or strangury, or agues, without suit, vexation, trouble,
penalty, or the loss of their goods ...[9]

Moreover, it seems very clear from the frequent and often
detailed references to medical lore in the plays of Shakespeare and his
contemporaries, not only that the dramatists themselves had a good
working knowledge of medical theories, but also that they expected
similar knowledge in their audiences. The physician's diagnosis of
Tamburlaine's sickness in Marlowe's play is puzzling to the modern
reader but would have been perfectly comprehensible to all but the
most ignorant of the play's hearers:

I view'd your urine, and the hypostasis,
Thick and obscure, doth make your danger great:
Your veins are full of accidental heat,
Whereby the moisture of your blood is dried:
The humidum and calor, which some hold
Is not a parcel of the elements,
But of a substance more divine and pure,
Is almost clean extinguished and spent;
Which, being the cause of life, imports your death:
Besides, my lord, this day is critical,
Dangerous to those whose crisis is as yours:
Your artiers, which alongst the the veins convey
The lively spirits which the heart engenders,
Are parch'd and void of spirit, that the soul,
Wanting those organons by which it moves,
Cannot endure, by argument of art.

Accidental, as opposed to essential, heat, the 'divine substance' of the calor, the notion of a crisis, the spirits engendered in the heart which are the organons of the soul, all this would be clear and familiar to a sixteenth-century audience, even to the poorer groundlings.[10]

The wealth of medical allusions in analogies and metaphors, together with the occasional detailed descriptions of medical ideas which appear in Shakespeare's plays have led a number of literary scholars to claim that he must have had privileged knowledge, learned perhaps from his physician son-in-law, John Hall. A simple tally of medical references in the plays clearly shows, however, that Shakespeare's knowledge of medicine pre-dated his acquaintance with Hall.[11] Furthermore, with regard to medical references at least, Shakespeare is just one playwright among many. There can be little doubt that medical knowledge was widely diffused through Renaissance society. On the face of it, this fact may seem rather remarkable but it is not too difficult to see why this came about.

First, we simply have to remember that we are dealing with a time before medicine was established as the specialist preserve of a professional elite. A major prerequisite for professionalization is the recognition, outside the nascent profession, that a particular body of knowledge is beyond the reach of the majority, either by dint of its technicality, the amount of time needed to sufficiently master it, or (a historically later development) its irrelevance to 'getting on' with the daily grind of life. This was not yet the case with medical knowledge. A knowledge of medicine was regarded as an important element in the make-up of a cultured gentleman or gentlewoman, while lower down the social scale a basic level of medical knowledge was regarded as an essential for daily life. This derived from the conviction that everyone was ultimately responsible for their own health. Moreover, the medical theory underlying these social assumptions was couched in such a way that any individual was more likely to be a better judge of their own health than a doctor.

All disease was held to be caused by an imbalance in the four humours which were all present in every individual, indeed the humours effectively constituted the body in the same way that all things in the physical world were made up of the four elements. The humours were called choler or bile, blood, phlegm and melancholer or black bile. To cure disease the balance of these four humours had to be restored, usually by drawing off an excess of one of the humours, say blood by bleeding, or phlegm by administration of an expectorant.

However, each individual person was endowed with their own particular 'temperament' or constitution. This meant that normal, healthy individuals displayed a range of different proportions of each humour. Generally speaking, it was held that one of the four humours would tend to be dominant in any healthy constitution. Individuals were, accordingly, choleric or bilious, sanguine, phlegmatic or melancholic (the phlegmatic type was held to be fat, dull of understanding, sleepy; the sanguine were muscular, amiable, merry, bold, lecherous and ruddy in complexion; the choleric were held to be hasty, envious, covetous, subtle and cruel, lean of build and yellowish or fair in colour; the melancholic were solitary, fearful, curious, heavy in build and dark in colour). It would obviously be dangerous if a physician, unaware of the normally bilious nature of one of his patients, tried to remove what he mistakenly took to be excess bile in an attempt to restore the patient to health. Assuming that the patient knew his own constitution, he would automatically have the advantage over any doctor who was called in to treat him for any disease. Here then is the origin of two old proverbs (no longer current since our system of medicine has made them inapposite): 'Every man is his own physician', and 'Every man is a fool or a physician'. The point of the latter is that everyone knows their own temperament or constitution (and, *ipso facto*, is a physician) unless they are foolish.[12]

Knowledge of the temperaments, the humours, and the humoral imbalance theory of pathology, was common intellectual currency, therefore, throughout the Middle Ages and the Renaissance. What is more, one of the ways in which people came to learn of these things was in every clinical encounter with a healer during the course of their lives. There were no stethoscopes, much less any X-rays, to give a doctor information not known to the patient himself. The doctor could only judge by the symptoms he saw and by asking the patient to describe his condition as best he could. Inevitably, therefore, doctors and patients shared a common language. The doctor was to a large extent dependent on the patient's claim that he was melancholic or phlegmatic. Before medicine could be fully professionalized doctors had to be able to free themselves from this reliance on the patient (or the patient's close family), to develop instruments such as the thermometer and stethoscope, which could be claimed to be more revealing of bodily states than the personal experience of the patient, and to develop a professional language, a jargon, which emphasized the specialist nature of the doctor's knowledge. None of these things could happen while a pathology based on notions of humoral imbalance persisted.

For those patients who were puzzled by, or felt at a loss in, their brief engagements with doctors during times of illness, there were numerous popular books to draw upon. The most successful of these in England were Sir Thomas Elyot's *The Castel of Helth,* first published in 1536 but running into many editions, Andrew Boord's *The Breviary of Helthe* (1547), and the *Regimen sanitatis salernitatum* ('The regimen of health of Salerno' – Salerno being a famous medical school in southern Italy which pre-dated university medical schools), first printed in 1484 but in circulation in manuscript long before. This too went through many editions. It would probably be a mistake to suppose that such works circulated at the lowest levels of society, but it seems safe to assume that much of the knowledge they contained did diffuse down to the illiterate classes. The *Regimen salernitatum* evidently flourished, at least in part, in oral traditions before the advent of printing and much of its advice appears in proverbs and folklore. Most famous, perhaps, is the adage that 'the best physicians are Dr Diet, Dr Quiet, and Dr Merryman'.[13]

Indeed, medical precepts filtered down into popular consciousness even from university texts like the Hippocratic writings. Some of these entered so successfully into popular thinking that they survive even today. Although many people nowadays believe that *ars longa, vita brevis* refers to the longevity of a masterwork of painting, sculpture or architecture, virtually everyone in the pre-modern period knew that the art in question was the art of medicine. This famous phrase is, in fact, part of the opening admonition in the Hippocratic collection of *Aphorisms.* Another of these medical aphorisms which survives to the present is, 'Desperate cases need the most desperate remedies'. Similarly, the familiar suggestion that someone in dire straits is 'clutching at straws' derives from a note in the Hippocratic treatise known as the *Prognosis* which says that it is 'a bad sign and portends death' if a patient's hands 'make grabs at the air, or pull the nap off cloth, or pull off bits of wool, or tear pieces of straw out of the wall'. Once again it is safe to assume that, although most people who use this phrase today are ignorant of its medical provenance, even those who used it metaphorically in the pre-modern period knew of its Hippocratic origin and meaning. Furthermore, it seems fairly safe to conclude that if phrases like these (and various other notions such as temperaments, the supposed link – now only jokingly proffered –

between virility and baldness or masturbation and bad eyesight) have managed to survive to our own day then the level of knowledge of even quite recondite medical ideas was once much more widespread.[14]

This diffusion of medical knowledge through all levels of society ensured that there was always someone in any locality who was capable of treating the sick, whether it were a parson, a well-read gentleman or lady, a concerned grocer, an adroit barber, tailor or sow-gelder, or a village 'wise' or 'cunning' woman. Indeed, it is perfectly possible that such 'irregular' medical practitioners were as capable of bringing about a cure as the most learned physician or surgeon. But, more to the point, even if some properly trained medical practitioners would have been a better bet than local autodidacts, they were not very often to be found. There is no shortage of evidence to suggest, and no good reason for denying, that most people's experience of physicians was thoroughly unsatisfactory. Proverbial expressions such as 'Physicians are worse than the disease', and 'Physicians kill more than they cure' sum up countless clinical encounters all over pre-modern Europe. Christopher Langton, an MD of Cambridge and author of *An Introduction into Phisycke (c.* 1550) has 'Physycke' herself speaking to 'her mynisters and Phisicions' thus:

> I praye you tel me, why doth every man now utterly abhorre my company: trumpyng eftsones in my waye, who is wylling to dye, let him goo to Phisyke: who is wylling to be robbed of his money, let hym go to Phisycke, yf I be not a thefe & a murtherer: But what sayd I, did I cal you my servauntes: no, I knowe ye not, & moche lesse ye know me: wo shall be unto you, which ravyshing me agaynste my wyll, hath thus brought me in captivitie & bondage, wo shalbe unto you, which for youre own lucre and advauntage, hath made me an instrument of mischefe: woo shall be unto you whyche have sclaundered me with the death of so many thousandes.

Later, speaking for himself, he wrote: 'I coulde prove that there is as moch iugglyng and deceyvyng of the people now a dayes amongest our phisitions ... as ever was amongest the Popysh preestes'.[15] Small wonder that many preferred to treat themselves or rely upon family and friends to care for them whenever they fell ill. As Geoffrey Chaucer advised in 1386, 'If thou hast nede of help, axe it of thy freendes, for ther nys noon so good a phisicien as thy trewe freend'.[16] 'Kitchen physic', a proverb reminded everyone, 'is best physic'.

Now, it is important to be clear about popular beliefs with regard to medicine. There was not a popular belief system of alternative medicine: the wise women and wise men called upon by villagers, as far as we can tell, held to roughly the same set of beliefs about humoral pathology, and used the same kinds of treatments, herbal remedies and manipulations as any physician or surgeon. Although it seems reasonable to suppose that their thinking on these matters was far less sophisticated than a learned practitioner's, we cannot conclude that their practice was less efficacious and successful. Even today occasional scientific reports vindicate the validity of folk remedies, and in the eighteenth century doctors made a point of testing folk remedies and found many of them to be superior in their effects to their standard drugs.[17] There is one very famous case of a learned sixteenth-century doctor taking advice from a wise woman. Ambroise Paré, the greatest surgeon of his day, met an 'old country woman' in an apothecary's shop while he was buying medicines for a burn. The old woman persuaded him to try raw onions beaten with salt. Fortunately for Paré, his patient was only a 'Kitchin boye', 'a greasy scullion', and so, Paré tells us, he was able to experiment on him. Fortunately for the kitchen boy, the old lady was right.[18]

The popular belief, then, was not that the system of medicine was wrong but that doctors themselves did not take their calling and its duties seriously enough. Here again it is important to notice that such criticisms were not confined to charlatans, quacks, and the like, who undoubtedly flourished in the 'medical market-place', but were clearly also aimed at the trained, the would-be professional, medical men. We can gain an interesting insight into the relationship between university medicine and the medicine of 'old wives' by considering the art of uroscopy. That late classical Greek source for much of western medicine, Galen, had advised physicians to note the colour, clarity and consistency of urines, and made some remarks on hypostases (or sediments). Based on the humoral imbalance theory of disease, Galen's original ideas were 'refined' and added to by the Arabs and subsequently by European university physicians. Uroscopy became extremely subtle and complex and was held to be extremely important for diagnostics. One measure of its importance can be seen in the speed and frequency with which treatises on uroscopy were produced by the printing presses. One of the earliest was Gilles de Corbeil's *De urinis* of 1484. Uroscopy was such an important element in medicine that, in illustrations, a flask of urine held in the hand was the badge or symbol of a doctor until well into the seventeenth century.

Complex though this art was, its precepts evidently filtered down to the less exalted levels of medical practice. When one of the characters in Shakespeare's *Twelfth Night* fears for the (mental) health of Malvolio he cries: 'Carry his water to th' wise woman'.[19] Uroscopy was part of popular beliefs about medicine: if a wise woman was supposed to know about medicine she would have to at least pretend that she could diagnose from urine. Once again we have to suppose that wise men and women learned to do this at least as well as many physicians. We can see this from a fascinating text attributed to one of the most renowned medieval physicians, Arnald of Villanova (1235?-1311). In a short treatise entitled *De cautelis medicorum* ('On the precautions that physicians must take'), Arnald tells his fellow physicians how to avoid making a fool of oneself, or being made a fool of, when trying to diagnose from urine. There are no guide-lines here about what a particular colour or kind of sediment actually signifies, instead the reader is told of certain subterfuges, usually ways of engaging the one who brought the urine sample in discussion about the case, with a view to reaching a conclusion from this conversation. Significantly, the one who brings the urine is usually referred to as an old woman. Perhaps that is how the wise women learned their skills. But what skills were they learning on these visits to the physicians? Consider these examples:

> There is a seventh precaution, and it is a very general one; you may not find out anything about the case, then say that he has an obstruction in the liver. He may say: 'No, sir, on the contrary he has pains in the head, or in the legs or in other organs.' You must say that this comes from the liver or from the stomach; and particularly use the word, obstruction, because they do not understand what it means, and it helps greatly that a term is not understood by the people ...
>
> The ninth precaution is with regard to a woman, whether she is old or young; and this you shall find out from what is told you. Should she be very old, say that she has all the evils that old women have, and also that she has many superfluities in the womb. Should she, however, be young say that she suffers from the stomach, and whenever she has a pain further down, say that it comes from the womb or the kidneys; and whenever she has it in the anterior part of the head, then it comes from the spleen; whenever to the right, then it comes from the liver; and when it is worse and almost

impedes her eyesight, say that she has pains or feels a heaviness in the legs, particularly when she exerts herself.

Lamentable though this may be (and Kurt Sprengel, an eminent nineteenth-century historian of medicine, judged Arnald to be an appalling charlatan), the fact remains that uroscopy was an important aspect of diagnostics until the seventeenth century.[20] People had come to expect a healer, of whatever stamp, from the highest level of practice to the lowest, to examine a patient's urine and, for the most part, healers obliged. Moreover, just as self-medication was common, so people routinely inspected their own urine to keep an eye on their state of health. There is an extensive body of literature on uroscopy and virtually all of it is written in an entirely rationalistic way, equating different states with different diseases, with no hint that any of this is bogus or fraudulent. It would surely be wrong to conclude that there was a conspiracy among all doctors to systematically fool their clients. It seems much more likely that doctors did believe in the validity of uroscopy and its relevance to determining the state of the patient's humours. However, all doctors knew the aphorism of Hippocrates, 'Life is short, the art [of medicine] is long; opportunity is elusive, experiment is dangerous, judgement is difficult'.[21] The difficulty of judgement made it necessary from time to time to take extra precautions, most doctors were probably very grateful for Arnald's warnings and unconsciously drew upon his advice more often than they ever knew.

So, it seems clear that popular belief about medicine can be characterized in two ways. First, with regard to the intellectual system of medicine, popular belief was essentially congruent with, though almost certainly simpler or vaguer than, the system that was propagated in medical faculties. Secondly, doctors themselves, particularly self-professed doctors or healers, that is to say all those who made or tried to make a living exclusively by treating the sick, were regarded as discomforting and even downright dangerous, frequently ineffective and, to say no worse, gave poor value for money. Before medicine could become professionalized this public image had to change.

3   Attempts to change the image of Renaissance medicine

Doctors' efforts to change public perceptions of medicine can be grouped under two major headings. First, there were defences of 'correct' medicine and good medical practitioners, as distinct from those which give medicine a bad name. These efforts are little more than rhetorical exercises in so far as they do not try to reform medicine, merely to show that in the right hands medicine is perfectly sound. Secondly, there are a number of different efforts to refine or even to reform the old system to make it more beneficial to mankind. In practice, there was a lot of overlap between these two approaches.

One commonly used means of trying to improve the physician's stock, and to separate him from the crowd of lesser healers, was to emphasise his learning. In the prologue to his *Breviary of Helthe* (London, 1547), Andrew Boord deplores the fact that many people 'wyll enterpryce to smatter & to medle to ministre medecynes, and can nat tell how when, and what time the medicine shuld be ministred'. Such people feel qualified merely because 'they have ben in the companye of some doctoure of physicke, or else havynge an Auctour of phisicke [i.e. a medical book], or Auctours and wyll ministre after them, and can nat tel what the auctour ment in his ministration.' This will not do, Boord goes on to insist, because what is lacking is the wealth of learning that is a *sine qua non* for sound physic:

> O lord what a great detriment is this to the noble science of phisicke that ignorant persons wyll enterpryce to medle with the ministration of phisicke, thus Galen prince of phisicions in his Terapentyk doth reprehend and disprove sayeng: Yf phisicions had nothing to do with Astronomy, Geomatry, Logycke, and other sciences, Coblers, Curryars of ledder, Carpenters and Smythes and such maner of people wolde leve theyr craftes and be physicions, as it appereth nowe a dayes that theyr be many ...[22]

According to Christopher Langton (1550), the physician must be 'exercysed, even from hys tender age, in dialect, arithmaticke, and mathematicke, he must also be very payneful, setting his mynde on nothyng but only on learnyng, and conferryng his studyes alwayes with the best, and ought alway to be a very diligent searcher of the truthe'.[23] Similarly, John Securis (1566) wrote that the physician 'must have and get his learning of the best learned men of his time, who diligentlye, even from his childhode must instructe him with these sciences: grammer, Logick, musicke, Astronomie, and chiefely (as Plato consayleth) Arithmetick and geometrie, and also Philosophie'.[24]

Grammar, of course, meant Latin grammar to enable one to read the ancient texts (all of which were translated into Latin from Greek or Arabic – a number of doctors prided themselves on a knowledge of Greek too). A command of logic (or dialectics – 'dialect' as Langton puts it) was considered essential to enable one to reason properly and so understand why a particular treatment worked (or why it did not, as the case may be). This was meant to distinguish the true physician from the mere 'empiric', the practitioner who was guided only by a limited experience. Having been successful once, an empiric would try the same treatment on other patients with other diseases without rhyme or reason. These claims for the importance of grammar and logic seem straightforward enough, if not terribly convincing, but what use is arithmetic, geometry or astronomy to the study of medicine?

The clue to understanding this is given in Boord's lament that ignorant practitioners cannot tell 'when, and what time the medicine shuld be ministred'. This takes us beyond the realms of mere rhetorical claims and into a real refinement of medieval and Renaissance medicine. Astrology was an important element in the skills of the learned practitioner. The complex techniques of drawing up a horoscope for a particular individual and his or her disease enabled the doctor to reach safer conclusions about the temperament of the patient (which was held to be linked to their ruling star-sign or planet), and the optimum timing for therapeutic intervention. Expertise in astrology was dependent upon expertise in astronomy and geometry and these mathematical sciences were regarded as important prerequisites for the study of medicine. They were usually taught in the Arts faculties of universities and all aspiring physicians began their training in Arts before moving to the 'higher' faculty of medicine.

Needless to say, astrology at this time was regarded as a legitimate art. Although there were a number of eminent thinkers throughout the Middle Ages and the Renaissance who objected to many of the conclusions of astrology, to its uncertainty, and therefore its futility, no one has ever come to light who actually dismissed its legitimacy. Everyone was agreed, at least in principle, that the heavenly bodies somehow affected events on earth; disagreement was confined to whether we could ever know precisely how and in what ways. Astrology was a useful and noble adjunct, therefore, to the art of medicine and it was frequently pointed to by learned physicians as a realm of learning which raised them above the competition in the medical market-place. But this was only true to a partial extent. There is suggestive evidence pointing to a surprising amount of knowledge about astronomy and astrological lore at the lower levels of society. But, and this is perhaps more important, one prominent kind of irregular medical practitioner was the astrologer. Astrologers, no doubt of varying quality, can be found practising their art at every level of society and a substantial proportion of the problems they are called upon to solve are medical problems. To digress into the eighteenth century for a moment, it can be seen that when trained medical practitioners began to succeed in establishing themselves as professionals, they did so by separating themselves from the ranks of such astrologers as much as from other irregular healers. Nevertheless, we can see from the quotations above, and countless others that could be cited, that in the sixteenth and seventeenth centuries astrology was still regarded by physicians as one of their more noble and learned distinguishing features.

Arithmetical expertise was also invoked in another refinement upon basic Galenic medicine. According to classical medicine, an individual drug (a so-called 'simple') had a complexion which could be analysed in terms of two dominant qualities, one active and one passive. There were four fundamental qualities: hot and cold, which were held to be active, and dry and wet which were passive. The strength of the simple, which might be, say, predominantly hot and wet, was quantified on a scale from temperate (neutral) to the fourth degree. In the thirteenth century Arnald of Villanova, drawing on Arabic ideas, extended these notions to work out the degree of a compound medicine, a drug made up of a mixture of simples (compound medicines were far more commonly used than simples during the Middle Ages and the Renaissance, so much so that the period has been called the age of polypharmacy). In the Arab system it was assumed that equal weights of all medicines were equally effective but Arnald developed another refinement. He argued that drugs had their own characteristic doses, so that a dram of sandalwood, say, which is cold in the second degree, would be neutralized not by a dram of honey which is hot in the second degree, but only by a proper dose of honey, which in this case was held to be two ounces. A sick patient could also be said to be hot in the third degree and so on, thus making an arithmetical problem for the physician to work out. There was some controversy in the universities about the details of this system. A drug was temperate by virtue of a 1:1 ratio of the active, or passive qualities, and the first degree was caused by a 2:1 ratio (2 wet to 1 dry, or 2 hot to 1 cold). There was some discussion about whether the subsequent

three degrees should proceed arithmetically (3:1, 4:1, 5:1) or geometrically (4:1, 8:1, 16:1). The mathematics of ratios was designated in the
medieval universities as part of the theory of music and that is why
John Securis included a knowledge of 'musicke' along with grammar,
logic, and astronomy as prerequisites for the physician.

Unlike astrology, the mathematics of the strength of drugs does
seem to have separated the medical sheep from the goats. In practice,
however, it was obviously insufficient to give university physicians an
advantage over their goatish rivals in the market-place. Judging from
the flourishing nature of irregular medicine throughout the Renaissance, and the fact that medicine did not really begin to be seen as the
exclusive province of a professional before the second half of the
eighteenth century, the same must obviously be said about all the
efforts of the learned doctors to establish their superiority. The claim
of a learned physician like Langton that the best form of medicine
'ioined experience alwayes with reason' and took great care to consider
'very diligently the causes' behind both diseases and cures, did not cut
much ice.[25] For too many people the physicians' talk of causes was so
much hot air. Molière's little skit summed up the impression that many
people had of university physicians:

> Mihi a docto doctore
> Domandatur causam et rationem quare
> Opium facit dormire?
> A quoi respondeo
> Quia est in eo
> Virtus dormitiva,
> Cujus est natura
> Sensus assoupire.

(Written by Molière in a Frenchman's dog Latin, intended to be
reminiscent of a blustering, jargonizing physician, the joke goes
something like this: 'To me by a learned doctor is asked the cause and
reason why opium produces sleep? To which I reply because there is
in it a dormitive virtue, whose nature it is to send the senses to
sleep.')[26]

Similarly, the frequent attempts of learned doctors to intimidate
people into dealing only with properly trained physicians and surgeons failed to counter the public's experience of the relative success
rates of regular and irregular doctors. Simon Harward warned in his
Phlebotomy: Or, A Treatise of letting of Blood (London, 1601) that there

are many in country towns 'which doe practize the opening of vaynes' without 'any learned counsaile'. As a result 'there often insueth more hurt and danger, then ease and succour' because, he explained,

> great ... are the harmes which may ensue by letting of bloud, if the same be rashly and unconsiderately attempted, the spirits and blood are spent and wasted, the naturall heat is pluckt away and dispersed, the principal parts are made overcold, and utterly lose their strength, old age is hastened on,... and many also are brought to an unrecoverable destruction of their health and life.[27]

In spite of many such strictures, most people continued to think that learned physicians were just as lethal. It is perhaps worth pointing out, by the way, that Harward, although a learned man, was not a regular physician himself. He was a minister of the Church, and a school master as well as a local healer for his community.

It would seem, then, that neither rhetoric nor a repertoire of specialist technical refinements were successful in persuading the public of the superior efficacy of properly trained and qualified medical practitioners. No amount of rhetoric was able to overcome the everyday experience that irregular practitioners were, generally speaking, just as successful as the regulars. And this common experience must also have contributed to popular scepticism about the learned doctors' technical niceties. After all, the fundamental system of medicine, based on the ancient notions of humoral pathology, had remained essentially unchanged through the Middle Ages and, over the centuries, knowledge of this system had become part of everyone's mental furniture. Before medicine could be recognized as a profession, the details of medical knowledge had to cease to be a part of popular belief, and the notion that doctors' were in possession of specialized knowledge had to enter into popular belief. But medical knowledge depends not merely upon theory but also on practice. Both theory and practice had to be changed before the common man would give up his claim that he, or at the very least local wise men and women, knew medicine as well as doctors did, or, the more temperate claim that various untrained individuals could be trusted to know as much as the doctors did.

Not even the otherwise supremely successful innovations of the mechanical philosophy could help medicine out of this impasse. The mechanical philosophy sought to explain all physical phenomena in terms of the motions and interactions of the parts of matter. Since many physical phenomena (for example, organic decomposition) did not seem to depend on the motion of matter, the mechanical philosophy had to entail a belief that matter was composed of invisible corpuscles or atoms (decomposition, therefore, was merely rearrangement of the atoms of the decomposing thing). The mechanical philosophy was very quickly embraced by leading natural philosophers and led, ultimately to the successes of Isaac Newton's *Philosophiae naturalis principia mathematica* (London, 1687). The precepts of the mechanical philosophy were similarly embraced by a number of innovative physicians to give rise to a movement known as iatromechanism. This new medical philosophy drew principally upon the many new discoveries in anatomy and physiology made during the sixteenth and the first decades of the seventeenth century and endeavoured to see the body as a complex machine. To a certain extent this also entailed a new way of regarding disease. To take a simple example, the blood-purifying function of the kidneys was explained in the mechanical philosophy by analogy with a sieve. Kidney disease could be explained, therefore, by supposing that the holes in the seive had become too large or too small to function properly. This immediately raised a new problem: So what? What could an iatromechanist physician do about a kidney-sieve in which the holes were the wrong size to be useful to its purpose?

The answer is that he must treat kidney disease the way doctors had always treated it. After all, the traditional therapies were tried and trusted and so they must work even if the mechanical philosopher could not explain how they work. The upshot of this was that the mechanical philosophy made no significant impact on the practice of medicine. This can be clearly seen, for example, in the casebook kept by Thomas Willis who is regarded by historians as one of the foremost iatromechanists and who was one of the most successful medical practitioners of his day. In spite of his many important publications in which the principles of iatromechanism are clearly expounded, the casebook reveals an extremely old-fashioned way of dealing with disease. Willis often records the temperament of the patient ('of sanguine temperament', 'of melancholy temperament', etc.), his diagnoses and his therapies are virtually all based on humoral therapies and the medicines he prescribes are nearly always traditional. Consider, for example, the case of Widow Wise:

> The Wise son came to me bringing the urine of his sick mother. The urine was red, thick and turbid; I suspected it was tinged with menstrual blood; but there was no one who could tell me for sure. He related that she was complaining strongly about pain in her stomach and head. I decided a light vomit was to be prepared, and that afterwards I must prescribe a dietary and await the result.

(Inducing vomiting was the standard way to remove an excess of yellow bile from the system.) In the case of 'the small daughter of F. Bodily' which Willis dealt with in October 1650, a complication arising from measles was exacerbated by the fact that the breast-feeding baby's mother was stricken with melancholy. 'It was also inevitable', Willis wrote, 'that the milk taken by the infant was infected with a melancholy juice or black bile'.[28] Willis's practice is entirely representative of the way iatromechanists dealt with disease and it is easy to see that, as far as patients and their relatives were concerned, there was no difference between what iatromechanists did and what the more familiar kinds of medical practitioner did. Accordingly, iatromechanism completely failed to change the patients' perceptions of learned doctors and learned medicine. It was unable, therefore, to change the popular belief that unlearned but experienced healers could do as good a job as elite physicians, or to persuade the public that medicine should be the special preserve of a professional body.

If iatromechanism failed to reform medicine because its new theories did not dictate new practices, we must look elsewhere to understand the failure of the other major development in Renaissance medicine – Paracelsianism. This system derived originally from the work and writings of a Swiss surgeon who called himself Paracelsus, by which he meant to imply he was the equal of the greatest ancient Roman writer on medicine, Celsus (*fl.* AD 50). Paracelsianism, which was supplemented and altered by a number of disciples, did imply a new practice as well as a new theory of medicine.

Paracelsus believed that all things were endowed by God with their own *archeus*, a 'divine spark' which defined their true nature. This was applied by Paracelsus not just to what we would think of as living creatures but to all creation. There was no distinction between animate and inanimate since all things were endowed with *archeus*. Paracelsus even thought of diseases as having their own *archei*. He was, therefore, one of the earliest subscribers to what is called the 'ontological' concept of disease, the belief that a disease has a real existence of its own (as opposed to being a mere epiphenomenon of an imbalance in the bodily humours). Illness occurred, according to Paracelsus, when the *archeus* of a disease set itself up in the body in opposition to the *archeus* of the healthy body. The doctor's role was to help the body to combat the *archeus* of the disease. Fortunately, God had ensured that there was always something in creation, a plant or a mineral, whose *archeus* was antagonistic to that of any given disease. In practice, therefore, the Paracelsian doctor would not be endeavouring to restore the balance of the humours by administering drugs whose effect was merely to remove one of the humours or to heat or cool the body. Instead he would try to discover and administer a 'specific' drug whose *archeus* was directly inimical to the *archeus* of the patient's disease.

The Paracelsian system was far richer and more complex than can be adequately discussed here and in the hands of his followers there were so many amendments and refinements that they have often appeared to historians as mere inconsistencies. Even so, it remains true to say that Paracelsian medical practice was potentially simpler than that of full-blown Galenism. Consequently, many aspects of Paracelsianism, in spite of being many centuries younger than Galenism, were quickly absorbed into the public's medical awareness. Such absorbtion was made easier by the fact that Paracelsianism drew heavily upon magical traditions which were already dominant aspects of popular belief. Furthermore, although Paracelsus did have his followers and admirers among academic physicians, for the most part Paracelsianism was confined to unlicensed or irregular medical practitioners. It tended to contribute, therefore, to the fragmentation of the institution of medicine, rather than to its unification as a profession.

Indeed, it is surely significant that Paracelsianism in the seventeenth century was associated with reformist movements which opposed medical monopolies as detrimental to the public welfare. In England, for example, Paracelsianism was vigorously promoted during the Interregnum by a number of radical Puritans. Nicholas Culpeper proclaimed in his *Physicall Directory* of 1649 that 'The Liberty of our Common-Wealth' was 'most infringed by three sorts of men, Priests, Physicians, Lawyers'.[29] Noah Biggs, writing just two years after the execution of Charles I, called upon Parliament to topple the Royal College of Physicians which he pointedly described as the 'Palace Royal of Galenical Physick' where Galen was 'Monarch thereof'.[30] One way of helping to break the alleged (though, as we have seen, by no means actual) monopoly of the College, reformers believed, was to provide the general public with English translations of medical writings. Medical reformers presented themselves as good Protestants, extending Martin Luther's 'priesthood of all believers' to enable everyman to be his own physician. As Culpeper remarked, 'Papists and the College of Physicians will not suffer Divinity & Physick to be printed in our mother tongue'.[31] The popularity of Paracelsianism, at least among the reformers themselves, is proved by the large number of Paracelsian works that emerged from the English presses at this time.[32]

In France, Paracelsianism played a major part in the medical reforms of Théophraste Renaudot, a royal physician whose ideas on poor relief earned him a position as Commissaire Général des Pauvres in 1618. Renaudot established the Bureau d'Adresse, which started as a public registry of goods and services for buyer and seller, employer and employee, but which soon became a more active institution. His interest in educational reforms led Renaudot to establish a popular academy within the Bureau. This took the form of 'conferences', that is 'experiments, courses, lessons, debates, and lectures', held every Monday afternoon and open to all comers. The conferences frequently dealt with medical matters and many of these were concerned with Paracelsian issues: 'If there are remedies specific to each malady', 'If it is useful to use chemical remedies' ('chemical' medicines, as opposed to drugs derived from herbal or animal matter, were characteristic of Paracelsianism), 'The principles of chemistry', 'If maladies are cured by similars or opposites' (homoeopathy was sometimes advocated by Paracelsus), and even a recondite topic such as, 'What did Paracelsus mean by the Book "M"?'.[33] Furthermore, by 1641 free medical treatment was available at the Bureau on five mornings a week. In September 1640 Renaudot had been granted permission to establish chemical laboratories at the Bureau and the medical practitioners employed there were predominantly Paracelsian. Renaudot's medical reforms did not end there. In 1642 he published *La présence des absens*, a fully indexed 60-page pamphlet consisting of diagrams of the human body upon which a sick person unable to visit the Bureau could mark the location of their trouble, together with lists of symptoms which could also be selected by the reader as those closest to his or her feelings. The filled-in booklet could then be sent to the Bureau from any part of France for assessment. Written in clear, simple language, the pamphlet was intended for anyone who could read, including 'simple peasant women and their children'.[34]

Paracelsianism, therefore, might well have become absorbed into popular culture even more effectively than traditional humoral pathology, thanks to the publishing and educational reforms of radical Puritans in England or the Bureau d'Adresse in France. This was not to be, however. The forces of conservatism were not routed for long. With the restoration of the monarchy in Britain (1660) and the death of Richelieu in France (1642 – he was Renaudot's patron) the schemes of the reformers came to nothing or, in the French case, were curtailed. Nevertheless, the popularization of Paracelsian ideas by vernacular publications, the public conferences and clinics of the Bureau d'Adresse and so on, could not be reversed. If nothing else, the increased awareness of Paracelsian alternatives to traditional medicine made it harder for bodies like the Royal College of Physicians or the Medical Faculty in Paris to insist upon their exclusive claims to sound physic. Medicine showed no signs yet of becoming a system of knowledge exclusive to a professional elite.

4   Magical healing

A final and extremely important aspect of popular belief about medicine which must be mentioned is magical healing. Curing disease by magical means was always popularly regarded as being at least as effective as any other kind of treatment. Among the irregular classes of healer, therefore, magical operations and medicaments were frequently resorted to. Once again it has to be said that magical traditions and beliefs were richly varied and complex, extending from general views about the nature of the world system to highly localized beliefs about the power of a particular place or a particular herb, and even to beliefs that were little more than family superstitions. Very little can be said here about the nature and extent of these beliefs but it is important for our theme to indicate the interaction of these popular magical beliefs with the learned medicine of trained practitioners.

The first thing to bear in mind is that a belief in magic, throughout the Middle Ages and up to the end of the seventeenth century, was not confined to the ignorant and cranks. Belief in magic, at all levels of society and among all intellectual groups, was not the exception but the rule. Each of the three most important aspects of the magical tradition – natural magic, demonology, and semiotics or symbology – had a great deal to be said for it. Natural magic was based on the not unreasonable assumption that many natural things affected other natural things by 'hidden' or 'occult' means. The moon has a secret power over water to affect the tides; the dandelion has a secret power to make the body generate excess urine. Demonology was based on the assumption that angels and demons really existed. This was virtually impossible to deny at a time when Christian beliefs were so dominant and, for the most part, so unquestioned. Once it is accepted that such creatures do exist it is reasonable to suppose that they are more powerful than men or women and that they might be able to be summoned if the right method can be found. As William Perkins wrote in his *Discourse of the Damned Art of Witchcraft the* devil has

> Exquisite knowledge of all natural things, as of the influences of the starres, the constitutions of men and other creatures, the kinds, vertues, and operations of plantes, rootes, hearbes, stones etc. which knowledge of his, goeth many degrees beyond the skill of all men, yea even those that are most excellent in this kind, as Philosophers and Physicians are.[35]

Similarly, the belief that physical events and objects can be affected merely by manipulating symbols or by using a particular combination of words (a spell) is perfectly reasonable given that one has been convinced that priests do this at every mass when they turn wine and wafers into the blood and body of Christ, and that the priests' ability to change objects by the power of words in this way was merely imitative of God's creation of light by saying 'Let there be light'; in the beginning, after all, 'was the Word'.[36]

The second point to grasp is that magic was not usually regarded as a *supernatural* art in the pre-modern period. Our own idea of the supernatural is somewhat wider in its embrace than that of medieval and Renaissance thinkers. As the above quotation from Perkins's treatise on witchcraft implies, if the Devil could perform seemingly wonderful and awesome effects it was only by dint of his superior knowledge of natural phenomena. It was generally acknowledged that no demon or angel, mere creatures like mortal men, could perform a supernatural act. Only God could operate supernaturally. So, if Satan himself brought about a particular state of affairs he did so by virtue of the natural power God gave him at the Creation, or by manipulating natural phenomena in the way that an alchemist, a philosopher, or a physician might do (though much more successfully because his knowledge of physical phenomena was more complete). The power of spells was a more moot point. Those who believed in the power of spells were obliged to suppose that God had given words a *natural* power to command and effect physical phenomena (this was usually justified by pointing to Genesis 2: 19 where Adam gave names to all the creatures). Others, however, insisted that only God could command the physical world by his word and that the priest during the mass was endowed by God with a genuinely supernatural power over the Eucharist.

Magic, then, was a practical art which used the natural powers of things to achieve certain desired effects. This was equally true whether the thing employed by the magician was the supposed natural power of foxglove to make the heart race and quicken the blood or the supposed natural power of a demon to raise strong winds. So, if a wise woman or a witch provided, say, a charm or amulet that made an insomniac able to sleep it did so by a natural power. The belief was that by manufacturing the charm in a particular way (no matter how superstitious it may seem to us – under a full moon, borne by the seventh son of a seventh son who must not utter a word, and the like) it somehow captured a natural power to make the wearer sleep. In this sense the amulet was precisely like opium, both had an 'occult' power to promote sleep, a 'dormitive virtue'. Certainly, the Church (and others) could and did distinguish between healing by a natural, God-given, herb and an artificial charm and could condemn the latter, but the condemnation arose from the belief that knowledge of how to produce such powers artificially could only have been acquired by entering into compact with the devil. Indeed, there seems to have been a clear distinction between learned and popular attitudes to witches. For the Inquisitions, and other representatives of learned authority, what mattered in accusations of witchcraft was whether a witch had had commerce with demons or not. For the accusers, usually neighbours of the accused, the issue was simply whether she had used her knowledge of natural phenomena to deliberately blight her neighbours or their property. For the former witchcraft was a crime against God and religion, for the latter it was just like a special case of assault or burglary. But one thing that united elite and popular culture was the conviction that magical effects were as natural and as real as any other kind of physical phenomena.

If a belief in the efficacy of magic in the natural world was not confined to the vulgar crowd we might expect to see elements of it, at least, in learned medicine as well as in the medicine of the so-called 'cunning folk'. This is indeed the case. We have already noted that astrology was a recognized part of university medicine. The power of the heavenly bodies to affect things on earth was as undeniable as the power of the moon on the tides. Similarly, physicians were used to the notion that drugs had 'occult' powers to affect the body in particular ways. The effects of some drugs, it was believed, could be analysed in terms of their sensible properties, their hotness, coldness, moistness or dryness, but there were increasing numbers of drugs whose operations could not be made sense of in this way. They were supposed to operate in an occult way on 'the total substance' of the body rather than on one of its humours. This made the notion of a specific drug for a specific disease (as opposed to a drug which merely made the body hotter, colder, more moist or dryer and therefore could be used for any appropriate imbalance in the humours) much more plausible among academic doctors. In so doing academic medicine was drawing closer to folk medicine in which there were a number of remedies which were supposed to act directly to cure a particular ailment. This paved the way for the Paracelsian emphasis on 'specifics'. Indeed, a large proportion of Paracelsianism can be seen to have derived from magical traditions.[37]

Once again, then, it would seem that there was considerable overlap in the Renaissance between popular beliefs and the medical theories and practices of properly trained physicians and surgeons. Like the other areas of shared belief mentioned above, this tended to prevent the true 'professionalization' of medicine. When medicine did begin to establish itself as the preserve of a professional elite in the late eighteenth century it did so, not because it became more successful in dealing with sickness and injury, but because it developed theories and practices which were not, and could not easily be, part of popular consciousness. That story, however, is part of the history of the so-called 'Enlightenment'. The Renaissance, which is characterized by so many changes in other fields, including fields closely related to medicine such as anatomy and physiology, saw no major or lasting changes in the institutions of medicine.

Notes

1 John Securis, *A Detection and Querimonie of the daily enormities and abuses committed in Physick, Concerning the three parts thereof: that is, The Physicians*

part, The part of the Surgeons, and the arte of the Poticaries, London, 1566, sig. Aiii[r]. The Pythagorean provenance of the oath is discussed in Ludwig Edelstein, *The Hippocratic Oath: Text, Translation and Interpretation,* Supplements to the Bulletin of the History of Medicine 1, Baltimore, 1943. In spite of their traditional designation as 'Hippocratic', this famous collection of ancient medical texts was written by a number of writers from different medical sects over a period of a century or more (mostly between 430-330 BC but some later).

2 Andreas Vesalius, *De fabrica humani corporis* ('On the structure of the human body'), Basel, 1543, preface.

3 For a fuller discussion of the Renaissance attitude to the past see the article by Stephen Pumfrey in this volume.

4 William Salmon, *Pharmacopoeia Londinensis; or, the New London Dispensatory,* London, 1678, the preface, sig. A3[r].

5 Letter, Thomas Browne to Henry Power, 1646, in T. Browne, *Works,* edited by Geoffrey Keynes, 4 vols., London, 1964, iv, p. 255.

6 Margaret Pelling and Charles Webster, 'Medical practitioners', in *Health, Medicine and Mortality in the Sixteenth Century,* edited by Charles Webster, Cambridge, 1979, pp. 165-235, see p. 188.

7 William Clowes, *A Brief and Necessarie Treatise Touching the Cure of the Disease called Morbus Gallicus, or Lues venerea, by Unctions and other approoved Waies of Curing,* London, 1585, fol. 8r, quoted from *Selected Writings of William Clowes 1544-1604,* edited by F. N. L Poynter, London, 1948, p. 77.

8 First recorded, as far as I know, by Gaspard Bauhin in his appendix to François Rousset, *Isterotomotochia ... Gallice primum edita, nunc vero C. Bauhini opera Latine reddita, multisque et variis historiis in appendice additis locupletata,* Basle, 1588, and now cited in many standard histories of medicine. See for example, Kurt Sprengel, *Histoire de la médecine, depuis son origine jusqu'au dix-neuvième siècle,* 7 vols., Paris, 1815, iii, p. 414; and Cecilia C. Mettler, *History of Medicine,* Philadelphia and Toronto, 1947, p. 952.

9 34⁰ and 35⁰ Hen. VIII, chapter 8, 'An Acte that persones being no coen [*sic*] Surgeons maie mynistre medicines outwarde', *The Statutes of the Realm,* 11 vols., London, 1810-1828, iii, p. 906

10 Christopher Marlowe, *Tamburlaine the Great,* Pt. 2, Act V, Scene 3.

11 See R. R. Simpson, *Shakespeare and Medicine,* Edinburgh and London, 1959, and F. N. L. Poynter, 'Medicine and public health', in *Shakespeare in His Own Age,* edited by Allardyce Nicoll, Cambridge, 1976, pp. 152-66.

12 The attributes of the different temperaments listed here are taken from John Guerot, *The Regiment of Life, whereunto is added a Treatyse of the Pestilence ...,* London, 1546, fol. 2[r-v], but they frequently differ from author to author. Each temperament was held also to be constitutionally liable to different diseases. All the proverbs cited in this article are drawn from *The Oxford Dictionary of English Proverbs,* 3rd edn, edited by F. P. Wilson, Oxford, 1970, but there are many alternative compilations. One aspect of learned physicians' attempts to persuade the public that medicine should be left in the safe hands of trained and duly licensed practitioners was to attack such proverbial 'wisdom' as 'vulgar errors'. That 'every man is his own physician' was said by doctors to be the most erroneous of all medical

proverbs. But then, they would say that wouldn't they? On this see Natalie Zemon Davis, *Society and Culture in Early Modern France*, London, 1975, pp. 227-67, especially pp. 258-67.

13 This was so successfully absorbed into 'proverbial wisdom' that *The Oxford Dictionary of English Proverbs* (see note 12) gives no indication of its original source.

14 *Aphorisms*, I.1, and I.6. *Prognosis*, section 4. The main site of phlegm, the watery principle, in the body was thought to be the brain which looked rather phlegm-like. Male semen also looked rather like phlegm and, as pubic hair did not grow until the age of virility, it was assumed that phlegm/semen in some way promoted hair growth. Accordingly, a highly sexually active man, by draining phlegm (i.e. seminal fluid) out of his body was depleting the supplies in his head, so resulting in hair loss through insufficient nourishment. Likewise, eyes were clearly watery organs and so depended on adequate supplies of the body's watery principle, phlegm. Excessive sexual indulgence could, therefore, lead to poor eyes. At some point this became used as a 'naturalistic', as opposed to religious or moral, argument against masturbation. See Aristotle, *Problems*, translated by W. S. Hett, 2 vols., London, 1936, bk. IV, sections 2, 3, 4, 18, 32, pp. 108-13, 122-3, 132-3. This work is now thought to be pseudo-Aristotelian.

15 Christofer Langton, *An Introduction into Phisycke*, London, no date, *c.* 1550, fol. v[r-v], fol. x[v].

16 Geoffrey Chaucer, *The Tale of Melibee*, line 1306. Quoted from *The Works of Geoffrey Chaucer*, edited by F. N. Robinson, 2nd edn, Boston, 1957, p. 176.

17 See R. H. Shryock, *The Development of Modern Medicine: An Interpretation of the Social and Scientific Factors Involved*, Madison, Wis., 1936, pp. 74-8. For a medieval example consider John of Gaddesden's treatment of Edward II's son for smallpox. He instructed that everything around the sick-bed should be red in colour. One modern writer has compared this with the Finsen red-light treatment: Wilfred Bonser, *The Medical Background of Anglo-Saxon England: A Study in History, Psychology, and Folklore*, London, 1963, p. 219.

18 For a full discussion see Henry E. Sigerist, 'Ambroise Paré's onion treatment of burns', *Bulletin of the History of Medicine*, 15, 1944, pp. 143-9, reprinted in *Henry E. Sigerist on the History of Medicine*, edited by F. Marti-Ibañez, New York, 1960, pp. 177-83.

19 William Shakespeare, *Twelfth Night*, Act III, Scene 4.

20 A complete translation of *De cautelis medicorum*, from which these extracts were taken, and fuller discussion can be found in Henry E. Sigerist, 'Bedside manners in the Middle Ages: The treatise *De Cautelis Medicorum* attributed to Arnald of Villanova', in *Henry E. Sigerist on the History of Medicine* (see note 17), pp. 131-40. For Kurt Sprengel's opinion see his *Histoire de la médecine* (see note 8), ii, p. 443.

21 Here, once again, we have part of the famous first aphorism of Hippocrates which begins: *ars longa, vita brevis. Aphorisms*, I.1.

22 Andrew Boord, *The Breviary of Helthe for all Maner of Sycknesses and Diseases the whiche maybe in Man or Woman doth folowe. Expressynge the obscure terms of Greke, Araby, Latyn and Barbary, in to Englysh concerning Phisicke and Chierurgye*, London, 1547, fol. ii[r-v].

23 Langton, *Introduction into Phisycke*, fol. vi^r.

24 John Securis, *A Detection and Querimonie of the daily enormities and abuses committed in Physick*, sig. Avi^v.

25 Langton, *Introduction into Phisycke*, fol. ix^r.

26 Molière, *Le Malade imaginaire*, Third Interlude.

27 Simon Harward, *Phlebotomy; or, a Treatise of Letting of Blood, fitly serving, as well for an Advertisement and Remembrance to well minded Chirurgions, As also to give a Caveat generally to all Men to beware of the manifold Dangers which may ensue upon rash and unadvised letting of Blood*, London, 1601, sig. A2^v-A3^r.

28 Quoted from *Willis's Oxford Casebook (1650-52)*, edited by Kenneth Dewhurst, Oxford, 1981, pp. 128, 141.

29 Nicholas Culpeper, *A Physicall Directory; or, A Translation of the London Dispensatory*, 2nd edn, London, 1650, sig. B^{r-v}, quoted from Christopher Hill, 'The medical profession and its radical critics', in his *Change and Continuity in Seventeenth-Century England*, London, 1974, 157-78, p. 160. Hill's article has much to say about the 'professionalization' of medicine in Interregnum England.

30 Noah Biggs, *Mataeotechnia Medicinae Praxeos. The Vanity of the Craft of Physick. Or, A New Dispensatory...*, London, 1651, sig. a4^v.

31 Culpeper, *Physicall Directory*, 3rd edn, London, 1651, sig. A2^v.

32 On the publishing of Paracelsus in translation and other Paracelsian works during the Interregnum see Charles Webster, 'English medical reformers and the Puritan revolution: A background to the "Society of Chemical Physicians"', *Ambix*, 14, 1967, 16-41, pp. 40-1; and by the same author, *The Great Instauration: Science, Medicine and Reform 1626-1660*, London, 1975, especially pp. 275-81.

33 These quotations are from Howard M. Solomon, *Public Welfare, Science, and Propaganda in Seventeenth Century France: The Innovations of Théophraste Renaudot*, Princeton, 1972, pp. 64, 79-80, 81-2; on the general prominence of medical issues in these conferences see pp. 75-80.

34 *Ibid.*, p. 176. For a fuller discussion of *La présence des absens* see pp. 175-7 and 237-8 of Solomon's study of Renaudot.

35 William Perkins, *A Discourse of the Damned Art of Witchcraft*, London, 1608, pp. 11-12.

36 Genesis 1:3; John 1:1. On the magic of the Christian Church, both before and after the Reformation, see Keith Thomas, *Religion and the Decline of Magic*, London, 1971, pp. 27-206.

37 On Paracelsus and magical traditions see Charles Webster, *From Paracelsus to Newton: Magic and the Making of Modern Science*, Cambridge, 1982. On occult qualities in medicine see Brian P. Copenhaver, 'Astrology and magic', in *The Cambridge History of Renaissance Philosophy*, edited by Charles B. Schmitt and Quentin Skinner, Cambridge, 1988, 264-300, pp. 283-4 and 286; and Linda Deer Richardson, 'The generation of disease: occult causes and diseases of the total substance', in *The Medical Renaissance of the Sixteenth Century*, edited by A. Wear, R. K. French, and I. M. Lonie, Cambridge, 1985, pp. 175-94.

10

The rational witchfinder: conscience, demonological naturalism and popular superstitions

Stuart Clark

1   The natural science of demons

In the modern world witchcraft beliefs are usually assumed to refer to things supernatural. Almost instinctively, we locate the deeds of devils and witches outside the realm where natural laws operate, or totally redescribe them until they can be accounted for naturalistically. As C. S. Lewis wrote: '... such creatures are not part of the subject matter of "*natural* philosophy"; if real, they fall under pneumatology, and, if unreal, under morbid psychology.'[1] This means that we also tend to regard demonology – the academic study of demonism, magic and witchcraft undertaken by Renaissance scholars and intellectuals – as an 'occult' or 'pseudo' science. Usually, reasons other than those intrinsic to it are sought for its popularity throughout much of the sixteenth and seventeenth centuries; it was the product, so it is supposed, of lingering superstition, of irrationality, or worse still, of collective derangement. But the history (as well as the anthropology) of science shows that the perceived boundary between nature and supernature, if it is established at all, is local to cultures, and that it shifts according to changes of taste. The one now generally in force among the tribes of the West is only as old as the scientific production that goes with it. Before the Enlightenment and the coming of the 'new science', things were different, metaphysically speaking, and nature was thought to have other limits. In fact, views of the demonic were entirely the reverse of today's. In Renaissance Europe it was virtually the unanimous opinion of the educated that devils, and, *a fortiori*, witches, not merely existed in nature but acted according to its laws. They were thought to do so reluctantly and with a good many unusual manipulations of natural phenomena, yet they were always regarded as being inside the category of the natural. This was a matter of principle for demonologists; for them, *not* to accept it was superstitious. We, on the contrary, have to set aside some of our most automatic assumptions to grasp its significance.

The idea was neither novel nor restricted to the sphere of natural philosophy. The demands of medieval Christianity itself were that the devil should be strong in relation to men and weak in relation to God, a power differential of which St Thomas Aquinas's discussion 'Whether angels can work miracles' was the model account. Angels did seem capable of miraculous actions, as did demons (whose fall from angelic grace had not deprived them of their knowledge and powers, the skills they imparted to demonic magicians being an obvious example). But (Aquinas argued) something could only be properly called a miracle if it took place entirely outside the natural order, of which the powers of all angels, being creatures, were necessarily a part. Its causation was thus the prerogative of the Creator alone. God might work miracles at the request of angels or through their ministry (only the latter in the case of demons). Otherwise, all presumed examples of angelic agency beyond nature must be attributed to human mistakes about just where the natural limits of actions lay. This was the case with demonic magic; it was thought to exceed nature but was in fact worked entirely through the natural powers of demons, and only seemed miraculous by comparison to the natural powers of men and women ('... relatively to us').[2]

Aquinas's point, to be endlessly elaborated in the sixteenth and seventeenth centuries, was that Satan worked not miracles (*miracula*) but wonders (*mira*). Demonic phenomena went beyond what would normally have been expected from the ordinary flow of natural effects from natural causes, and their strangeness made them seem impossible. They were, after all, the product of great intelligence and power, largely unimpaired by the devil's original fall from grace and exceeding what most men could muster. But this did not make them supernatural, only extraordinary – they were, in the terminology of Renaissance scholars, 'preternatural'. As the Zwinglian professor Peter Martyr insisted, the devil had only two 'bridles': one was the will of God, the other the boundary of nature. The Lutheran Paulus Frisius agreed that even his most astonishing feats should be classed as physical events within nature and not beyond it. A Calvinist expert, the pastor and then Genevan academic Lambert Daneau, put the point most bluntly. 'Sathan', he wrote in a witchcraft treatise of 1574, 'can doo nothing but by naturall meanes, and causes... As for any other thing, or that is of more force, hee can not doe it.'[3]

All this has important implications for the way we read the literature of witchcraft. Generally, it was much more a function of a particular view of nature, and much more intimately involved in debates about that view's validity, than has often been realized. The Spanish theologian Pedro Ciruelo spoke for all demonologists when he insisted that the existence and agency of incorporeal spirits were truths of 'philosophy' – that is to say, natural philosophy – as well as of theology.[4]

This means that we can treat Renaissance conceptions of the witch in much the same way as Michel Foucault has treated pre-modern conceptions of the medical patient – that is, in relation to historically particular ways of constituting objects of knowledge. Witches became 'visible', so to speak, when the conditions governing what could be seen and described in nature changed in their structure.[5] They remained 'in view' as long as the vision of demonologists continued to be directed by the perceptual codes of pre-modern, Thomistic natural philosophy. And they were 'lost to sight' when, in the course of the 'long eighteenth century', the world of objects to be known by natural philosophers was radically reconstituted.

But in particular, witchcraft writings were marked by outright naturalism. Despite a reputation for intellectual incoherence and inflammatory rhetoric, one of their main aims was to demystify demonic phenomena by subjecting them to careful, and essentially negative, scrutiny – sorting out, as the magistrate Henri Boguet did in his study of the confessions of the witch Françoise Secretain, just where the limits of demonic efficacy through 'second and natural causes' were reached.[6] Again, in so far as his preternatural powers as an agent were concerned, demonologists often compared the devil to an astonishingly knowledgeable and adept scientist. They portrayed him as someone who, in addition to having mastery over local motion, specialized in manipulating the secret properties of nature and the hidden causes of things. In fact, he was seen quite precisely as an expert in what Aristotelians called 'occult qualities'. This, however, had little to do with what is meant by the 'occult' today, since one of the most widely held assumptions in Renaissance natural philosophy was that the qualities in question were real qualities which, even if they could not be perceived, were capable of causing real effects – among which were gravitation, magnetism, and many forms of purgation. In contemporary terms, then, demonology was certainly concerned with the occult; yet demonologists were never exponents of inefficacious science. Indeed, it is one of the ironies of the history of witchcraft beliefs that the modern notion of the occult, which stresses inefficacy, applies not to the things that witchcraft theorists accepted as true but (as we shall see) to the things they themselves rejected as false. This is another reminder that decisions of this sort are always licensed by historically particular conceptions of reality.

We can see that, although they were theological in inspiration, Aquinas's arguments issued at every stage in questions about the precise workings of the natural – and, especially, the preternatural – world. These questions preoccupied the demonologists but they were also crucial to Renaissance natural philosophy as a whole. Although the principle that only the Creator could break his own laws was unshakable, the correct classification of individual phenomena proved to be increasingly difficult and contentious. The requirement that true miracles remain 'beyond the order of the whole created nature' tied their authentication, along with the exposure of demonic look-alikes and the identification of prodigies that were nevertheless natural, to an exact understanding of where nature's boundaries actually occurred. And in these matters of categorization and epistemology, Renaissance natural philosophy was often in turmoil. It experienced what might be called 'frontier' problems – problems about how to allocate phenomena lying along the borders between different classes of events. Demonology was very much continuous with the debates that resulted. For example, in 1646, the English preacher John Gaule said that it would take someone 'very learned' in natural philosophy to distinguish between the trickery of magicians in manipulating natural effects artificially and the spontaneous 'Mirables [i.e. wonders] of Nature, in her occult Qualities, Sympathies, Antipathies, and apt conjunction of Actives to Passives'. Robert Filmer agreed that what could and could not be done by the power of nature was an issue for the 'admirable or profound Philosopher'; it would, he said, have taxed Aristotle himself, yet 'there be dayly many things found out, and dayly more may be which our Fore-fathers never knew to be possible in Nature.' Both these remarks were made in books about witches.[7]

Above all, describing the devil as a worker in occult causes and wonderful effects established an epistemological (though not, of course, a moral) equivalence between demonic agency and the subject matter of what was known as 'natural magic' – one of the most enduring enthusiasms of Renaissance natural philosophers. This is because natural magic too was defined as the pursuit of nature's secrets, especially those based on affective natural actions of 'sympathy' and 'antipathy', and, again like demonism, was admired for its capacity to produce *mira* that seemed impossible to the uninitiated. Bert Hansen has shown how the study of nature's normal processes (*scientia naturalis*) and the study of preternatural or artificial marvels (*magia naturalis*) complemented each other in later medieval scholasticism.[8]

This continued to be true of much of the sixteenth and seventeenth centuries, not least because, in addition to the many enthusiasts for natural magic in the so-called 'Hermetic' tradition, and the support of an influential figure like Francis Bacon, it was routinely included in the teaching curriculum by the Aristotelian natural philosophers who still largely dominated academic physics.[9] In epistemological terms, natural magic was demonic magic's essential, and (it must be stressed) very orthodox, point of reference. Equally, the causation and efficacy of demonic phenomena were often discussed in natural magical contexts – for example, in collections of natural secrets, or by specialists in natural magic like Georg Pictorius, Girolamo Cardano and Gottfried Voigt. The two fields of knowledge rose to prominence together, and ceased to be taken seriously, or were resolved into other disciplines, when the same changes of scientific taste made them both seem equally implausible.

The best, and, hopefully, not now paradoxical, way to sum up this initial point is by saying that we would do better to associate demonology with the advancement of science than with its stagnation or decay. The modern assumption has been that witchcraft beliefs were somehow inimical to the welfare of science – that they were consistent only with an ossified Aristotelianism and were quickly and inevitably overhauled as soon as the pace of scientific change quickened. Since the Enlightenment, Thomistic puzzles of the devils-and-miracles type have always seemed to stem from ignorance, not knowledge. As more and more came to be known about nature, so (the argument has been) it was less and less likely that witchcraft would be accepted as a real thing.

Yet if the devil was a part of Renaissance nature, then demonology was a part of Renaissance science; there was thus no reason for it to remain intellectually inert. Implicit in the medieval inheritance was a programme for further inquiry, an agenda of discussable issues in metaphysics, epistemology and physics, which, far from being threatened by any improved understanding of natural causation, actually presupposed this. If anything, the issues became more and more pressing. The need to reconsider the validity of preternatural phenomena of every kind, and the rules for categorizing them, became especially urgent. It is as though the existence, during much of the sixteenth and seventeenth centuries, of a conceptual free-for-all added intellectual allure to subjects like demonology and natural magic which, precisely because they existed at the borders between causal categories, exposed the categorizing process itself for inspection. Here lay the scope for a genuinely natural and genuinely scientific exploration of the demonic.

2 Demonology and the reform of popular technique

We can see in more detail how naturalism was put to work in Renaissance demonology if we concentrate on the issue of causal efficacy. More importantly, this will allow us to recognize the crucial importance of demonology for religious and cultural, as well as scientific change – in particular, its significance in the widespread campaign to improve lay spirituality. It was one of the central arguments in demonology that the ostensible causal connections exhibited by witchcraft were invariably non-existent. For the most part, they were attempts to link entities that had no bearing on one another in nature; indeed, it was this that made them demonic and not merely what they would be for us – bad science. Boguet, for example, spent the central chapters of his *Discourse on Witches* examining the supposedly maleficent operation of rituals, powders, ointments, breath, words, looking, touching, magic wands, and images of wax – all of these allegedly in use by early modern witches. In every case, efficacy was said to be either natural or spurious, with the devil intruding his own agency to secure the *maleficium* or resorting to outright deception. Of course, because of its capacity to produce *mira* this agency seemed to go well beyond the nature known to men and to be in this sense 'unnatural'. But, as we have seen, in the end the devil too could only work with the causal connections actually available between natural things. He could, for instance, bring hailstorms because they had natural causes and because all natural phenomena were at his command. But no accompanying ritual could physically effect this; it merely symbolized the demonic entanglement of the witch who performed it in the belief that it could. If, as in the case of actual poisons, powders possessed natural properties to harm, then they could certainly be used to kill or injure, and obviously by humans. Otherwise they too were only signs. But afflictions 'caused' by witches breathing or blowing on their victims, charming them with maledictions, looking at them, touching them by hand or with wands, and performing sympathetic magic on their likenesses, were all in reality caused directly by the devil stepping in to procure the right effects when such 'causes' failed to do so. Boguet's preference was always for a demonic (but still ultimately natural) efficacy, with the ostensible means reduced to a merely symbolic accompaniment.[10]

These arguments are found throughout the literature. The French physician Jacques Grevin wrote that there were only two things to consider in cases of witchcraft – what was natural and what was not. Daneau said that the overriding criterion for judging the deeds of witches was whether 'the accomplishing and trueth therof, plainly repugneth against the course of nature'.[11] It was generally assumed that, in and for themselves, witches had no greater capacity to effect things than other human beings; all alike were constrained by the same natural limits to the powers of created beings. Every other thing in creation had also been given its 'nature' – its attributes, virtues and properties. It followed that any effect lying beyond these various capacities could only be achieved, or even be hoped for, if some agent with the ability to substitute alternative efficacies was also involved. In witchcraft this could plainly be neither God nor some good angel. It was devils, then, who made good the causal *lacuna* that opened up whenever the intentions of witches exceeded the limits of natural efficacy. The basic principle was stated succinctly by Thomas Erastus, a physician who wrote medical works against Paracelsus and (in 1578) a demonology in reply to Johann Weyer: 'Whoever tries with natural instruments to do things which surpass the strength of nature, using neither the help of God nor the help of good Angels, is necessarily appealing for demonic aid by means of an open or secret pact.'[12] This is not the remark of a man whose thought allowed for no distinction between genuine and empty causation – between what would now be called scientific and occult knowledge. And it is not the remark of a man who believed that witchcraft *per se* was anything but spurious.

These technical questions concerning efficacy remained the subject of intense academic debate throughout the early modern period. Demonology became rather like what Bacon called a 'prerogative instance' – that is to say, a subject-area with the potential to lay bare especially important empirical and conceptual issues. It therefore interested natural philosophers for reasons that had nothing to do with witch hunting. But efficacy in nature was also brought to bear vitally on broader questions of cultural reform; it became a social issue. The debate about how to categorize and correctly identify phenomena according to a 'grid' of causes came to divide not merely the theologians and philosophers from each other but the learned from the unlearned; getting these things wrong became a sign of cultural unregeneracy and not just an intellectual error. This is because, contrary to our expectations, a great deal of the literature of demonology was not addressed to the relatively specific problem of the classic, maleficent, devil-worshipping witch – the sort of witch represented by Boguet's Françoise Secretain. It was directed instead at the everyday behaviour of ordinary men and women. In particular, it was aimed at what it is best to call 'popular technique'. By this I mean the enormous repertoire of techniques, recipes and rituals for good health, healing and fertility, for securing good fortune and preventing misfortune, for detection and divination, or simply for making decisions, that existed alongside official religious observance and constituted the practical dimension of early modern popular culture.

Chronicled extensively by historians under the heading 'popular magic', these practices seem to have little to do with demonic witchcraft and *maleficium*; those whose practices they were evidently thought so too. In societies that accepted *maleficium* as simply one of the risks of day-to-day life they were often directed *against* the possibility of witchcraft or in response to it and were generally considered to be beneficent in character. Those who wished to control and improve popular belief – essentially, the clergy – saw things very differently. They were more concerned, indeed, in an age of reformation, preoccupied, with the spiritual predicament of anybody who, faced with a practical need or misfortune of any kind, turned not to the Church and the approved remedies, but to the resources of popular tradition. This showed lack of patience and trust in God. It treated as contingencies those things that divines called 'Providence', and ignored the need for repentance and other benefits of affliction. Worse still, it placed a faith amounting to idolatry in the protective, curative and revelatory properties of 'creatures' – whether persons, places, times or things. Worst of all, it was grounded in a faulty view of causation. Like witchcraft itself, popular techniques were spurious in nature – that is, the nature known to clerics. In the eyes of the latter they were 'magical', in the sense not of an allowable *magia naturalis* but of a very unallowable *magia daemonica*. This is the reason why 'magic' is a difficult word for historians.

In so far as essential services like healing, divination and counter-witchcraft were professionalized in the hands of 'cunning' or 'wise' men and women, the Church was even challenged by a rival institution. In the European countryside, in particular, the priest and the 'magician' confronted each other as rival therapists for the community's afflictions. The recent cultural history of Renaissance Europe has shown beyond doubt that popular technique was a major target of the Protestant and Catholic Reformations but that it was remarkably resilient in the face of attack. Even after a century and a half of 'improvement' ordinary people still had ideas about the origins of fortune and misfortune, the causes of events, and the relative benefits of religion and 'magic' that could be radically at odds with those of their clerical reformers. This was the context for much of the demonology of the period.

3   Witchcraft and the concept of superstition

The extent to which demonology was taken up with the condemnation of popular technique, rather than *maleficium* as such, is especially clear in the case of Protestant authors. Contemporary English experts on witchcraft like George Gifford, William Perkins, James Mason and Richard Bernard – all, except Mason, clergymen – were outspoken in their attacks on the cunning men and women of Tudor and Stuart England, and thought that the services they offered their clients were much more subversive of God's spiritual order than any material damage inflicted by witches intent on evil. The Swiss reformer Heinrich Bullinger spent most of his witchcraft tract discussing the ordinary household use of charms and exorcisms and the astrology and soothsaying of professional diviners. Bernhard Albrecht devoted one of the most substantial German treatments of witchcraft and magic largely to the blessers, exorcists and astrologers who drove away evil and illness and plotted propitious and unpropitious moments for their clients' activities. In Saxony, his fellow Lutheran, the preacher Johann Rüdinger devoted ten published sermons to maleficent witchcraft but ten more to soothsaying, 'the observation of days', augury, healing by charms, astrology, and the interpretation of dreams. The most important Danish contribution to the witchcraft debate, by the Copenhagen theologian Niels Hemmingsen, concentrated likewise on those arts which ordinary people thought of as useful and beneficial, notably divination and charming. It is clear from the ethnographic detail in these texts, and many others like them, that they were meant to embrace a very considerable number of the arts and techniques on which lay people relied for their material and psychological welfare.

Catholic treatises were often more inclined to dwell on the circumstantial details of devil worship and *maleficium* – the sorts of things that come first to mind when 'witchcraft' is spoken of. But the greatest Jesuit treatment of the subject, Martín Del Río's *Six Books of Disquisitions on Magic*, included a good deal of advice to priests and confessors regarding the 'vain observances' of the laity and a whole section enumerating them. Another major contribution, Francisco Torreblanca's *Demonology*, opened with thirty chapters on divination, including augury and astrology, the interpretation of dreams and omens, and the observation of days. Members of the French orders, like the Minim Pierre Nodé, and even Catholic lawyers like Pierre Massé, wrote demonologies which turn out to be concerned mainly with charming and divination. Jacob Vallick, the Catholic priest of Grosen (or Groessen) in Gelderland in the Netherlands, published a dialogue *Of Magicians, Witches and evil Spirits* which exactly parallels *A Dialogue Concerning Witches and Witchcraftes* by the Calvinist minister of Maldon in Essex, George Gifford. Even Boguet found space to attack those who healed with charms.[13]

Besides, precisely because clerical demonology had this very much broader frame of reference, we should not be looking only at those texts normally taken to be illustrative of learned witchcraft beliefs and the desire to hunt witches. Evidently, demonology was often the context for castigating lay views concerning fortune and misfortune, affliction and redress. But equally, there were many other attempts to proscribe these views and substitute clerically approved versions which draw on demonological arguments. The important point is that conceptual continuities make it impossible to separate the literature of witchcraft from the literature of religious reform. This applies particularly to two major areas of preaching and publishing, one devoted to 'cases of conscience', the other to the specific sin of 'superstition'.

The importance of casuistry to the evangelical ambitions of the two Reformations can hardly be over-estimated. The improvement of lay behaviour presupposed that ordinary men and women could learn to treat their actions as moral cases resolvable in terms of general rules of conduct. The clergy were there to offer the necessary guidance, with or without the confessional, and they in turn could be provided with model guides to the workings of perfect consciences. Edmund Leites has recently said that the Catholic Church's commitment in this area was 'massive' but that casuistical literature was also vital to Lutherans in Germany and Calvinists in England. Between the thirteenth and seventeenth centuries, he writes, 'church-sponsored casuistry was a culturally dominant force in the West'.[14] Practical moral theology on this scale could hardly avoid cases arising from daily needs and routine afflictions and, in particular, the seemingly constant popular demand for techniques that would ward off misfortune in the present and ensure some knowledge of and control over the future. The ground shared by casuistical demonology and demonological casuistry was thus extensive. Discussions of demonic magic and witchcraft that exactly paralleled those in specialist demonologies were available to Protestant audiences all over Europe via this different textual and evangelical route – for example in Friedrich Balduin's *Treatise of Cases of Conscience*, William Ames', *Conscience with the Power and Cases Thereof*, and Ludwig Dunte's *One Thousand and Six Cases of Conscience*.[15] The Catholic contribution was also voluminous, especially after the Council of Trent. It included the great collections of moral questions by Jesuits like Domingo de Soto, Thomas Sanchez, Francisco Suarez and Leonardus Lessius; manuals on cases of conscience addressed to priests, like the much reprinted *Instruction of Priests* by Cardinal Toledo, the Benedictine Gregory Sayer's *Key to the Palace of Priests*, or Martin de Azpilcueta's *Manual, or Guide for Confitents and Penitents*; and many, many instruction books for inquisitors.[16]

These indispensable aids to the reformation of the laity contain some of the most substantial and influential discussions of demonic magic and witchcraft available anywhere in Renaissance Europe. But in every case, the sin against which consciences were warned was that of superstition. This means that the enormous literature dealing with superstition has also to be taken into account if we are to appreciate the extent to which demonology was implicated in the clerical campaign to reform popular culture. Jean Delumeau has rightly argued that Protestant attitudes to superstition are unintelligible if separated from the demonology that went with them.[17] In addition to the examples he gives from the major Protestant theologians, this is also amply illustrated from the German lands by the writings of Lutheran pastors like Conrad Platz, Christoph Vischer and Abraham Scultetus.[18] But it is equally true of Catholic culture. The devil is so important in works like Pedro Ciruelo's *A Treatise reproving all Superstitions* or in Jodocus Lorichius' *Superstition* that it is impossible not to read them as intellectually continuous with the texts that defended witch hunting. Yet Ciruelo was, once again, chiefly concerned with practices popularly thought to be beneficial to material life, and Lorichius' divisions of superstition were all aspects of popular technique.[19]

'Superstition' is one of the key concepts in the history of early modern culture. This is not, of course, because in the past some historians have themselves used the term 'superstitious' to describe aspects of early modern beliefs and behaviour. It is because the notion was immensely important to contemporaries and informed a great many of the things they said and did. It was important because it embraced the three things with which this essay is mainly concerned. First, it defined what religious reformation, seen in the broadest moral terms, was actually about. Following St Augustine and (again) Thomas Aquinas, the clerics who led the two great reform movements thought of superstition as among the most serious (often *the* most serious) of religious transgressions. It was the very antithesis of true religious worship and seen as a fundamental obstacle to success. In Catholic casuistry it was 'religion's opposite' and always classed as a sin against the First Commandment. This made it, with blasphemy and impiety, the greatest of all moral vices. The First Commandment, wrote Ciruelo,

> is also of greatest worth and sacredness. And the virtue which God commands us to practice by it is the most perfect among the moral virtues. It is the most pleasing to God... On the contrary, the sins man commits against this commandment and this virtue of religion are the most hateful of all, because they are most displeasing to God and most harmful to men.

When reading passages like this we need to bear in mind John Bossy's recent suggestion that, from the sixteenth century onwards, the Ten

Commandments replaced the Seven Deadly Sins as the West's moral system. 'The main substantive difference between the two systems', he writes, 'was the greater importance given in the Decalogue to offences against God'. Among these, superstition was one of the most serious.[20]

Secondly, the concept of superstition was a cultural weapon directed by churchmen mainly at the populace at large (although also at their clerical competitors). It was a form of proscription in terms of which many of the routine material practices of pre-industrial, rural cultures, together with the categories and beliefs that shaped ordinary people's experience, were denounced as valueless. It was not entirely this, of course. Adopting distinctions laid down in Aquinas' thirteenth-century theological *summa* the casuists allowed for two broad types of superstition, one of which consisted in service to the true God but in an incorrect manner, the other in service to a false god but in the manner due to the true. The usual explanation, given, for example, by Toledo, Lessius and Azpilcueta, was that the first of these occurred when God was honoured either with invalid ceremonies, like those derived from Mosaic law or based on false relics and miracles, or with superfluous ceremonies, like those which went beyond the official liturgy by, for example, multiplying its rituals.

In these contexts, superstition meant irrelevant or excess worship, provoked (so it was said) by ignorance and fear. On the other hand, we know that laymen and women often associated the use of such para-liturgical formulas and the mechanical repetition of pieces of church ritual with material benefits; to call these 'superstitions', then, could, even here, connote the broader proscription of popular culture. The second broad type of superstition, that of according the right service to the wrong god, was also invariably sub-divided into species. Of these, idolatry and (in Toledo) *maleficium* referred to the outright devil-worship of the evil witch – that is to say, they referred to sins which, although real enough in the minds of the clergy, were hardly as significant in the daily lives of their parishioners. But alongside them were categories like 'magic', 'divination' and (most inclusive of all) 'vain observation', all of which embraced aspects of popular technique which were both very widespread and very real. It is noticeable that the casuists spent far more time analysing the latter group than the former.

Finally, superstition was demonic. 'Superstition', Aquinas had said, 'includes not only sacrifices offered to the demonic powers in idolatry, but also invoking their help in order to know or to do something'. 'All types of superstitions', repeated Ciruelo, 'come from evil spirits... the devil has discovered and taught men all superstitions'.[21] To indulge in them was to transfer to him the spiritual allegiance promised at baptism to God, to become a religious traitor and apostate. At the general level, this is scarcely a surprising view. Behind the greatest sin must lie God's greatest antagonist; all manner of errors and excesses were thus attributable to him. But what exactly made countryside healers and diviners worshippers of a false god? What made everyday decisions based on propitious or unpropitious moments acts of spiritual treason? Why was the simplest charm or the most common blessing a sign of entry into a pact with Satan? On the status of such routine aspects of practical life hung the demonologizing of popular culture. But, beyond the fact alone that they *were* proscribed actions (from the Old Testament onwards, it was argued), it is not immediately obvious why they should be demonic at all.

4 Inefficacy and its implications

The answer lies in their inefficacy. Supporting the moral theology of superstition, and thus the entire campaign to improve popular cultural habits, was the naturalism of Renaissance demonology. The essential criterion of a superstition in the spheres of knowledge and practical activity was that it too was faulty in nature (and, as a consequence of this, faulty in religion). The natural incapacities insisted upon in the case of the evil witch were, necessarily, those of all human beings; *anyone*, we recall Erastus insisting, who used 'natural instruments' for purposes beyond their natural strengths, and who had neither divine sanction nor the help of good angels in doing this, was in effect appealing to the devil to bring about the required effects. Yet again, this was a principle that Renaissance authors found codified in Aquinas:

> When things are used in order to produce an effect, we have to ask whether this is produced naturally. If the answer is yes, then to use them so will not be unlawful, since we may rightly employ natural causes for their proper effects. But if they seem unable to produce the effects in question naturally, it follows that they are being used for the purpose of producing them, not as causes, but only as signs, so that they come under the head of a compact entered into with the demonic.

This was why superstitious practices depended on 'auxiliary demonic forces'.[22] In the age of witch hunting itself, superstitious persons were witches in another guise, their real sin merely disguised by their good intentions. Without knowing it, they too had entered into a pact with the devil, a pact which should be considered 'tacit' or 'implicit', in order to distinguish it from the 'open' or 'express' agreement into which witches intent on evil were thought to enter. All the texts we are considering insisted on the reality of tacit demonism in cases of inefficacy. It occurred, according to the Ingolstadt theologian Gregorius de Valentia (another Jesuit), 'whenever anyone employs, as capable of effecting something, such means as are in the truth of the matter empty and useless'.[23]. They also took up Aquinas's distinction between a cause and a sign. Superstitious means, because they caused nothing, were tokens of another kind of efficacy; they were, indeed, sacraments, pledges of a covenant with the devil. Casuists, then, like demonologists, were committed to a conception of nature, a covert natural philosophy, which it was their aim to impose on lay consciences. This is the reason why it is impossible to separate the two sets of arguments.

Protestant casuistry on the subject is neatly summarized in Ames's *Conscience with the Power and Cases Thereof* at the point where he considers the case of consulting the devil. This occurs not merely 'by a direct petition, or by an expresse compact' but 'by a silent and implicite compact' whenever 'those meanes are used, either for the knowing or effecting of things which have no such use by their owne nature, nor by the ordinance of God; and no extraordinarie operation of God with them can bee expected by Faith'. In such cases, 'the Devill is the author both of the operations, and significations which doe depend on such meanes, and ... is consulted with by them that doe expect any thing in such waies'.

The sins involved, accordingly, are infidelity to God and, chiefly, giving honour and worship to Satan. Ames considers the two examples of astrology and the 'vaine observations' of 'the simple, ignorant, and credulous common people'. The stars do influence inferior things in a general way and it is, therefore, within nature to make general predictions from their movements (just as it is in the cases of predictions 'taken from the elements, from the frame of the members of mans body, from dreams, from prodigies, &c'). But particular predictions concerning future contingencies and, especially, the voluntary actions of individuals are not warrantable in nature and are, therefore, demonic. Likewise, the demonism of the 'common people' lies in associating causes and effects that do not belong together in nature; for example, conjecturing 'some joyfull, or sad events, upon some accidentall words or deeds aforegoing', counting saints' days as lucky or unlucky 'to beginne any worke in', attributing material efficacy 'to certaine formes of prayer, and to conditions annexd to them, for the procuring of this, or that singular thing', employing 'Figures, Images, Characters, Charmes, or Writings' to drive away diseases, or ascribing effects to 'Herbes, and other Medicines, not as they are applyed in a naturall way, but as they be charmed, or as they bee used in some certaine forme and no other'.[24]

These distinctions between efficacy and inefficacy in nature were the intellectual foundation of Protestant attitudes to 'superstition' in whatever context they were expressed – whether in the encouraging of magistrates to hunt witches, the 'demonologizing' of communal beliefs and techniques, or simply the guiding of individual consciences. For the seventeenth-century preacher John Gaule, whose *Select Cases of Conscience Touching Witches and Witchcrafts* covered the whole ground, the crucial relationship was between the *genus* superstition and the *species* witchcraft: 'Because in this main Act, superstition and Witch-craft both agree; to apply to the Creature as means unto those ends and uses unto which it is neither apt by its own nature, nor thereunto ordained by divine Institution'. They differed, he said, only in degree, 'for superstition is witchcraft begun, and witchcraft is superstition finished'. James Mason, author of *The anatomie of Sorcerie* also argued that the activities of countryside healers and blessers ('commonly called wise and cunning men, and women') were demonic because they ignored the fact that God had given everything in the world 'a severall nature, vertue, and property, to be wrought, or to worke this or that effect (if it be rightly used and applyed) the which it is not possible for any creature in the world to alter, or change, but onely for the creator...'. In another book devoted substantially to 'the Curing Witch', the minister of Batcombe in Somerset begged his churchwardens to present to the church courts not only the local healers and charmers but their clients too. Anybody who used words and prayers superstitiously, that is, believing in their natural efficacy, was invoking the devil in principle; if effects resulted, the devil was at work in fact, for there was simply no other way they could have come about.[25]

In Germany, in sermon collections like those of Abraham

Scultetus (on divination and astrology) and Bernhard Albrecht (on all the popular techniques), it was said that magic and witchcraft occurred

> when anyone uses something in God's creation, such as herbs, wood, stones, words, times, hours, gestures and the like, or seeks to bring about some effect, other than God has decreed, with the assistance and support of devils, either to reveal hidden or future things, or to obtain unnatural things, supposedly to help a neighbour or more likely to bring harm and loss.

What Albrecht meant by God's decree was that each created thing had been given 'its nature, virtue, power and efficacy [*verrichtung*]'.[26] Specialist treatments of individual superstitions, like the Schmalkalden superintendent Christoph Vischer's *A Brief Report Against Blessings, by which Men and Beasts are Thought to be Helped* simply elaborated the general principle. It was magic, said Vischer, to use words alone to cure illnesses, because this was 'against their natural efficacy (*Wirkung*)'. It was superstitious, agreed Johann Rüdinger, the pastor at Weyra in Saxony, to regard days or hours as lucky or unlucky, since 'there is no reason for this in nature'. Caspar Peucer's hugely successful mid-sixteenth-century study of divination rested in part on the distinction between the legitimate predictions of effects from the natural properties with which things were endowed ('natural divination') and the superstitious abuse of the natural order in supposing it to contain causal relationships that were in fact spurious ('diabolical divination').[27]

Catholic acculturation was likewise based on a particular view of nature and natural processes – in this case, transmitted from scholasticism via authoritative academic pronouncements on superstition (like that of the Paris theology faculty in 1398) or by the writings of earlier reformers) like Joannes Gerson, Jacobus van Hoogstraten and Martin de Arles). It was often expressed in terms of formal commentary on Aquinas as, for example, in the cases of Lessius and De Valentia and, in addition, the Italian Dominicans, Silvestro Mazzolini da Prierio and Tommaso de Vio (Cardinal Cajetan). What he called 'the general rule' on the subject was stated by Jacobus Simancas in one of the most frequently cited of all inquisitors' manuals, his *Institutiones Catholicae*: 'Those things that cannot naturally bring about the effects for which they are employed are superstitious, and belong to a pact entered into with devils'.[28] Toledo was typical of the Tridentine casuists in defining each of the three specific superstitions that embraced popular techniques in terms of inefficacy. 'Magic' was any attempt, unassisted by God's miraculous agency, to do 'that which is above nature'. The devil was tacitly invoked in any intention 'to do something by such means as are neither by themselves nor by supernatural power able to produce such effects'. 'Divination' was any attempt, unassisted by divine revelation, to know things 'that cannot be known naturalistically *[per naturam]*'. The devil was again invoked whenever methods of acquiring knowledge assumed causal connections in nature which did not in fact exist. And 'vain observance', like magic, was the employment of means which had no natural ability to bring about the required effects.[29] In the standard accounts by Sanchez and De Valentia it comprised attempts to know future contingencies from 'reading' present ones, and to use spoken or written prayers for material purposes like healing. The various superstitions might therefore vary in their aims – their final causes – but their defficient causes were identical.

Ciruelo's rule for the identification of 'vain, superstitious, and diabolical' works was another influential application of the inefficacy principle. It was originally offered as a help to the taking of confessions and, in Catalonia a hundred years later, still recognized by the Jesuit Vincente Navarro and the jurist Pedro Jofreu as invaluable to those 'who deal with the Forum of the conscience... and with matters which concern the Inquisition':

> The rule is this: in every action which man performs to bring about some good or avoid some evil, if the things that he uses or the words that he employs possess neither natural nor supernatural power to bring about the desired effect, then that action is vain, superstitious, and diabolical; and the effect which is produced comes from the secret operation of the devil.[30]

Like all who wrote on the subject from an essentially Thomistic theological and natural philosophical perspective, Ciruelo recognized three types of causation: natural, supernatural and preternatural. Superstitions were not based on the first; they differed, for example, from the practices of orthodox physicians. And they were not based on the second; God might arbitrarily endow certain of the priest's words with miraculous efficacy, but plainly not those of the magician. They were thus preternaturally effected, and not by the good angels but by that consummate natural magician, the devil, who (as Jofreu noted) 'through his knowledge of nature and his wide range of experience, knows all the secrets and qualities of things...'[31]

Such also was the view of those Catholic authors who offered a set of five rules for the identification of superstitions. This was commonly in use by the later sixteenth century, when Peter Binsfeld, the Bishop of Trier, included it in his much reprinted *Manual of Pastoral Theology and Doctrine Necessary for Priests Having the Care of Souls*. There is also a version in the Freiburg professor Lorichius' *Superstition*, a work which consists largely of lists of popular forms of divination, astrology, charming, healing by orthodox means, rituals performed in excess of those required by the church, and the 'observation of times'. Of all these instances, Lorichius writes, there is one sure rule (the first) for judging them to be superstitious. It is simply 'whether an object is associated with an effect which it has neither in nature nor from the consecration of the Church'.[32]

One consequence of this bi-partisan insistence that it was nature that helped very largely to determine what was superstitious was that the categories 'witchcraft', 'superstition' and 'magic' became completely interchangeable. In consequence, the beliefs and activities that could be considered demonic were multiplied enormously. The notion of the implicit pact was particularly influential in this respect, turning even the simplest cultural transgression into a rejection of God's most important commandment. John Gaule remarked that for every witch 'so become after an explicit manner of Covenanting, more then ten of them are guilty after the Implicite and Invisible way onely'. The latter comprised not only the clients of the much-attacked diviners and healers of popular culture, but any private individual who indulged in the same techniques. William Perkins, in his *The Damned Art of Witchcraft*, expressed this exactly – once again reminding us of the criterion at work: 'By Witches we understand not those onely which kill and torment: but all Diviners, Charmers, Juglers, all Wizzards, commonly called wise men and wise women; yea, whosoever doe any thing ... which cannot be effected by nature or art'.[33]

5   The magical power of words

The themes of this essay have been the existence in early modern natural philosophy and theology of a particular scheme, different from our own, for classifying the causes of things, and its role in the promotion of cultural change by would-be reformers of popular beliefs and behaviour. It has not been necessary to say what, in substance, the view of nature *was*, or where the boundary between 'natural' and 'superstitious' causes was actually located in particular instances. Instead, I have concentrated on the way the principle that nature had ascertainable limits was, itself, a central aspect of acculturation. All the same, a reliable knowledge of where these limits were reached was obviously crucial to the application of the classificatory scheme to particular cases. From the resolution of private consciences to the making of public accusations of witchcraft lay a whole epistemological terrain filled with difficult choices – choices which the casuists often had to explore in natural philosophical detail. What made the remedies of the physician work and the amulets of the magician fail? Where exactly was the boundary between natural divination (licit) and judicial astrology (illicit)? What was the difference between the efficacy of utterances and the inefficacy of charms? Deciding these matters – other than on grounds stipulated simply by the professional interests of physician or clergyman – depended on a detailed understanding of the workings of nature in concrete instances.

Since issues of substance were therefore crucial for cultural reform, and since demonology contributed to the resolving of them, it may be worth looking finally at how one of them was tackled. Demonology was, of course, undertaken by men with pre-modern ideas concerning what was possible and impossible in nature. When, for example, the Danish theologian Niels Hemmingsen distinguished between properly natural predictions and magical divination, he included in the *former* the forecasting of rain or drought from the flight of birds, and of storms from the state of the human body, together with the tracing of moral inclinations in physiognomy.[34] However, there is one case which shows the demonologists adopting a philosophical position that was not only important to later medieval and Renaissance science but consonant with the values of the 'new' philosophy that eventually supplanted it. This will be another reminder, then, of the point sketched at the outset of this essay concerning the contribution of demonology to change and innovation in early modern scientific circles.

The case is that of the efficacy of words – or, more strictly, of the words, figures, characters and images in which superstitious causality was so typically embodied. It was obvious to all that magicians and witches intent on evil used linguistic and symbolic instruments, and assumed by many (especially their victims) that these had, as Perkins again put it, 'a miraculous efficacie to bring some extraordinarie and unexpected thing to passe'.[35] Incantations, spells and curses appeared to produce physical changes in objects and persons; actions to images were apparently conveyed to the things they depicted. But the supposed efficacy of signs was not only a basis of *maleficium* and the evil arts; it was also indispensable to what I have called popular technique. Referring to the latter, the Protestant casuist Balduin complained that, 'Among all the ceremonies of magicians and witches, nothing is more commonly used than words'. In Lutheran Germany, for example, it was reported in the 1590s that 'the use of spells is so widespread among the people here that no man or woman begins, undertakes, does, or refrains from doing, desires or hopes for anything without using some special charm, spell, incantation, or other such heathenish medium'.[36] Everywhere in the literature of conscience, in sermons and books on lay beliefs, and, of course, in treatments of witchcraft and demonism, the popular belief that signs could act transitively on the objects they referred to or the persons who used them was condemned as outright superstition. Reflection on the nature of language and the power of words was thus an important component of religious reform, as well as the occasion for theological and liturgical contention between the faiths.

But in natural philosophy too the problem of language and its uses was crucial and controversial – crucial to the ways in which knowledge was defined and validated, and controversial enough to become a focus of intellectual allegiance. It is one of Foucault's suggestions, for example, that the discontinuity that separated Renaissance knowledge systems from those associated with the 'new' science was marked by a fundamental shift in the perceived relationship between words and things. Until the early to middle part of the seventeenth century the order discerned in things was itself linguistic. Language, he says, was 'interwoven' with the world and resided in its forms as its 'prose'. It was not something arbitrary whose importance lay simply in its ability to convey meaning. Rather, its capacity stemmed from the fact that it was related by analogy to the things it depicted; in effect, knowledge consisted in relating one form of language to another by similitude. The Renaissance commonplace of nature as a 'text' was not merely a metaphor for natural knowledge but referred to a genuine act of reading natural signs and manipulating the properties they signified. Some contemporary theorists of language even thought it likely that 'before Babel, before the Flood, there had already existed a form of writing composed of the marks of nature itself, with the result that its characters would have had the power to act upon things directly, to attract them or repel them, to represent their properties, their virtues, and their secrets'.

From the seventeenth century onwards, what became important about language was its ability to mirror nature, not resemble it. It entered what Foucault calls its 'classical' period, 'a period of transparency and neutrality'. Representation became its essential task, a task discarded only by modern structuralism. The man-made nature of signs was now insisted on. They could no longer exist as natural entities independently of being known, and natural resemblance could no longer enter as a third term between signifiers (phonetic sequences) and signifieds (mental concepts). It came to be thought that precisely because language was artificial men should be able to devise names for things with ever-increasing faithfulness to the sense impressions they received of them. Language was seen as essentially nomenclative, and theorists spoke with absolute confidence about its capacity to follow reality. In this way, linguistic conventionality became the servant of philosophical realism.[37]

In fact, Foucault badly underestimated the extent to which these two accounts of signification, the one 'natural' and the other 'cultural', had always been available as alternatives. But it does look as though the first received most attention in Renaissance Europe from those whose Neo-Platonism and 'Hermeticism' encouraged them to see an equivalence between words and things. The conventionality of language, on the other hand, drew emphatic support from many who associated themselves with the newer styles of scientific thought that gained momentum from the seventeenth century onwards – writers like Mersenne, Sennert, Van Helmont, Hobbes and Locke. This has led scholars like Brian Vickers and (in a different vein) Richard Waswo to continue to use the language issue as a criterion of epistemological, and hence natural philosophical allegiance. Vickers, for example, cites as typical of the coming scientific orthodoxy Locke's description of the relationship between the concepts and the names of things as an 'appropriated connexion' without which the latter would be 'but empty sounds'. It was as true for Adam as for any other speaker that signification was 'by a perfectly arbitrary Imposition.' All that was needed for communication was that words should 'excite in the Hearer exactly the same Idea they stand for in the Mind of the Speaker.' The new demands of communication in natural philosophy were that men should by convention exchange absolutely clear and distinct meanings, not that they should expect these to arise directly from nature itself.[38]

On which side of the argument were the demonologists? To a man they defended the cultural account of signification. Here, for example, is Perkins arguing for the natural inefficacy of the spells and charms of witches:

> That which is in nature nothing but a bare signification, cannot serve to worke a wonder, and this is the nature of all words; for as they be framed of mans breath, they are naturall, but yet in regard of forme and articulation they are artificiall and significant, and the use of them in every language is, to signifie that which the author thereof intended; for the first significations of words, depended upon the will and pleasure of man that framed and invented them. Beeing therefore invented onely to shew or signifie some thing, it remaines that neither in nature nor proper use, they can be applied to the producing of wounderfull and strange effects.[39]

If Hobbes could say that signifiers functioned only because of an agreed relationship with their signifieds, not because of any natural link with their referents, and that a name was merely 'a word taken at pleasure to serve for a mark' to recall a thought, Thomas Erastus could say likewise that words were only 'marks and images of our thoughts, and in themselves do not have any property other than what they signify according to the common assent and intention of people'.[40] Mersenne and Martín Del Río were similarly in agreement in conceding that God might originally have called things according to their essences. For mortals, however, names served only to signify what they wished to say to others; someone who lived alone would have no need of them. In his witchcraft treatise of 1616 Alexander Roberts wrote that words had 'no vertue, but either to signifie and expresse the conceits of the minde, or to affect the eares of the Auditors'. For James Mason words only signified, 'neither can characters doe or effect any thing, but onely represent'; they were 'not things naturall, but artificially made'.[41]

Given the wider role and relevance that I have sketched for demonological naturalism it is not surprising to find this theory of signs repeated by the casuists and disseminated throughout the literature of cultural reform. In early sixteenth-century Spain, Ciruelo was expressing the sentiments of John Locke and asking confessors to pass them on to sinners. 'The meaning of words in any language' he wrote, 'is willingly assigned to them by the men who speak that language. [They] signify nothing except those things which men who speak the language wish to designate by them. And it is the same with other languages. If, then, the meaning of either spoken or written words is a matter of choice and not of innate quality, one cannot find in the meaning some natural power to create some natural effect in mankind, either to cure or hurt them.' A century later, in Protestant Wittenberg, Balduin was also insisting that words were signs of things but not because they had any natural power over them. Words and things were different in kind. The first were only sounds made by beating the air, or characters written on paper, and they signified only arbitrarily (*voluntate*).[42]

It is not that these were anticipations of Modernism in the philosophy of language. The Thomism of the writers means that they were only drawing out the implications of Aquinas's own view that 'we make signs only to other intelligent beings.'[43] Nevertheless, in this particular area of the debate about natural efficacy, the writers on demonology I have been considering were not merely trying to reform popular notions of causality in a fundamental way – in itself, of considerable significance for the history of cultural change in Renaissance Europe. Although dealing with subjects now thought to be incompatible with science, they were handling them in a manner consistent with contemporary scientific values – indeed, with those values that were becoming important for the 'new' science. They were also disseminating these on a very wide front, extending across the entire programmes of two Europe-wide religious reform movements – with all that this implies in the spheres of education, preaching and publishing. The years that separated Ciruelo's book on superstition and Balduin's on the conscience saw probably the most sustained attempt there has ever been to alter ordinary people's perceptions of the natural world. Demonology had a vital part in one intended cultural revolution, and, in its long and consistent advocacy of language as a human artefact, helped in addition to sustain the knowledge system to which the fortunes of a second revolution – the revolution in science – were tied.

Notes

1 C. S. Lewis, *Studies in Words*, Cambridge, 1960, p. 67 (author's italics).

2 St Thomas Aquinas, *Summa theologiae*, Blackfriars edition, 60 vols., London, 1963, xv, pp. 15-17 (Part Ia, Quaestio 110, article 4).

3 Pietro Martire Vermigli, *Sommaire des trois questions proposées et resolues*, in Ludwig Lavater, *Trois livres des apparitions*, Geneva, 1571, p. 285; Paulus Frisius, *Von dess teuffels nebelkappen, das ist: Ein kurtzer begriff, den gantzen handel von der zäuberey, belangend, zusammen gelesen?*, Frankfurt am Main, 1583, reprinted in *Theatrum de veneficis*, Frankfurt am Main, 1586, p. 221; Lambert Daneau, *De veneficis*, Geneva, 1574, translated by Thomas Twyne as *A dialogue of witches*, London, 1575, Sig. lii^v.

4 Pedro Ciruelo, *Reprobación de las supersticiones y hechicerías*, Alcalá, c.1530, translated by Eugene A. Maio and D'Orsay W. Pearson as *A Treatise Reproving all Superstitions and Forms of Witchcraft*, London, 1977, p. 83.

5 Michel Foucault, *The Birth of the Clinic*, tr. by A. M. Sheridan, London, 1973, p. xii.

6 Henri Boguet, *Discours des sorciers*, Lyon, 1602, translated by M. Summers as *An Examen of Witches*, London, 1929, p. 107.

7 John Gaule, *Select cases of conscience touching witches and witchcrafts*, London, 1646, p. 98; [Robert Filmer], *An advertisement to the jury-men of England, touching witches*, London, 1653, p. 8.

8 Bert Hansen, 'Science and magic', in *Science in the Middle Ages*, edited by David C. Lindberg, Chicago, 1978, pp. 483-506.

9 J. L. Heilbron, *Elements of Early Modern Physics*, London, 1982, pp. 1-89, especially pp. 11-22.

10 Boguet, *op. cit.* (n. 6), pp. 62-87.

11 Jacques Grévin, *Deux livres des venims*, Anvers, 1567-68, p. 34; Daneau, *De veneficis*, Sig. Gvii^v.

12 Thomas Erastus, *Repetitio disputationis de lamiis seu strigibus*, Basle, 1578, translated into French anonymously and republished in Johann Weyer, *Histoires, disputes et discours, des illusions et impostures des diables*, Geneva, 1579, ii, p. 499.

13 Martín Del Río, *Disquisitionum magicarum libri sex*, Lyons, 1608 (several other editions); Francesco Torreblanca (Villalpandus), *Daemonologia*, Mainz, 1623; Jacob Vallick, *Von Zäuberern, Hexen, und Unholden*, reprinted in *Theatrum de veneficis*, cit., but first published in German in 1576.

14 Edmund Leites, 'Casuistry and character', in *Conscience and Casuistry in Early Modern Europe*, edited by Edmund Leites, Cambridge, 1988, pp. 119-20.

15 Friederich Balduin, *Tractatus de casibus conscientiae*, Frankfurt, 1654 (also published Wittenberg, 1628); William Ames, *Conscience with the power and cases thereof*, n.p., 1639 (also published in Latin in 1630); Ludwig Dunte, *Decesiones mille et sex casuum ... Das ist Kurz und Richtige Erörterung Ein Tausend und Sechs Gewissens Fragen*, Lübeck, 1636.

16 Francisco de Toledo, *Summa casuum conscientiae*, Cologne, 1599, in later editions entitled *Instructio sacerdotum*; Gregory Sayer, *Clavis regia sacerdotum*, Venice, 1615; Martín de Azpilcueta (Navarrus), *Enchiridion, sive manuale confessariorum et poenitentium*, in his *Commentaria et tractatus*, 3 vols.,

Venice, 1588.

17 Jean Delumeau, 'Les reformateurs et la superstition', in *Actes du Colloque l'Amiral de Coligny et Son Temps*, Société de l'Histoire du Protestantisme Français, Paris, 1974, pp. 451-87.

18 Stuart Clark, 'Protestant demonology: sin, superstition and society, c. 1520-c.1630', in *Early Modern European Witchcraft Centres and Peripheries*, edited by Bengt Ankarloo and Gustav Henningsen, London, 1990, pp. 45-81.

19 For Ciruelo see note 4 above; Jodocus Lorichius, *Aberglaub*, 2nd edn, Freiburg im Breisgau, 1593.

20 Ciruelo, *op. cit.* (n. 4), p. 77; John Bossy, 'Moral arithmetic: seven sins into ten commandments', in *Conscience and Casuistry, cit.* (n. 14), p. 216; cf. John Bossy, *Christianity in The West, 1400-1700*, Oxford, 1985, pp. 35-8.

21 Aquinas, *Summa, cit.*, XL, p. 41 (IIa IIae, *Quaestio* 95, art. 2); Ciruelo, *Reprobación, cit.*, pp. 88, 91.

22 Aquinas, *Summa, cit.*, xi, pp. 75, 9 (IIa IIae, *Quaestio* 96, article 2; *Quaestio* 92, article 2).

23 Gregorius de Valentia, *Commentariorum theologicorum*, 4 vols., Ingolstadt, 1591-97, iii, col. 1985.

24 Ames, *op. cit.* (n. 15), pp. 28-31.

25 Gaule, *op. cit.* (n. 7), pp. 39-40; James Mason, *The anatomie of sorcery*, London, 1612, pp. 2-3, 17; Richard Bernard, *A guide to grand jury men, divided into two bookes ... In the second, is a treatise touching witches good and bad*, London, 1627, Sig. A6ʳ, pp. 120-1.

26 Bernhard Albrecht, *Magia; das ist, christlicher bericht von der zäuberey und hexerey ins gemein*, Leipzig, 1628, pp. 10-11, 22; cf. Abraham Scultetus, *Warnung für der warsagerey der zäuberer und sterngücker*, Neustadt, 1608, pp. 5-6.

27 Christoph Vischer, *Einfeltiger bericht wider den ... segen, damit man menschen und viehe zu helffen vertmeinet*, Schmalkalden, 1571, Sig. Dviiᵛ; Johann Rüdinger, *Decas concionum secunda, de magia illicita*, Jena, 1635, p. 110; cf. Caspar Peucer, *Commentarius de praecipuis divinationum generibus*, Wittenberg, 1553.

28 Jacobus Simancas, *Institutiones catholicae*, Valladolid, 1552, f. ccxᵛ.

29 Francisco de Toledo, *Instructio sacerdotum ac poenitentium, in qua absolutissima casuum conscientiae summa continetur*, Douai, 1622, pp. 592-601.

30 Ciruelo, *op. cit.* (n. 4), p. 94 (and for Navarro's judgement, p. 27).

31 *Idem*, p. 101.

32 Peter Binsfeld, *Enchiridion theologiae pastoralis et doctrinae necessariae sacerdotibus curam animarum administrantibus*, Trier, 1594 and many other editions; Lorichius, *Aberglaub, cit.*, p. 15.

33 Gaule, *op. cit.* (n. 7), p. 73; William Perkins, *A discourse of the damned art of witchcraft*, Cambridge, 1610, p. 255.

34 Niels Hemmingsen, *Admonitio de superstitionibus magicis vitandis*, Copenhagen, 1575, Sigs. H1ᵛ-2ʳ.

35 Perkins, *op. cit.* (n. 33), p. 131.

36 Balduin, *op. cit.* (n. 15), p. 541; Gerald Strauss, *Luther's House of Learning: Indoctrination of the Young in the German Reformation*, London, 1978, p. 304.

37 Michel Foucault, *The Order of Things: An Archaeology of the Human Sciences*,

London, 1974, pp. 17-67.

38 Brian Vickers, 'Analogy versus identity: the rejection of occult symbolism, 1580-1680', in *Occult and Scientific Mentalities in the Renaissance*, edited by Brian Vickers, Cambridge, 1984, pp. 95-163, quotations from Locke at pp. 110-14; cf. Richard Waswo, *Language and Meaning in the Renaissance*, Princeton, 1987.

39 Perkins, *op. cit.* (n. 33), pp. 136-7.

40 Vickers, *art. cit.* (n. 38), p. 103; Erastus, *op. cit.* (n. 12), p. 421.

41 Marin Mersenne, *La verité des sciences, contre les septiques* [sic], Paris, 1625, pp. 69-72; Martín Del Río, *op. cit.* (n. 13), pp. 26-7; Alexander Roberts, *A treatise of witchcraft*, London, 1616, p. 69; Mason, *Anatomie, cit.*, p. 22.

42 Ciruelo, *op. cit.* (n. 4), p. 334; Balduin, *op. cit.* (n. 15), p. 543.

43 St Thomas Aquinas, *Summa contra gentiles*, in *Basic Writings of Saint Thomas Aquinas*, edited by Anton C. Pegis, 2 vols., New York, 1945, ii, p. 208 (Book III, part 2, chapter CV).

11

Astrology, religion and politics in Counter-Reformation Rome*

Germana Ernst

1 Animal superbum

Early in the year 1586 Pope Sixtus V promulgated his bull *Coeli et terrae* against divination in all its guises.[1] The opening solemnly declared that knowledge of future events is the exclusive preserve of God, the creator of heaven and earth. Man may, with great efforts, use his reasoning powers to explore higher forms of knowledge, but he must never venture beyond the proper limits and above all he must never claim to have raised himself to a point where he knows the future. That is the prerogative of God, before whose eyes the entire sweep of time past and future is eternally present and every single thing stands 'open and exposed'.

Any doctrine which aspires to knowledge of this kind is fallacious, continues the bull, because they are based on the cunning of evil men, or on the intervention of demons who regularly take a hand in the divinatory arts because they desire nothing so much as the ruin and damnation of men.

On the strength of these premises the bull outlaws a multitude of 'curious' arts. It relates the condemnation of the Tridentine Index which had forbidden all works of divination and magic. As far as astrology was concerned only the natural observations useful to navigation, agriculture and medicine were allowed. Unlike the Index, the Sistine Bull does not confine itself to a simple list, but presents us with a vivid impression of the outlawed practices and superstitions.

It observes, for instance, that many impious men, obsessed by an insane desire for knowledge of the future, have gone beyond such well-

* I am very grateful to Angus Clarke for translating this article from Italian and to Paolo L. Rossi for help and advice.

known arts as geomancy, chiromancy and necromancy to try casting dice, grains of wheat and even beans. They have entered into binding pacts with death and hell. Others, thirsting to know secret things, to find hidden treasures and to perform acts, have invoked the devil himself by drawing circles and hieroglyphs, by lighting candles, by offering sacrifices, incense and all manner of sacrilegious rituals. They manufacture rings, mirrors and phials in which to enclose the demons they have invoked. They interrogate demons when they find them shut up in the bodies of people who are possessed. They – and this is especially true of superstitious old wives – pretend to worship the 'pure and holy' angels but actually invoke the devil in jugs full of water, or in mirrors lit by holy candles, or in the nails and palms of the hand which sometimes they smear with oil. And they persist until the devil appears before them and answers their questions.

The Sistine edict is eloquent proof of how widespread even the most peculiar superstitious practices were, regardless of prohibitions, and it warns that no matter how bizarre and fantastic, they are all equally impious and those who devote themselves to such vanities are ensnared and deceived by the devil.

Among those who so mortally offend God with their claim to know the future, continues the bull, are the astrologers. Indeed, astrologers, indiscriminately called 'genethliacs' (i.e. casters of nativities), 'mathematici' or 'planetarians', have the dubious honour of heading the catalogue of superstitious practitioners. Basing their claims on what is in fact a false and worthless science of the stars, they pass judgement on the past, the present and the future. They brag with intolerable audacity that they can know in advance of events as they will unfold in the order laid down by God. They presume, on the basis of natal horoscopes, to predict the principal events of a person's life – honours, wealth, children, journeys, friends and enemies, sickness and death. Simple people believe their predictions because they express their fondest wishes, and this leads to a perilous marriage between the falsity of the masters and the credulity of their followers. Moreover, they ascribe to the stars matters which depend on free will, thus overthrowing the true relationship between men and the heavenly bodies: man is not made for the stars, the stars are made for man. How then could the stars 'machinate and act' against the protection of the angels into whose stewardship God, in His goodness, has entrusted his rational flock?

Guilty of aspiring to knowledge which is the exclusive preserve of God, astrology is degraded to popular fortune-telling. Far from being a legitimate and useful science, it is one of the most dangerous by-products of man's pride. Instead of striving to raise his mind to levels of awareness which are beyond its reach, man should fear God and, prostrate on the ground (*humi stratus*), he should revere the immensity of God's majesty.

2 Os sublime

In the introduction to the bull there is only cursory reference to the notion of man as being made in the 'image and likeness' of his creator – while humanists had appealed to this biblical reference to glorify the 'dignity' of man. On the contrary, the greatest emphasis falls on the profundity of the abyss which separates the omnipotence of God and the deficiency of his creatures. Man is warned most severely to bow his head in reverence instead of aspiring to know higher things. He is enjoined to fear rather than to know.

This assault on man the 'proud animal' is the cornerstone of the condemnation of the celestial sciences. It is the reply to astrology's attempt in the preceding decades to substantiate its nobility and excellence with the *topos* of the *os sublime* ('the face directed upwards'). In their writings astrologers were wont to quote a passage from Ovid which declared that the special difference between men and other animals lies in the fact that 'while the other creatures on all fours look downwards, man was made to hold his head erect in majesty and see the sky and raise his eyes to the bright stars above'.[2]

This image – a commonplace from patristic writing through to the Renaissance – was employed to emphasize a divine quality in man, his ability to pass beyond the confines of the natural world and his irresistible tendency towards that upper realm from which he came.

Astrologers appropriated the image in order to refute opponents who accused astrology of being too close to pagan idolatry, emphasizing, on the contrary, that contemplating the regularity and harmony of celestial dynamics and the *machina mundi* made men aware of the existence of a sovereign ordering intelligence.

Hand in hand with this celebration of astrology's 'excellence' went an insistence on its *pietas* (meaning both morality and respect of the Divinity). Girolamo Cardano, in the Dedication of his *Commentary on Ptolemy's Tetrabiblos*, gave impassioned voice to a number of basic motifs which were widespread in astrological texts. He opened by declaring that the highest and most arduous knowledge – that of the future – and the contemplation of the greatest works of creation are the twin elements which combine to constitute astrology's pre-eminence:

> Nothing comes closer to human happiness than knowing and understanding those things which nature has enclosed within her secrets. Nothing is more noble and excellent than understanding and pondering God's supreme works. Of all doctrines, astrology, which embraces both of these – the apotheosis of God's creation in the shape of the machinery of the heavens and the mysterious knowledge of future events – has been unanimously accorded first place by the wise.[3]

This conception of astrology fits coherently into a general notion of nature as divine epiphany. Each natural thing, however paltry, manifests God's wisdom, consequently nothing is meaningless – everything has been ordered towards predetermined ends. If even the most insignificant blade of grass does not grow by chance, as the German astronomer and astrologer Johann Garcaeus states in his *Astrologiae methodus*, how much more certain is it that the stars, made with such skill and regulated by such unvarying laws, 'do not shine uselessly in the firmament, but each one has its own precise effects'?[4]

In his *Admonition Concerning the True and Licit Use of Astrology*, Hieronymus Wolf, a classical scholar with astrological interests, uses the familiar image of the heavens and nature as a representation of God's language to stress the fact that the peculiarity of man's awareness consists in the progressive deciphering of the meaning of natural signs, by contrast with the animals which live immersed in their senses:

> He has inscribed the heavens with a marvellous script and if we could read it perfectly we would be perfectly wise. He has written men's intelligence, habits and fortune in their faces, on their hands. But we either neglect or fail to understand His writing, and so we behave like the animals which are guided only by their senses, while the excellence and divinity of the mind is so great.[5]

From this point of view the mind's capacity to go beyond the immediacy of sense data blurs ambiguously with its capacity to 'foresee the future', however fraught with difficulties that particular ability may seem to be. Of course only God has a complete knowledge of the future and man's achievements are like 'a grain of sand' or 'a small coin' compared with the treasures of a king. However imperfect and circumscribed such foreknowledge may be, it is neither futile nor forbidden – rather, it is one of the highest peaks of human experience.

The astrologers reject their adversaries' contention that this impulse towards the heavens and the future derives from human pride – or fuels it. On the contrary, contemplating the stars sharpens men's awareness of their limitations and the fragility of their condition. As Cardano puts it, in a passage of some pathos, when gazing on the stars

> these things come to mind: the proof of eternity, the fragility of our condition, the vanity of ambition and the bitter memory of wrongdoings. Whence comes contempt for so short a life – even if it lasts a hundred years what is it compared with the immensity of eternity? Is it, perhaps, a point compared with a circle? What is all this human happiness? Is it as if someone had tasted it, perhaps even you, and found it no more than wind, smoke and dreaming?[6]

As this rapid survey shows, astrologers met accusations of impiety by deploying a conception of nature as the reflection of God's wisdom and this lends their *apologia* an intense natural religiosity.

After Pico della Mirandola's fierce attack on divinatory astrology in his twelve books of *Disputations*[7] in the last decade of the fifteenth century it was generally agreed by astrologers that their art was in a condition of serious decay and needed to be purged of corrupting superstitions.

On occasion, we find apologists of astrology like Garcaeus adopting the *topos* which compares the purity of ancient wisdom with the corruption of contemporary wisdom; the decadence of their art becomes a symptom of the general ageing of the world, a process which accounts for a tarnishing of man's cognitive abilities. While the sharper and clearer gaze of the ancients could penetrate the secrets of celestial and elemental nature, contemporary minds were weak and contemporary eyes dim. Nature was not entirely changed though, and it was still possible to forge a science of the stars. To restore astrology's ancient dignity, it was necessary – declared Cardano – to liberate it from Arabian astrologers' disfiguring superstitions and re-establish it as 'a conjectural part of natural philosophy'. It should have nothing to do with oracles and haruspices, divination and magic. Once the irrational trifles introduced by the Arabs had been eliminated, astrology would cleave to the three guiding principles of 'ratio, sensus et experimentum', on the basis of which it would formulate somewhat cautious judgements.[8]

Astrology, said Cardano and many with him, is an art, not a science. It is to natural philosophy as prognoses are to medicine in general. If it is permissible for physicians, sailors, farmers and metallurgists to make forecasts, is it not legitimate for the natural philosopher to work up conjectures on the basis of subtle observations of the stars?[9] Of course the more general the events predicted, the more reliable such predictions are. Astrology is an art of conjecture whose accuracy diminishes in inverse relation to the degree to which it descends to particulars. Let us not confuse astrology with divine and demonic oracles, warns Garcaeus while insisting on its veracity within a 'natural' dimension. Astrologers derive their arguments from signs and natural causes. On these they base predictions, which are dependable and have certain effects because of a physical rather than an absolute necessity.[10]

Apart from being true, it is an art which is useful to individuals and the collectivity, at both the psychological and the practical levels – contrary to the objections of those who repeating the argument of the second-century sceptical philosopher Favorinus claimed that the foreseeing of negative events increases anxiety, while that of positive events diminishes happiness, stressing its lack of utility. Astrology is true and useful, but also 'very difficult and laborious' by reason of the enormous variety of causes which combine to produce events – astral causes being by no means the only ones in play.[11] Whence the proliferation of charlatans – not in themselves a sufficient reason for condemning astrology out of hand, any more than heretics obscure the holy scriptures or incompetent physicians sully the good name of medicine.

Astrology is an intricate labyrinth in which it is easy to go astray, so it is necessary to proceed with caution and humility. Whatever the case, the conviction that astrology reinforces rather than diminishes *pietas* remains firm against the calumnies of those who would, in their ignorance, declare it to be idolatrous: 'Oh you fools, who would wish to abandon Christ and worship the stars?'.[12]

3   *Theologia quasi soboles astrologiae*

The 1586 bull sanctioned the divorce of astrology and theology but their relationship had been deteriorating since the 1560s. An almost unknown episode in the 1570s illustrates the head-on collision between them. The astrologer in question is Francesco Giuntini, a Carmelite from Florence. Giuntini was prey to heretical inclinations and as a result found himself compelled to leave Venice in a hurry. He sought refuge in the prosperous French city of Lyons and remained there till his death in 1590. From autobiographical notes which accompany his horoscope we learn that he entered the Carmelite order as a sub-deacon at 15. At 18 he celebrated his first high mass, before a large congregation, and in 1551, aged 29, he received his doctorate in theology at Pisa. However, at the age of 35 an unfortunate aspect of Saturn to the Sun brought him upon evil times, dishonour, imprisonment and torture. His thirty-ninth year was especially dire: a malign aspect of Mars to the Sun 'signified the wrath of a great prince, hatred from powerful people, and the loss of honour and worldly goods' as well as another spell in prison and exile as an alternative to yet worse sufferings.[13]

Abandoning theological controversies Giuntini now applied himself single-mindedly to astrology. The two monumental volumes of his *Speculum astrologiae* (Mirror of astrology) testify to his extraordinary dedication. Giuntini was primarily a compiler – he made no special claims to originality. The *Speculum* is essentially an encyclopedia: it contains the 'classic' texts of astronomy and astrology from Ptolemy through Sacrobosco to Peurbach, some with commentaries, as well as more personal works more deeply indebted to contemporary thinking. Broadly speaking, as far as Giuntini was concerned there was no conflict between astrology, in the proper sense, and theology. Indeed for him theology derived quite amazingly from astrology, and both arts stood at the summit of human experience: 'astrology and its near-progeny theology are the most excellent of all sciences'.[14]

What concerns us here, however, is a modest volume entitled *In Defence of Good Astrologers*. It was aimed at those who 'ignorant of astrology criticize it'. Giuntini opens by distinguishing between three different views about the influence of the stars on man. First is the view which champions a rigorous astral determinism: it ascribes every human action to the heavens which, as with the Chaldeans and the Babylonians, are identified with God and worshipped as such. The second view stands at the opposite extreme – it tolerates no relationship whatsoever between the stars and human vicissitudes and it holds that 'God governs everything through himself'. The most correct opinion is the third, midway between the first two, which allows a celestial influence on man but one which cancels neither God's freedom to intervene nor man's free will.[15]

Armed with this premiss it is easy enough to see off the objections of the 'calumniating moderns' who, muddling 'good' astrology with 'bad', fail to realize that scriptural, juridical and theological condemnations refer to the latter – espoused by Stoics and deterministic philosophers – while 'good' astrology is entirely legitimate, useful and founded upon the authoritative writings of Thomas Aquinas, who both in his *Summa theologiae* and *Summa contra Gentiles* admitted a direct influence of the stars upon the human body and an indirect one upon his mind and will, which could also be affected by bodily tendencies and passions.[16]

After dismissing a number of hoary old arguments against astrology – how to accommodate miracles, how to explain the fact that twins differ, how to explain massacres where many individuals, each no doubt with an entirely different horoscope, all die at the same time – Giuntini replies to those who accuse astrology of 'endangering the spiritual health of ordinary people', a line of argument which was destined to acquire considerable weight, as we will see. He restricts himself to the counter-argument that even conundrums like predestination or image cults can be misunderstood and so spawn heresies and idolatries.

Despite its moderation this slender *Defence* provoked some responses. Two years later when, translated into Latin, it was adopted as a general introduction to the first edition of the *Speculum* (dedicated to Catherine de' Medici, at that time regent of France after the death of Henry III), the *Defence* had almost doubled in size. The author tells us that he was forced to reply to the theological objections which had been raised against his text. The objections were gleaned from the Bible, from legal texts, from the proceedings of the church councils and from the Fathers of the Church in order to assert the utter incompatibility of astrology and theology.

Giuntini does not tell us his adversary's name, simply alluding to him as a 'certain slanderer'. But some clues suggest that this biting epithet might refer to Antonio Possevino, who was in later years to become one of the most prominent figures of the Jesuit order. In the 1560s, at the beginning of his career, he was active in France, spending long periods in Lyons, where the Huguenot 'threat' was strongest. Between 1571 and 1573 – the years in which the objections were made

– Father Possevino was in Lyons and he made Giuntini's acquaintance, as he himself tells us in his partly biographical *Bibliotheca selecta*, published twenty years later. In the chapters devoted to refuting astrology, when he speaks of 'modern' astrologers, Possevino mentions only two names. One is the Sienese Lucio Bellanti, whose defence of astrology against Pico's *Disputations* was exhaustively refuted by Pico's nephew Gianfrancesco in chapter five of his *De rerum praenotione* (Concerning prediction). The other is Giuntini – whom Possevino dismisses as unworthy of further consideration. He tells us that he knew Giuntini in Lyons, maliciously recalls his apostasy from the Carmelite order and his 'shipwreck' from the Catholic religion, and expresses strong reservations about his public return to the bosom of the Church. The main reason for Possevino's suspicion is Giuntini's fidelity to astrology. As no one can set his hand to the plough looking at the same time over his shoulder, he says, so Giuntini's refusal to repudiate his books on the 'impious art of divination' make him unworthy of the Kingdom of Heaven.[17]

Possevino adopted two especially effective tactics in his attack. First he denigrated his adversaries' morality. He insulted Bellanti viciously enough but also went so far as to insinuate that Giuntini had amassed an enormous fortune (which was nowhere to be found after his death) by dint of money-changing and usury, adding sarcastically that he was a fine example of religious piety and poverty. Possevino's second and more subtly effective tactic is to ignore the vast and serious debate for and against astrology which had spanned the century. This debate is, as it were, played down, squeezed into a minute and insignificant space. Possevino records only two people, besmirches their morals and buries the most famous authorities in total silence.

In addition to objections from Catholics Giuntini also records those of Calvin, though only to dismiss them as 'blunted weapons which cannot cut and awaken little fear'. Calvin had based his argument on miraculous events drawn from the Bible – Giuntini counters that there is no gainsaying the fact that super-natural events are also super-astrological and depend entirely on God. Obviously celestial influences have their limits and are not 'praetorian edicts' in so far as they are subject to divine rule, but it is absurd to elevate specific and exceptional cases to the status of general rules. In Giuntini's opinion astrology is being criticized by someone who quite simply does not know anything about it. There follows a lengthy and philosophically demanding theological excursus and, doctorate in theology notwithstanding, it is surprising that Giuntini ventures so confidently into such dangerous pastures. In fact, much of it consists of long passages transcribed word for word (complete with marginalia) from Pomponazzi's treatise *On Incantations*.[18] It is rather ironic, but not insignificant that Pomponazzi, who would be adopted by atheists and libertines as their spiritual father because he ascribed miraculous events to natural and astral causes, should be summoned to the rescue of an astrologer on account of his theological skills!

4 *Lux et tenebrae*

Giuntini's anonymous adversary's eighteenth objection was directed against Arabian astrologer Albumasar's doctrine which related the birth and the death of prophets like Mohammed and Christ to certain celestial conjunctions. When, in his *Defence* of astrology, Giuntini answers this question, he dwells at length on Christ's natal horoscope. He refers to Cardinal Pierre d'Ailly's *Elucidarium* and cites the well-known passages from Albertus Magnus's *Speculum Astronomiae* (Mirror of Astronomy) in support of the contention that 'perhaps it would not be heretical to allow that the birth of Our Lord Jesus Christ, in as far as He was a man, was in the heavens *ut in signo et ut in causa*' (as a sign and as a cause). But this section is omitted from editions after that of 1578.

The deletion of this passage proves that the topic was a delicate one. In Renaissance times Christ's horoscope was cast by Cardano, who published it in his *Commentary* on Ptolemy. Cardano claimed that the human element in Christ's nature was, like that of any other human being, subject to the influence of the stars: but this was not intended to mean that his divinity, his miracles and his religion were all dependent on the stars. Despite this cautious caveat, Cardano was accused of impiety immediately after the horoscope's publication, because he had dared subordinate to the stars He who had created them, and the horoscope, while present in the early editions of the *Commentary*, was omitted from the posthumous Basle edition of 1578 – despite the author's passionate self-defence in the preface where he declared that the nativity of Our Lord is yet another proof of the excellence of Christianity.[19]

The horoscope of Christ according to Pierre d'Ailly, followed by the passages from Albertus Magnus's *Speculum* which in turn cites the Arab astronomer Albumasar – introduces a powerful critique of astrology by the Dominican Sisto da Siena in his *Bibliotheca Sancta*, a book of erudition and biblical criticism which ran into many editions. To refute such views Sisto cites Augustine's condemnation of astrologers for regarding the stars as 'causes' rather than as 'signs' of events. Sisto maintains that astrology is 'false and futile': 'cold and impotent' is how he describes the 'excuses' of contemporary apologists for astrology, both when they say that the stars do not compel but only incline, and when, even more vaguely, they attribute only 'certain signifying and admonitory powers' to the stars. All these attempts are useless, and, as he points out, the Tridentine Index regards all kind of divination as impious and superstitious: it disregards all the astrologers' subtle distinctions and condemns all their writings indiscriminately.[20]

As Sisto himself tells us, his views have drawn protests from one Serravalle, otherwise 'a not unlearned theologian'.[21] His replies to these protests clearly display his reasoning. First he insists that the 'astromancy' which claims to foresee the unfolding of an entire life in a horoscope is not an art but an imposture, baseless and condemned by both human and divine law. As for the fact that astrologers do make numerous true predictions, this happens for four reasons: first, because of 'fate', taken in the correct theological sense as 'a hidden disposition of divine providence'; second, because of a demonic pact; third, because of the perspicacity and human wisdom of certain astrologers; and fourth, because of the 'stupid credulity' of astrologers' clients who, in the hope of a favourable prediction or anxious about a bad one, behave in such a way that events turn out to match their passions.

When, however, Sisto turns to answer the accusation that he is contradicting Thomas Aquinas, he exhumes a series of distinctions between 'natural' and 'sham' astrology. These can be distinguished by observation: one considers only the physical influences of the stars, the other concerns itself with stellar emanations and occult virtues in images invented by astromancers. They can also be distinguished by their application: the first believes that celestial force and energy express themselves exclusively in the body, and only in the mind *per accidens* and *indirecte* (through its 'accidents' and indirectly). Finally the two astrologies differ with regard to foreknowledge: natural astrology abstains from definite predictions of individual and contingent events, restricting itself to general and hypothetical predictions. Of course, Aquinas approves of physical astrology and is utterly opposed to 'sham' astrology.

Though steadfast, Sisto's opposition to genethliacal astrology allows a degree of discussion about the limits within which it should be contained. The Spanish theologian Benito Pereyra, who entered the Jesuit order in 1552 and taught at the Collegio Romano until his death in 1610, is more intransigent. In the 1590s he collected three of his writings in his widely-read treatise *Against wrongful and superstitious practices*. It attacked magic, dreams and astrology – doctrines all generated by the human soul's mad and unbridled desire to know the future, all capable of producing only illusions and deception. Pereyra deplored the fact that not only private individuals but also princes and republics could be ensnared by such pernicious fantasies – which still had deep roots despite persecutions and bans. They constitute an insidious evil and their practitioners are extremely dangerous.[22]

The attack on astrology is five-pronged. The first and most important point is to insist on the radical incompatibility of astrology and theology. The other points concern the astrologers' *imperitia* (lack of judgement), the oppositions between astrology and philosophy, the fact that the stars are neither causes nor signs of human events, and finally the reasons why certain predictions turn out to be true – this section copied entire and without acknowledgement from Sisto da Siena.

The book makes easy, if not particularly original reading and sets itself up as an elegant compendium of the anti-astrological tradition since Pico. All the canonical texts relating to the subject are wheeled out, including lengthy quotations from classical writers like Cicero, Seneca and Favorinus, from the holy scriptures and the Church councils, from such Neoplatonists as Plotinus and Porphyry, from the Church Fathers, Greek and Latin, headed by Augustine and followed by Ambrose, Basil, Origen and Jerome.

The foundation upon which the argument is constructed is Pereyra's unshakeable conviction of the incompatibility between astrology and Christian truth – they are antithetical and mutually exclusive. By ostentatiously disregarding all the caveats and distinctions which the astrologers have so carefully erected, and by taking astrology in its most rigidly deterministic guise – according to which every single human deed is celestially determined – it is easy for Pereyra to show how astrology annuls freedom, ethical principles, the immortality of the soul, mysteries, miracles and Christian prophecy. In this fatalistic universe good works are ignored, the passions are unleashed, all crimes are justified and the law, human and divine, is entirely useless.

With perfect consistency Pereyra shows no qualms about re-
pudiating an authoritative figure like Cardinal Pierre d'Ailly who, by
adopting astral explanations for miraculous events like the Flood and
the birth of Christ, destroys the realm of the supernatural. It is not
d'Ailly's astrological sympathies which Pereyra condemns but his
monstrous attempt to 'marry' astrology and theology, two things as
fundamentally antagonistic as falsehood and truth, darkness and light,
Satan and God.

Accordingly, Pereyra ridicules astrology's claim to infallible
predictions, the fragility of its theoretical foundations, its internal
contradictions and the airy vagueness of the distinction between
'causes' and 'signs'. He also replays all the standard objections – the
twin paradox (why, if the same stars presided over their birth, are their
lives different), the problem of mass deaths (how can so many people,
born at different times, share the same 'astral' destiny) – and refers to
the 'beautiful treatise' in which the Syrian thinker Bardesanes (second
century), converted from Gnosticism to Christianity, attacked fatalistic doctrines and claimed that a people's conventions and customs are
a consequence of common laws which that people freely make for
themselves, and not a consequence of individual horoscopes. The
broader-minded attitude of Aquinas and Albertus 'is veiled in a
discreet silence'.[23] A further argument against astrology is that all the
wisest men of the past have condemned similar impostures that are
believed only by the ignorant masses, the simple-minded and those
who crave after novelty – impostures that are practised by restless and
curious individuals who prefer money to truth and kindle social
unrest.[24]

In his *History of Magic and Experimental Science* Lynn Thorndike
seems to distinguish between a tolerance for magic and a rigorous
intransigence where astrology is concerned.[25] In fact while Pereyra
does appear to allow a distinction between natural magic and demonic
magic – regarding the former as 'the most secret and excellent'
element of natural philosophy and backing that belief with quotations
from Pliny to Cardano – he restricts 'human' natural magic to a rather
narrow area which extends from magnets and the pilot-fish or remora,
through men who can sweat, weep and move their ears and hair at will,
to the principle of Archimedes and the flying wooden dove made by
the Pythagorean Archytas.

Pereyra devotes much more space to the intervention of demons who can interfere in natural magic as well as in what is more properly demonic magic. In order to get round the naturalistic conclusions of writers like Peter of Abano, Pomponazzi and Cardano who 'in vain and inappropriately' explain marvels purely in terms of natural causes, Pereyra is quite happy to accommodate Neoplatonic demons within his exclusively Aristotelian framework. He does this not in any positive sense but rather so that he can unmask their intervention, define their powers exactly (eliminating trivial powers) and, above all, supply criteria for distinguishing demonic 'marvels' from divine 'miracles'.[26]

The most important chapter is the fourteenth where Pereyra asks whether 'the study and practice of magic are legitimate'. The smoothness of his prose does not disguise a resounding negative. There is no question that natural magic, as the noblest part of physics, medicine and mathematics, is most worthy of study. However, 'public discussion of natural magic may justly be banned if there is reason to fear that once it is popularized people (and we must not forget that humankind tends towards evil) will use it wrongly and to the detriment of their fellows.'

As for demonic magic, a theoretical knowledge of it might not be a bad thing, if only for unmasking demonic deceptions. However, since such studies might go astray because of excessive curiosity, the potential for misuse and the human weakness for superstition, they too are justly forbidden to Christians unless specifically authorized by their superiors. The chapter closes by quoting Philo Judaeus who declares that those who practise magical arts and study poisons should be eliminated as soon as possible, just as it is right to anticipate the natural malice of scorpions, vipers and other poisonous creatures by killing them before they can move.[27]

By ruling out any possible accord with theology and by ostentatiously ignoring the debate about astrology since Pico − or treating it as if were not even worthy of mention − Possevino and Pereyra, faithfully reflecting Tridentine thinking and the Sistine bull, tend to do away with all distinctions between the divinatory arts and condemn them all by insisting on demonic interventions. In the Jesuit writers therefore the debate shifts decisively away from the truth or falsehood of magic, to the question of whether it is 'dangerous'. It is no longer a debate (as to some extent it was for Sisto da Siena) but quite simply a liquidation, motivated by the desire for political and social control, in which any disturbing, confusing or destabilizing factors are to be eliminated.

5   *De siderali fato vitando*

An episode in Rome around 1630 dramatically illustrates the inten-
sifying conflict between astrology, politics and religion in the post-
Tridentine period. The chief protagonists are Pope Urban VIII
(Maffeo Barberini), Tommaso Campanella, the abbot Orazio Morandi
and, in the background, Galileo.

In July 1626, after nearly twenty-seven years in the dungeons of
Spanish-ruled Naples, Campanella was transferred to the cells of the
Holy Office in Rome. His fame and his virtuosity as an astrologer were
by no means irrelevant factors in his transfer. He had already written
six books on the subject with the declared intention – earlier adopted
by Cardano – of establishing a 'physical' astrology purged of Arabic
superstitions and in agreement with the Christian religion. For some
years the manuscript had been in the hands of the publisher Soubron
in Lyons who had never had it printed. When Campanella arrived,
Rome was rife with rumours that malign stellar influences were
threatening the health and even the life of the Pope, so he wrote *De
siderali fato vitando*, a pamphlet on 'how to avoid what is destined by
the stars'.[28]

He begins by saying that every ill has its remedy and it is the
goal of the arts – daughters of nature (*naturae soboles*) and rays of the
single light of the divine Logos – to protect human life, to help it to
obtain what is good for it and to avoid what is harmful. Of course it is
necessary to distinguish between the true arts and sciences – daughters of true learning whose fundamental principles are the causes,
effects and signs supplied by nature and by God, its creator – and the
false arts, daughters of mere curiosity which rely on purely conventional and artificial signs rather than natural ones.

If medicine, ethics, politics and economics allow us to tackle the
ills of, respectively, the body, the soul, the commonwealth and the
household, then the sidereal science is effective against possible harm
derived from the stars. In this short pamphlet, says Campanella, he
does not intend to dwell on the complex metaphysical and theological
problem, because he has already thoroughly explored it elsewhere, of
the connection between divine foreknowledge, second causes and
human free will. He restricts himself to pointing out that God is above
destiny, because He is its creator, and that it is permissible for us to
behave 'as if' we too were free from destiny ('tanquam fato liberi'). It
is not that ills are inevitable, stresses Campanella, but simply that the

arts required to avoid them have been neglected, especially by princes who, instead of favouring the 'absurd sciences of the ancients', should encourage the development of modern science. The universities are full of sophists unable to find anything new, and itching to persecute those who seek the truth.[29]

Thanks to the sidereal science it is possible, with suitable 'lures', to capture the benign effects of the stars, and to avoid malignant influences. Of course this science too has its time limits and occasionally such remedies do not work, as in the 'classic' cases of Oedipus and Plotinus (but Campanella confuses Plotinus' death with Aeschylus's, whose skull was broken by a tortoise dropped on him by an eagle in flight), or in the more recent cases of the palmist Cocles and the astrologer Nabod. The former, knowing he would die from a hit on the head, always wore a helmet. But a killer hired by Ermete Bentivoglio and disguised as a porter murdered him with an axe while he was unlocking his front door. The latter, fearing that he would die by the sword, barricaded himself into his house in Padua with provisions for a month. However, he was murdered by thieves who had broken into the house thinking that he had gone away.[30]

In any case, whenever a celestial aspect seems to be threatening it is wise to take all possible precautions. With eclipses, for example, if the threat is to the entire region the best remedy is flight and total separation from the afflicted area ('abscinde te igitur a toto'). If on the other hand the eclipse threatens a specific individual he should have recourse to procedures capable of counteracting the malign influences. Such remedies, described in detail in chapter four, were studied by D. P. Walker who demonstrated the connections between these pages of Campanella and the magic of Ficino.

Campanella recommended that the individual in danger should close himself in a room with the doors and windows tightly barred in order to avoid all contact with 'poisonous seeds' borne on the infected air. The air in the room should be purified by sprinkling perfumes and burning wood and herbs like laurel, myrtle, rosemary and cypress. The walls should be decorated with foliage, leaves and strips of light-coloured silk. In the sealed and isolated environment it is necessary to reproduce the heavens symbolically by lighting seven torches which represent the two luminaries (the sun and the moon) and the five planets. Walker puts it very well: 'The lights in the sealed room are ... quite simply a substitute for the defective, eclipsed celestial world outside; the real heavens have gone wrong so we make ourselves another little, normal, undisturbed, favourable heaven'.[31]

When, some two years after his arrival in Rome, Campanella was released, it was said that the Pope took this decision after frequent meetings with him 'taking an extraordinary enjoyment in talking with him'. The friar 'had conquered a high place in his heart' because he had administered to the Pope 'certain infusions against evil humours and melancholy', thus giving him to believe that he would live 'long and in great peace' – when most astrologers were convinced that either the lunar eclipse of 1628 or the solar eclipse of 1630 would prove fatal for him.[32]

However, the pamphlet's publication in 1629 provoked the violent wrath of Urban, who felt himself to be publicly exposed and involved in superstitious practices, and led to an irrevocable chill in their relations. Campanella did everything he could to exculpate himself by proving that the publication was unauthorized and clandestine, the fruit of a malevolent plot by two of his most powerful and jealous enemies – Nicolò Ridolfi, vicar-general of the Dominicans, and Nicolò Riccardi, master of the papal household – to prevent the Pope from nominating Campanella as *qualificator* of the Holy Office. Campanella promptly wrote an *Apologeticus* to show that the remedies suggested in his work should not be regarded as superstitious ceremonies, because they had nothing to do with demonic pacts, implicit or explicit, but rather as entirely natural and therefore permissible.[33]

6   Inscrutabilis

The Campanella affair was closely entwined with the 'trial of the astrologers', set in train by Urban VIII in order once and for all to throttle both the unfavourable predictions of his early death and the rumours and intrigues surrounding his succession.

Public opinion was rife with expectations and superstitions, as is shown by the reappearance – a common occurrence when the death of the pontiff was deemed imminent – of spurious prophecies about the papacy ascribed to Joachim of Fiore[34] and visions experienced by self-proclaimed 'saints' equipped with sacrilegious paraphernalia. On 9 June 1600 Rome witnessed a public abjuration by the rector of the church of San Carlo, a priest 'of venerable aspect and long beard' and 'favoured by many cardinals' who had been unmasked as 'a hypocrite of the worst sort and a great necromancer'. Convinced that the Pope was about to die, and believing that he would be made a cardinal by his successor, this priest tried to hurry matters along by practising necromantic rituals with wax statues and by celebrating black masses over the body of a woman who believed herself to be a saint. The priest was hanged in the Campo dei Fiori and – so runs one gruesome and laconic contemporary report – the nun who was his accomplice 'was whipped upon a donkey and then immured, whereupon she expired after two days'.[35]

The death of Urban's brother in February 1630 was accompanied by cynical gossip to the effect that Urban's grief was much mitigated by the thought that the malignant influences of the stars had been 'earthed' by striking another member of his family. The rumours of Urban's own imminent demise flourished. Indeed, they flourished so strongly that Philip III thought that Urban's death was imminent and sent the Spanish cardinals to Rome to attend the conclave to elect his

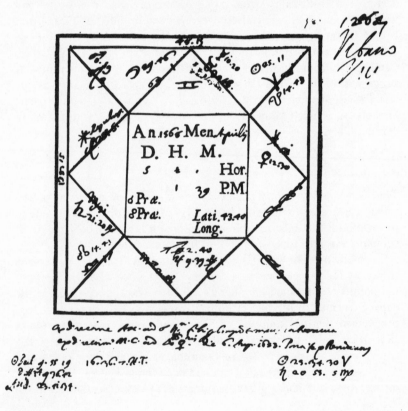

successor. The arrival of the French and German cardinals hard on the heels of the Spaniards was the last straw for Urban. He immediately ordered that the trial of Orazio Morandi, abbot of Santa Prassede, accused of practising judiciary astrology, composing 'political and maledictory' writings and owning forbidden books, should begin.

He instructed magistrates to search his monastery and the houses of other suspects as a suitable punishment for those who 'dared to study and practise judicial astrology with predictions of future successes in war, the overthrow of governments and princes, and the deaths of the same and of other private persons' and who 'in speech and in writing brazenly predicted such things'. Imprisonment was therefore the lot of the abbot – 'a widely-read man, beloved of many and of the Pope himself,[36] – and his presumed accomplices.

The Monastery of Santa Prassede was a well known and influential meeting place for a number of astrologer-politicians who,

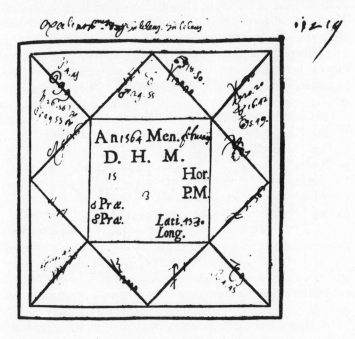

8   **Nativity charts of Pope Urban VIII and Galileo Galilei.**

by means of horoscopes, kept a close eye on and tried to influence the Italian and international policies of the Roman court. The thick folder of trial documents included dozens of horoscopes of popes and cardinals – effectively an astrological database for the Vatican (fig. 8). Campanella had his finger firmly on the pulse of all this activity and years later, in a letter written from exile in Paris on 3 April 1635, he reminded the Pope about the conference of astrologers at Santa Prassede in the autumn of 1629, and about the ensuing lobbying 'to move the souls of the cardinals to elect that pope who is suggested by the stars'.[37]

Apart from the predictions of his death the Pope was also annoyed by a number of astrologers who used the pretext of examining the horoscopes of cardinals who might be chosen to succeed to the papacy to find their vices and defects. According to the admissions of several copyists, Morandi had gone so far as to write a paper on the subject. He identified three factions within the Holy College and prepared a copy, embellished with portraits of the cardinals, for the Venetian ambassador.[38]

Once the trial was under way, the witnesses – astrologers, monks, booksellers and clerks – followed each other in quick and picturesque succession. One defendant justified his possession of astrological texts with the excuse that he 'fancied he might compose a modest criticism of astrology' and that he 'wished to confirm its falsity' – and by night he confessed to his cell-mate that if he managed to extricate himself from this 'squall' 'he did not want to play the astrologer ever again'. Another defendant disassociated himself from the charlatans and impostors, like certain Spaniards, who cast horoscopes for payment, and insisted that he was a serious scholar – 'I don't do this for a living, I have a doctorate in law and I have studied widely' – and a respectable astrologer, consulted by princes, cardinals and popes. Another still hoped to get off by denying everything including the fact that he had predicted the Pope's imminent death – which in fact he had done. By contrast, his astonishment at this sudden intransigence where previously there had been the greatest 'tolerance and permissiveness' was entirely unfeigned: everybody in Rome had horoscopes made for themselves and 'there is not a cardinal, nor a prelate, nor a prince who does not possess an astrological analysis of his birth horoscope with predictions of good fortune'.[39]

The monks of Santa Prassede tried everything to defend their abbot. They denied all charges until Teodoro Ameyden, the legal adviser of the monastery, let slip that some of them had tampered with the seals on a chest and had extracted the most compromising documents which they then hid or burned. After this Morandi's defence quickly fell apart.

The surprising inventory of Morandi's library, included with the trial papers, shows that he was a man of great learning and discrimination. Beside the works of Machiavelli and Tacitus (not to mention Erasmus and Paolo Sarpi) stood Jesuit catechisms and manuals about the techniques of power. Raymond Lull, Arnald of Villanova, Paracelsus, Khunrath and texts of astrology and the art of memory rubbed shoulders with Kepler and Galileo. Galileo also figures among Morandi's correspondents, and the five extant letters from him reveal his genuine admiration for the Florentine scientist. With the first, dated 6 July 1613, he encloses a letter from Francesco Sizzi in Paris which shows, Morandi hopes, that Sizzi now regrets the bumbling attack on Galileo in his *Dianoia Astronomica*. Morandi also shows a taste for typically Galilean imagery – 'truth is one' he insists, and he contrasts those who philosophize 'meditating on the beautiful and spacious book of nature' with the sophists who shackle science with 'the unworthy manacles of Aristotelian prejudice'.

From hints in letters from Francesco Stelluti we infer that early in March 1626 Morandi went to the trouble of sending his personal copy of Kepler's *Tychonis Hyperaspistes* to Galileo who was keen to know what was said about him in the Appendix. In April he replied to Galileo's letter of thanks. In the next two letters Morandi declared he would be glad to do Galileo any favour he could and considered himself honoured and happy to serve him in whatever way. The last, dated 24 May 1630, warmly invites Galileo to dine ('Please don't bother to reply, just make sure you come; I shall expect you without fail'), and Galileo, visiting Rome for a few weeks, accepted. Galileo must have reciprocated the abbot's friendliness despite his extreme caution and diffidence where occult sciences were concerned. When Don Orazio was suddenly tossed into jail Galileo enquired what was going on. He was told that the trial was extremely complicated and that it was proceeding 'with such secrecy that nobody could say anything for certain'. The position of 'our friend' is awkward; however, it was to be hoped that the Pope 'would not go so far as to take special steps against such a learned, capable and skilled person'. The following spring Galileo was still ignorant of Morandi's fate and he must have been very shocked when Benedetto Castelli told him that 'it is said that he has been dead three or four months and is buried in Santa Prassede'.[40]

In fact, Morandi had died on 7 November 1630. A monk from Santa Prassede was called to identify his abbot – a male cadaver, some 55 years of age, tall, chestnut beard shot with grey, wearing only a black habit with hood. The monk added that Morandi had died of natural causes around noon following a violent fever. The news was confirmed a couple of days later by the prison medical officer, no doubt in order to silence rumours that Morandi had been given poison in his food – this being the quickest way to close a rather delicate case while avoiding the dishonour of capital punishment.[41]

The following April Urban unleashed his ferocious bull *Inscrutabilis* against astrologers. It confirmed the bull of Sixtus V and added an absolute ban on predicting the death of a pope or any of his kin. Transgressors were threatened with the severest penalties possible – from expropriation of all possessions to death. There seems to be no substance to the rumours that Campanella himself had edited the bull ('so that he would be the only practising astrologer left').[42] In fact he seems to have been stunned by the severity of the bull and immediately penned a *Disputatio* which, on the pretext of defending the two anti-astrological bulls, tried to limit the damage by jettisoning the entire ballast of superstitious doctrines and allowing the power of the stars only the narrowest possible, and purely physical, scope. Campanella protested that astrologers were being shown far greater severity than heretics (and how was one supposed to answer the accusation that the Pope was more worried about his and his relatives' peace of mind than about the faith and the worship of God?) and still hoped it would be possible to at least avoid the absurdly restrictive view espoused by 'Padre Mostro' ('Father Monster'), the Dominican Nicolò Riccardi, so nicknamed for his 'monstrous' erudition and fatness, that it should be illegal even to name astrology, let alone refute it.

But Campanella had watched the whole affair from the ringside and he was fully aware that the situation was very serious and any decisions taken would be 'political'. Not only had Sixtus V, as *theologus doctus* ('the learned theologian'), forbidden false doctrines; as *princeps prudens* ('the prudent prince') he had also forbidden dangerous beliefs. Urban's prohibitions were also aimed at the causes of social unrest and dangerous explosions of human passions:

Everybody knows about the disorders last year caused by the empty gossip of astrologers and superstitious predictions against the life of the Pope and the state of the church. Every one knows how expectations were raised and dashed, how changes were thought about and plotted, and how the princes of the church came to Rome for this reason. Everyone knows how powerful men with evil intentions always get false prophets to stir up the passions of the masses and to cause sedition, war and schism.[43]

Sixtus V condemned the 'proud animal'. Pope Urban 'who banned others from astrology while being highly skilled in it himself,[44] – insisted on the social and political dangers of human arrogance:

The unfathomable depth of God's judgements does not tolerate the human intellect, locked as it is into the shadowy prison of the body, to rise above the stars. Not only does it dare in its sinful curiosity to pry into mysteries which are hidden in God's heart and are unknown even to the spirits of the blessed, it also presumes, arrogantly, to set a dangerous example and hawks mysteries as certainties, scorning God, agitating the state and endangering princes.[45]

Notes

1 This and other bulls referred to are to be found in *Magnum Bullarium Romanum*, Lyons, 1692, II, pp. 515-517.

2 Ovid, *Metamorphoses*, I, 84-84, trans. A. D. Melville, Oxford, 1986, p. 3.; on the topic of the dignity of man see Ch. Trinkaus, 'In Our Image and Likeness', in *Humanity and Divinity in Italian Humanist Thought*, 2 vols., London, 1970, esp. I, ch. 4.

3 G. Cardano, *In Cl. Ptolemaei IIII de Astrorum Iudiciis aut Quadripartitae Constructionis libros commentaria*, Basle, 1554, f. A2^r (henceforward *Comm. in Quadr.*).

4 J. Garcaeus , *Astrologiae methodus*, Basle, 1576, f. B2^v; for Garcaeus (1530-1575) see Lynn Thorndike, *A History of Magic and Experimental Science*, 8 vols., New York, 1923-58, VI, pp. 102-104.

5 H. Wolf, *Admonitio de astrologiae usu* in C. Leowitz, *Ephemerides*, Hapsburg, 1557 [pp. n.n.]; for Wolf (1516-1580) see Thorndike, *op.cit.*, VI, 112-14.

6 *Comm. in Quadr.*, ff. A2^{r-v}.

7 G. Pico della Mirandola, *Disputationes adversus astrologiam divinatricem*, edited by E. Garin, 2 vols., Florence, 1946-52.

8 See respectively *Astrologiae methodus, cit.* (n. 4), f. A4v.; *Comm. in Quadr., cit.* (n. 3) f. A4^v; G. Cardano, *Aphorismi astronomici*, I, 13, in *Opera Omnia*, edited by C. Sponius, 10 vols., Lyons, 1663, V, 29.

9   *Comm. in Quadr., cit.,* f. A4^r.

10 *Astrologiae methodus, cit.* (n. 4), f. B2^r.

11 See for example *Astrologiae methodus,* ff. B2^{r-v}.

12 *Comm. in Quadr., cit.,* f A5^r.

13 F. Giuntini, *Speculum astrologiae,* Lyons, 1581, I, pp. 136-8.

14 After the first edition in one volume (Lyons, 1573), the *Speculum* was reprinted twice in two volumes (Lyons, 1581 and 1583); see in particular the dedication to G. Arluno in the 1581 edition.

15 F. Giuntini, *Discorso in difesa dei buoni astrologi,* Lyons, 1571.

16 *Summa th.,* I, 115,4; II-IIae, 95, 5; *Contra Gent.,* III, 82-6; 104-6.

17 A. Possevino, *Bibliotheca selecta,* Rome, 1593, 1. xv, ch. xv, pp. 283-4.

18 P.Pomponazzi, *De naturalium effectuum causis, sive de incantationibus,* Basle, 1556 ; The plagiarism was pointed out by G. Zanier, *Ricerche sulla diffusione e fortuna del 'De incantationibus' di Pomponazzi,* Florence, 1975, pp. 90 ff.

19 *Comm. in Quadr., cit.,* pp. 163-6; Lyons, 1555, pp. 369-73. Deleted from the Basle edition of 1578, the horoscope was reintegrated in Cardano's *Opera Omnia,* Lyons, 1663, V, 221-2.

20 S. da Siena, *Bibliotheca Sancta,* Venice, 1566, pp. 684-90.

21 A short account on Clemente Serravalle in J. Quétif, J. Echard, *Scriptores Ordinis Praedicatorum,* 2 vols., Paris, 1721, II, p. 186.

22 In the dedication to Cardinal Camillo Caetani which prefaces the 1592 Venice edition.

23 Thorndike, *op. cit.* (n. 4), p. 412.

24 B. Pereyra, *Adversus fallaces et superstitiosas artes,* Ingolstadt, 1591, p. 221 ff.

25 Thorndike, *op. cit.,* p. 410.

26 *Adversus fallaces, cit.* (n. 24), p. 14 ff.; 26.

27 *Ibid.,* pp. 105 ff.; Philo Judaeus of Alexandria (*c.* 25 BC–*c.* AD40) had tried to combine religion with philosphical views mainly derived from Platonic philosophy.

28 *De siderali fato vitando* (henceforward *De fato*) was published as the seventh and last of the *Astrologicorum libri.* I quote from T. Campanella, *Opera latina Francofurti impressa annis 1617-1630,* repr. by L. Firpo, 2 vols., Turin, 1975, II, pp. 1318 ss.

29 *De fato, cit.,* pp. 1319, 1326. For an exhaustive treatment on the subject of *Fatum,* see *Metaphysica,* Paris, 1638, repr. by L. Firpo, Turin, 1961, part II, p. 186 ff.

30 On the atrocious death of Plotinus, in spite of all his precautions, see I. Firmicus Maternus, *Matheseos,* Leipzig, 1897, pp. 21-4; for Aeschylus, see Valerius Maximus, *Facta mem.,* ix, 12. For the death of Bartolomeo della Rocca, known as Cocles (1467-1504), see P. Giovio, *Elogia virorum illustrium,* Basle, 1577, pp. 106-7. Nabod died in Padua in March 1593, while the young Campanella was living there: see Thorndike, *op.cit.,* VI, p. 121; *De Fato,* p. 1339.

31 D. P. Walker, *Spiritual and Demonic Magic from Ficino to Campanella,* London, 1958, pp. 203 ff. (esp. p. 223).

32 L. Amabile, *Fra T. Campanella ne' castelli di Napoli, in Roma e in Parigi,* 2 vols., Naples, 1887, II, doc. 203, p. 148.

33 *Ibid.,* I, pp. 342-3; L. Firpo, *Ricerche campanelliane,* Florence, 1947, pp. 155

ff.; *Apologeticus ad libellum de siderali Fato vitando*, in Amabile, *op. cit.*, II, doc. 242, pp. 172-179.

34 The most famous doctrine of the abbot Joachim (twelfth century) dealt with the division of the whole of history into three ages, that of the Father, the Son and the Holy Spirit, in which the millenarian expectations of an age of peace and happiness would be fulfilled.

35 G. Gigli, *Diario romano* (1608-1670), edited by G. Ricciotti, Rome, 1958, pp. 112-13.

36 Gigli, *op. cit.*, p. 115; Amabile, *op. cit*, I, pp. 284-5. The records of the trial are to be found in the State Archives of Rome. The most significant of them were published by A. Bertolotti, 'Giornalisti, astrologi e negromanti in Roma nel secolo XVII', *Rivista Europea*, 5, 1878, pp. 466-514.

37 T. Campanella, *Lettere*, edited by V. Spampanato, Bari, 1927, pp. 287-8.

38 Bertolotti, *art. cit.*, pp. 480, 484, 485, 489, 494-5.

40 G. Galilei, *Opere*, edited by A. Favaro, 20 vols., Firenze, 1890-1909 (repr. Firenze 1968), XI, 491-3; XIII, 299-300; XIV, 134-5, 250.

41 Bertolotti, *art. cit.* (n. 36), pp. 497-8. For rumours on poisoning see also Gigli, *op. cit.*, p. 118.

42 Amabile, *op. cit.* (n. 32), II, doc. 203, p. 150.

43 *Disputatio in Bullas*, in *Atheismus triumphatus*, Paris, 1636, pp. 267 ff.

44 Gigli, *op. cit.* (n. 35), p. 253.

45 *Magnum Bullarium cit.* (n. 1), IV, 173 f.

12

Astrology in early modern England: the making of a vulgar knowledge

Patrick Curry

Any true detection should prove that we are the guilty party.

Umberto Eco

1   Introduction

Astrology in late seventeenth-century England experienced a dramatic transformation. There was an unprecedented flowering during the Civil War and Interregnum years (1640-60). Yet within a few decades of the Restoration of 1660, its reputation and influence seemed to fall from this earlier peak into a disgrace from which it never really recovered.

The purpose of this essay is to treat English astrology in this period as an historical case-study of what is sometimes called a 'pseudo-science'. My general approach is neither precisely that of the history of science nor of social history; it could perhaps best be described as the social history of ideas. After giving an overview of the subject, I shall discuss how our understanding of astrology's transformation has changed in recent years. In particular, I suggest that the 'missing key' to that episode is now clear; it lies principally in contemporary changes in the relations between social classes in England. As an example of how these affected astrology, I concentrate on the apparent disappearance of such beliefs from elite social and intellectual life. Finally, I explore some explanations for this development centred around the ideas of 'appropriation' and 'hegemony'. In order to be useful for other historical contexts and purposes, the focus here is historiographical: that is, on the ways the historical facts are approached and organized, rather than on the evidence itself.

In order to emphasize new developments, both descriptive and analytical, I am going to use as a starting-point the work of Keith Thomas. His pioneering *Religion and the Decline of Magic* (1971) first opened our eyes to this subject. Bernard Capp's later *Astrology and the Popular Press: English Almanacs 1500-1800* filled in the picture with much valuable new detail, but didn't substantially alter the approach or framework of Thomas's book. The latter continues to attract the description 'magisterial', and with good reason. He discovered a wealth of magical beliefs and practices in sixteenth- and seventeenth-century England. Rescuing them from the silence and sneers of other historians, he succeeded in bringing this cultural history to life in all its weird and wonderful variety.

Such phenomena are usually described as belonging to 'popular culture'. However, it is important to recognize that for most of the seventeenth century, astrological beliefs – defined pragmatically as those systematically relating earthly to cosmic events – existed virtually throughout English society. Illiterate and semi-literate labourers sowed crops, slaughtered animals and even cut hair and nails according to the sign and phase of the Moon. Along with their social superiors, by the 1660s they were buying an average of 400,000 annual astrological almanacs a year.[1] The authors of the latter were not the ubiquitous village 'cunning' men and women but more educated and/ or self-taught individuals (virtually all men) who had mastered the mathematical and interpretative intricacies of horoscopic 'figures' or 'maps'. For this skill they were consulted by a remarkably wide range of people, including many from the gentry and nobility. In the course of the English Civil War and Interregnum, these astrologers acquired unprecedented power as propagandists and prophets on both sides of the religious and political divide. Meanwhile, among contemporary intellectuals such as mathematicians, astronomers and natural philosophers, astrology was sometimes studied, sometimes not, but very rarely unequivocally condemned. English astrology in this period, in fact, was in its heyday – but also, paradoxically, its final flowering. By the turn of the century, such beliefs had apparently been eclipsed by the recognizably modern commercialism and scepticism of the succeeding decades.

In addition to considerations of historical detail, the material Thomas presented raises two general questions which he put as follows: first, why were such beliefs, which are (in his words) 'now rightly disdained by intelligent persons, taken seriously by equally intelligent persons in the past'?[2]

And secondly, why did they die, or at least so drastically decline? In what follows, I would like to start with the second question. Not only is it more accessible, but our search for answers will lead, I suggest, to a re-formulation of the first.

However, let us briefly review beforehand the answers Thomas himself proposed for the decline of magic – although actually, in the case of astrology, he wavered between descriptions of death and of decline (the former tends to predominate, with metaphors of funeral, grave, and *coups de grâce*).[3] In his final chapter, three sets of reasons are discussed. The first, 'Intellectual Changes', includes the undermining of the old Aristotelian distinction between terrestrial and celestial bodies, the hostility of Pierre Gassendi and Pierre Bayle, and the discovery by Edmond Halley and Isaac Newton of the orbits of comets, thus making it harder (Thomas suggests) to regard them as heaven-sent omens. At this point, however, he quite rightly admits that the change cannot simply be attributed to the Scientific Revolution. There were 'too many "rationalists" before, too many believers afterwards ...' A second reason is therefore introduced, namely that of 'New Technology'. This includes improved agricultural and fire-fighting practices, insurance, a postal service and newspapers. All these developments arguably boosted a more sophisticated literacy and diminished the conditions of uncertainty that (it is assumed) encourage recourse to 'superstitious' strategies. Without doubting the effects of such innovations, however, Thomas again notes, with characteristic honesty, that 'in England magic lost its appeal before the appropriate technical solutions had been devised to take its place'. This leads him to introduce a third cause, 'New Aspirations', according to which there was a new ideology abroad of self-help and progress, leading to a preference for practical new ideas over old. He concludes that these three considerations taken together may explain the decline of magic, including astrology, although 'the connection is only approximate and a more precise sociological genealogy cannot at present be constructed'.[4]

2   A new Look at an old problem

What has changed in our views of this subject? To begin with, since the publication of Thomas's book it has become increasingly untenable to regard the fate of early modern astrology as unitary, subject to a single verdict on its outcome. Certainly something *like* its 'death' took place for the educated and/or upper classes after 1660; which really amounts to saying that it was no longer considered a fit subject for 'Books and Talk'.[5] Indeed, by the turn of the century astrology as such had disappeared from the discourse of astronomers and natural philosophers. Yet the popular, largely lunar astrology of the labouring classes continued almost as if nothing had happened. Sales of the bible of popular astrology, Moore's *Vox Stellarum*, grew steadily throughout the eighteenth century, seeing it out at more than 300,000 copies a year.[6] And folklore collectors in the early nineteenth-century documented countless examples of centuries-old popular astrological beliefs throughout the British Isles. Meanwhile, somewhere between these two ends of the social pole, men of middling income and education continued to consult ephemerides and cast figures, albeit less frequently and more privately than before. This practice was sustained by a thriving community of judicial astrologers throughout the eighteenth century, mainly in the West Midlands.[7] In short, it would now take an unconscionable degree of arrogance to pronounce on this complex picture of English astrology after 1660 the single sentence of death, or even of a kind of life not worth bothering about.

In a sense, the reason is simple: astrology actually never was a single entity, with a fixed identity. Formally speaking, it was a set of ideas or propositions concerning the relation of cosmic and earthly events. These ideas were themselves rather banal, even empty of experiential meaning; however, they found various different social interpretations and uses. It is these that are crucial to consider, because astrology only acquired its equally various particular meanings and significance in the course of, and in the form of, such uses. As the account just outlined suggests, it is possible to distinguish three principal instances. The first was high or cosmological astrology, the affair principally of natural philosophers concerned with the Earth as a whole. The second was judicial astrology. 'Middling' both intellectually and socially speaking, this kind centred on the interpretation of horoscopes for the purpose of making predictions for specific individuals. Finally, popular astrology was an integral part of oral culture which addressed the practical and spiritual needs of labouring people. According to which kind is being considered, therefore, astrology after 1660 can be said to have disappeared or died, declined, and survived. Obviously, any one of these descriptions, taken alone and unqualified, would result in an unbalanced view of the matter.

Another and equally crucial point follows from this first: namely that the social geography of this picture, although complex, is far from random. Astrology progressively 'disappeared', socially speaking, from the top down. As E. P. Thompson noted in a typically fertile remark, early modern popular magical beliefs did not simply constitute an incoherent mass, rejected by their holders' betters; rather, 'discrete and fragmented elements of older patterns of thought become integrated by *class*'.[8] This insight is important, for properly followed up, it provides the missing key to the mystery of astrology's transformation. First and foremost, that event was a consequence of a crucial new development after 1660, widely confirmed by historians: namely the new and rapidly widening divide between patrician and plebeian sensibilities.[9] This split did not follow the traditional distinction between aristocracy-gentry and commoner; it fell between the self-defined 'better sort' – comprised of the gentry and the newly powerful (but still far from independent) commercial and professional classes – and the labouring masses. And exactly coincident was the banishment of astrology from polite written and spoken discourse – except as an occasional object of scorn – and its re-making as an almost exclusively plebeian knowledge.

I say 'plebeian' here rather than 'popular' because as records of the nature and consumption of astrological almanacs show, prior to the period beginning in 1660 individuals from the upper and middling classes had ready access to popular astrological ideas. But this was now no longer the case, and as readers from other classes deserted it, the leading almanac – Moore's *Vox Stellarum* – became a publication purely of and for the labouring and lower middle classes.[10]

The specific impetus for the process of patrician/plebeian division came from the Civil War and Interregnum years. This period was the high point of English astrology when, untrammelled by ecclesiastical and governmental censorship, astrologers like William Lilly attained a degree of popularity and influence unmatched before or since. Astrologers were the Parliamentarians' (in particular) most valuable propagandists, and highly visible advocates of democratic and antinomian ideas which disparaged religious and political authority alike. It is therefore hardly surprising that with the conservative Restoration, they were firmly identified with the turbulence of the mid-century political and religious sects. That opinion was strengthened by succeeding generations of outspoken astrologers with popular almanacs (notably John Partridge) in the context of the turbulent 1670s and 80s. The umbrella term of opprobrium was 'enthusiasm', and far beyond the immediate crackdown at the hands of Charles II's censor Sir Roger L'Estrange, astrologers became, for an entire social class, first 'enthusiastic', and then – as the social and sensible split deepened – 'vulgar'. By about 1700, therefore, astrology had been re-fashioned as virtually a purely vulgar phenomenon; indeed, it was now only 'common-sense' to see it so. Thus, to return to the question of the death or decline of astrology as a whole – in so far as that is still meaningful – it is now untenable to say that astrology simply disappeared. It would be more accurate to say, following a current Latin American usage, that it 'was disappeared'. But in any case, judicial astrology survived, and the popular form thrived, throughout the late seventeenth and the entire eighteenth centuries. Astrology's invisibility after 1660 was therefore due more to elite commentators' inability or unwillingness to perceive it than to a real absence; and the same is true of modern historians who maintain the fiction of the latter.

There was a still more general historical context for this development, namely what Peter Burke identified as 'the reform of popular culture'.[11] Beginning in the sixteenth century with the European Reformation and Counter-Reformation, this consisted of an attempt by members of the educated ecclesiastical elite radically to alter the values and beliefs of the mass of people. Their chief ideological weapon was 'superstition', which was employed against astrology with increasing effect throughout the seventeenth century. But it took the events of the Interregnum to convince sufficient of the English upper and middle classes that such beliefs were no mere theological offence, but a real danger. Those events coincided with the start of a second phase in the attempted reform of popular culture, now led primarily by the intellectual laity – especially, in this case, natural philosophers.

Of course, that phase did not abruptly take over from the first. There was considerable overlap, both chronologically and biographically. Robert Boyle was both an influential 'Latitudinarian' or Low-Church Anglican and the leading Restoration natural philosopher. His colleague Henry More, the Cambridge Platonist philosopher and Anglican divine, devoted considerable effort in 1661 to a 'solid Confutation of Judiciary Astrology'. In line with a standard rhetorical position among European intellectuals sceptical of astrological claims, More denied any influence to the stars other than heat and light, and attributed correct predictions (which he admitted) to the intervention of demons. In view of its post-war associations, More was particularly anxious to distance himself from astrology. He had good reason, for in 1666 the Royalist High-Churchman Samuel Parker (later Bishop of Oxford) published a swingeing attack on the 'Enthusiasticke Fanaticisme' of 'Platonicke Philosophie'. He was especially concerned about the work of the astrologer John Heydon, whose 'Pestilential Influences' Parker maintained were obvious from the recent Civil War. In books like *The Rosie Crucian* (1660), Heydon peddled a kind of populist magical astrology to enable readers (in the words of its subtitle) to 'know all things past, present and to come'. Parker described such ideas as 'contrary and malignant to true knowledge', that is, in both religious and natural philosophical senses. Despite their other differences, More and Parker agreed on that (both were also Fellows of the Royal Society).[12]

I noted earlier that the early scientific community was often ambivalent on the subject of astrology. Earlier historians' discussions of this subject have tended to degenerate into attempts to demonstrate that all progressive or 'real' scientists rejected such beliefs. This attitude is sometimes mirrored, without any sense of irony, by amateur historian-astrologers who maintain that the entire early Royal Society was permeated by occultism, up to and including Newton. Such arguments, which dragoon various historical figures into supporting an *a priori* position, are unlikely to advance the cause of understanding. It is an historical fact that leading English astronomers like Jeremy Shakerley, Vincent Wing and Thomas Streete were committed judicial astrologers, as were such early Fellows of the Royal Society as Elias Ashmole and John Aubrey. It is also the case that while Boyle was sympathetic to the subject, Robert Hooke and John Flamsteed were clearly not. Much more interesting, however, is what these men actually said and wrote when it is taken in context. Examination reveals, for example, that when attacking astrology they largely failed to draw a clear distinction which has been too often assumed by modern historians: namely between social–political criticisms, on the one hand, and on the other philosophical–epistemological (including what we would now call 'scientific') objections.

Thus, for example, the polemical *History of the Royal Society* (1667) by Thomas Sprat, FRS and future Bishop of Rochester, railed against 'this melancholy, this frightful, this Astrological Humor', maintaining that it was 'a disgrace to the Reason, and honor of mankind, that every fantastical humorist should presume to interpret all the secret Ordinances of Heven'. In terms whose resonance after 1660 is unmistakable, he argued that astrology 'withdraws our obedience, from the true Image of God the rightfull Soveraign'; nothing could be more injurious to 'mens public, or privat peace' [13]

John Flamsteed, writing in 1674, objected in the same breath to astrologers' 'pernicious predictions of ye Weather *and* State affaires' – pernicious because of their esteem among 'the superstitious Vulgar'. 'Of what ill consequences their predictions have beene,' he wrote, 'and how made use of in all commotions of the people against lawfull and established sovereignity ... our own sad experience, in the late Wars, will abundantly shew the considerate; how they have erred in their Judgements, the *same* experience will informe us ...'.[14]

David Gregory, Savileian Professor of Astronomy, wrote in 1686: 'we prohibit Astrology from taking a place in our Astronomy, since it is supported by no solid fundament, but stands on the utterly ridiculous opinions of certain people, opinions that are so framed as to promote the attempts of men tending to form factions' – this being a clear reference to the Civil War sects which Gregory, a Jacobite, detested.[15] And in a tiny but revealing incident, Robert Hooke – certainly no friend to astrology, even if he did share London's coffee-houses with several of its leading practitioners – recorded in his diary the death of Bishop Wilkins, on 19 November 1672. His laconic entry is accompanied by the note, 'a conjunction of Saturn and Mars,' and (in large letters) 'Fatall Day'.[16] The fact that this was written in the space normally reserved for meteorological reports graphically symbolizes the extent to which astrology – even for its critics – still shared 'natural' and 'social' categories. The point is not that the former was merely, at bottom, the latter; nor that a distinction between them was unavailable at the time. I merely want to point out that it is poor historical practice automatically to privilege one over the other, when the historical subjects themselves did not.

The last example points to another distinction – one which contemporary natural philosophers seemed keenly aware of, but which has been neglected by historians. That was between public and private discourse. In a manuscript written in 1665 or 1666 and published in 1686, Boyle expressed concern about those 'enemies of Christianity' who explain the New Testament miracles by reference to 'coelestial influences'. Only in his posthumously published essays and letters, such as one addressed to Samuel Hartlib, do we learn of Boyle's high hopes for astrology, based on his conviction that notwithstanding all the objections against it – such as 'the superstition and paganism incident to this kind of doctrine' – the celestial bodies 'may have a power to cause such and such motions, changes, and alterations ... as the extremities of which shall at length be felt in every one of us'.[17]

Flamsteed, writing to Richard Townley on 4 July 1678, commended the astro-meteorological research of John Goad, 'whose conjectures I find come much nearer truth than any I have hitherto met with'. He added, 'You know I put no confidence in Astrology, yet dare I not wholly deny the inferences [sic] of the stars, since they are too sensibly impressed on [us]'. Yet in a meeting of the Royal Society the following year and subsequently published, Flamsteed criticized Goad's predictions severely, maintaining that he had found 'not one of three true'.[18]

Most consistently of all, we find a process of re-naming, whereby hitherto astrological ideas become 'safe' through being grafted on to the body of Christian natural philosophy. The virtuoso John Evelyn, FRS., for example, abused 'knavish and ignorant star-gazers' while musing that several recent comets – a centuries-old staple subject for astrologers' speculation – 'may be warnings from God, as they commonly are forerunners of his animadversions'.[19] It is interesting to trace this process after 1700, using Boyle's argument of distinct planetary effects on the atmosphere, and hence on the weather, epidemics, and so on. The theory appears in John Harris's *Lexicon Technicum* (1704), which simultaneously lambasts astrology as 'a ridiculous piece of Foolery'. Ephraim Chambers' popular *Cyclopedia* (1728) has it under 'Natural Astrology', carefully bracketed from 'Judicial Astrology'. Yet the basic idea – that every planet has 'its own proper Light ... embued with its specific Power,' and 'the Powers, Effects, or Tincture, proper to each, must be transmitted hitherto, and have a greater or lesser effect on sublunary things' – had always been integral to the intellectual basis of judicial astrology. By the time of the first *Encyclopedia Britannica* (1771), astrology as such receives five lines, against 66 pages on astronomy; significantly, however, the latter now included the cosmological and eschatalogical functions of comets.[20]

The transformation of comets from astrological into astronomical entities has already been treated elsewhere by Simon Schaffer.[21] I only want to stress here that far from breaking with the millennia-old astrological attributes of comets, the success of Newton and Halley's programme, beginning in the 1680s, enabled them to retain just such functions for comets without attracting unwelcome attention as astrologers. (This was implicitly recognized in the later German astronomer J. H. Lambert's sarcastic description of his colleagues as 'authorized prophets'.) Indeed, Newton believed that in addition to replenishing the vitality of the cosmos, the sacred power of comets extended to sweeping away all idolatrous corruption on the Earth, *including* 'astrologers, augurs, aurispicers ... [and] such as pretend to ye art of divining'.[22] These ideas were taken up and developed still further in the popular lectures and books by Samuel Clarke and (especially) William Whiston, who inferred 'the mightyest effects and Influences' from the Sun, Moon and comets, while simultaneously deploring 'that mighty quoil and pother so many in all Ages have made about the Conjunctions, Oppositions, and Aspects of the Heavenly Bodies ...'.[23]

Even at this elite level then – where the changes were the most dramatic and complete – the so-called death of astrology was neither simple nor straightforward; and it certainly didn't simply vanish in the light of 'the truth'. It is not my brief here but, as I have already suggested of its other forms, horoscopic astrology survived until its flowering again in early Victorian England; while popular lunar astrology – although subject to some subtle qualitative changes – carried on among 'the bootless and unhorsed' throughout the eighteenth and arguably the nineteenth centuries, virtually as if the Enlightenment, scientific or otherwise, had never occurred.[24]

3 Implications

To discover and describe historical 'facts' without trying to explain them in terms of theories is not just undesirable, it is impossible. The apparently least theoretical account conceals selection and interpretation; it also makes itself harder to think about and learn from. I would therefore now like to turn to some ideas which have a direct bearing on the transformation of early modern astrology – ideas which are both suggested by the material and help to illuminate it. They have developed mainly in continental Europe, where intellectuals, including historians, are traditionally less wary of theory (or more enthusiastic, perhaps) than in England. Indeed, it is worth noting that the common English distaste for ideas as such – 'We are plain quiet folk and have no use for [them]. Nasty disturbing uncomfortable things!' – to adapt the words of one literary representative[25] – stems precisely from the reaction which we have been considering to the exotic profusion (including prophetical astrology) of the Civil War years. That suspicion, subsequently given a huge boost by the events of the French Revolution, unfortunately still pervades English culture, intellectual as well as popular.

Particularly relevant for our concerns here are the work of Roger Chartier, an historian of early modern France, and recent developments in the tradition of the Italian Marxist, Antonio Gramsci. Chartier combines his knowledge of this period with a thorough grasp of the complex historiographical issues embodied in the *Annales* school and its 'history of mentalities'. The latter arose in an attempt to consider phenomena long neglected by historians: the mental world-views of human historical subjects, and the social determinants, or at least correlates, of those views. This concern produced pioneering work on (for example) attitudes to death, childhood and the family, and popular religion.

Chartier's work is important for confronting the problem that faces anyone engaged in this kind of historical practice. That is, how to relate *social* practices to *cultural* (including mental and ideological) representations. The specific difficulty is to avoid two opposing dangers: on the one hand, that of reducing cultural phenomena entirely to social (a tendency among both Marxists and *Annalistes*); on the other, that of detaching them completely (the traditional vice of conservative and idealist intellectual historians). Chartier encourages us to recognize the intimacy of mentalities with issues of social structure. At the same time, however, he argues cogently against the attempt to carve up people's mentalities, or cultures and artefacts, into separate categories like 'popular' and 'elite'. In particular, such a tendency smothers any awareness of complex interrelationships, and the dynamics of change over time. Instead, Chartier suggests the concept of culture as *appropriation*, whereby what really matters is 'the different ways in which groups or individuals make use of, interpret and appropriate the intellectual motifs or cultural forms they share with others'. Indeed, he argues that far from bearing a fixed stamp of identity, written or spoken discourse acquires its sense and significance 'only through the strategies of interpretation that construct its meanings'.[26]

Astrology is a clear case in point. As I have already noted, its basic ideas, centring on the intimacy of celestial and terrestrial phenomena, were both simple and very widely shared. As we have also seen, however, different social groups adopted and adapted them for very different purposes; and therein lies their meaning and interest. Take, for example, the appropriation of astrological ideas by natural philosophers after 1660, which I have discussed in some detail. Riding on the back of class revulsion against popular enthusiasm, it was part of a strategy intended to establish their own authority as a Christian proto-scientific elite in the interpretation of cosmic and celestial phenomena, while discrediting that of radical populist astrologers, whose 'Almanacks are Oracles to the Vulgar'.[27] This involved taking over and re-interpreting considerable chunks of hitherto astrological discourse, particularly respecting comets and astral influences on weather and disease, in such a way as specifically to exclude astrologers. Meanwhile, the latter – in their almanacs and text books – were proclaiming not only their own right and ability to 'interpret the secret Ordinances of Heven', but (in most cases) those of everyone else, potentially, to do so for themselves. In the hands of these different groups, then, astrological ideas acquired radically different meanings.

As this description implies, such ideological (that is, simply, socially interested) appropriations of part of a set of common ideas are usually in direct conflict with one another. In other words, their particular proponents are engaged in a hegemonic struggle – a struggle to make their interpretation alone prevail among as many social groups as possible. Here I would like to turn to some closely convergent work inspired by Antonio Gramsci. A brilliant innovator in the Marxist tradition, Gramsci was critical of Lenin's assumption that the working class – or rather its self-selected representatives – had a privileged relationship to the truth, whereas that of the other classes (e.g. the bourgeoisie) was characterized by 'false consciousness'. Like the supposed privileged access of scientists, which this view parallels and from which it partly developed, such a position relies on teleology; that is, on some individual or group being in a unique position to know the final truth. Gramsci found it difficult to maintain the orthodox Marxist (and scientistic) view that there is truth on the one hand, and mere beliefs on the other; or that cultural beliefs were so directly or simply determined by class and, in turn (since class is supposedly the result of a particular position in the 'relations of production') by the economy.

None the less, Gramsci tried to address the problem of how certain ideas, although more in the interest of some classes than others to believe, become (through what he called 'moral and intellectual leadership') wide-spread 'common-sense' throughout society. By common-sense is meant a view which, while rarely questioned or thought-out itself, informs more conscious judgements; a current example might be the popularity of British royalty. In its least conscious aspects, such a view is a 'mentality'; at its most conscious, an ideology.[28]

In several ways – admitting that ideas cut across classes, reducing epistemological privilege, and encouraging a more sensitive look at social/cultural connections – Gramsci's idea of hegemony constitutes a real advance. It resembles our historical picture of early modern astrology, for example, a great deal more closely than either 'vulgar Marxist' or scientistic fairy-tales. By recognizing that it was in the elite's interest to accelerate, cement and exploit astrology's 'decline' it also offers a sharper and clearer picture of the 'precise sociological genealogy' of that process than Thomas's empirical social history. Yet problems remain.

First, the so-called decline of astrology occurred well before the appearance of mature classes resulting from the Industrial Revolution; yet Gramsci assumed that hegemony could only proceed from the social base of a fundamental class like the proletariat. Secondly, he also seemed to believe that there could only be a single successful hegemonic centre, whose 'common-sense' would dominate virtually all society. But we found that the hegemonic success of the campaign against astrology, despite its apparent conclusiveness, was patchy: while conquering virtually all of the upper and middle classes, it left the resistance of labouring people and their plebeian astrology contained but uncracked. Thirdly, that campaign was conducted not by a single social class, but by a cross-class alliance composed of the gentry and aristocracy and the fledgling middle classes. The resulting bloc was neither purely gentry – the dominant partner, and the one originally most affronted by the enthusiastic excesses of the Civil War astrologers – nor professional and commercial – the younger and lesser partner, but no less anxious to discourage sectarian instability. Indeed, in some respects, such as aristocratic privilege and excess, and in matters of religion, individuals of the latter origin were often highly critical of the former. Yet on other subjects, such as astrology, they spoke as one. In other words, the hegemonic attack on astrology was led not by a single class, but by a strategic alliance of classes, based on a contingent and unstable agreement of interests.

The political theorists Ernesto Laclau and Chantal Mouffe have recently developed Gramsci's work in a way that resolves these anomalies. Their principal contribution is the idea of hegemony as 'a logic of articulation and contingency'[29] – articulation not in the sense of expression, but that of joining-up; and contingency not in the sense of randomness, but that of final unpredictability. According to this approach (which is similar to that of Chartier), the human meanings of ideas are determined neither by their social origins, nor by any intrinsic or essential characteristics, but by the ways in which they are used and connected to other accepted ideas, embedded in the particular historical circumstances of the time and place. The more widely and convincingly such articulation takes place, the greater the degree of hegemonic success. Such success is contingent because it is not predetermined by any necessarily privileged considerations or 'inevitable' outcomes. That includes the traditional Marxist star, class struggle – although the latter *may* prove crucial, as indeed it did in the case of early modern astrology. But any one of class, race, gender, nation, or religion – all the ways in which we define ourselves and each other – may predominate, or interact in complex ways to produce an unpredictable outcome.

In the context of British historiography, a flexible concept of hegemony breaks through a sterile confrontation which has developed in recent years between two schools of thought, and which reflects the problem mentioned earlier facing the historian of ideas. On the one hand, there is social 'structuralism': usually Marxist, often reductionist and implicitly European, that is, 'foreign'. On the other hand, there is what proponents of the latter persuasion term 'culturalism': often (although not necessarily) conservative and native, that is, 'English'. Gramsci's concept of hegemony, as a synthesis of both concerns, was a decisive new departure; Laclau and Mouffe have now strengthened and clarified its promise.

That promise remains largely unexplored by historians but not, I hope, for long. Certainly it is borne out in this case-study. The importance of class (not closely defined in economic terms, but rather in social terms, against other classes), even in apparently unrelated and eccentric matters of belief; the real but limited success of the hegemonic campaign against astrology; its cross-class character, and the contingent alliance of interests that sustained it; the discrediting of astrology by linking it with the ignorance of 'the vulgar Mobb', while connecting natural philosophy with Anglican sobriety and erudition – all these things are explicable in terms of Laclau and Mouffe's concept of hegemony. In short, the early modern transformation of astrology involved the ascendancy (albeit partial) of a set of hegemonic ideas, or rather uses of ideas; just as its survival involved resistance (albeit limited) to those uses.

I want to give the final word here to Jonathan Swift. Along with Joseph Addison, Swift led another phase of the hegemonic campaign which there is no room to discuss. But the words Swift put into the dying astrologer John Partridge's mouth, in his mock almanac of 1708, lay bare the nature of its triumph. Swift has the astrologer say,

> 'I am a poor ignorant fellow, bred to a mean trade, yet I have sense enough to know that all pretences of foretelling by astrology are deceits, for the manifest reason that all the wise and learned, *who alone can judge* whether there be any truth in this science, do unanimously *agree to laugh at and despise it*, and that none but the ignorant vulgar give it any credit'.[30]

4   Conclusions

What has permitted us to see the transformation of astrology in seventeenth-century England – not only the various forms astrology took, but the social dynamics missing in earlier accounts – more clearly in recent years? I think the answer is two-fold. The first is simply a closer acquaintance with the mass of historical evidence, of which few people were aware before Thomas's book. The second reason is a sea-change in attitude: one that is both a cause and a result of the kinds of ideas we have just been discussing – perhaps especially as seen in recent work in the history and sociology of science, whose effects on an often straitened and dispirited discipline have been exhilarating. The new attitude enables us to recognize and refuse the voice that whispers, 'Astrology is really just rubbish, now and always'. That voice is a product of the positivism and teleology that, under the banner of truth, has bedevilled the history of science for so long, aided by the anti-intellectualism of English culture which I have already noted, and which takes the form, in the discipline of history, of 'antiquarian empiricism'.[31]

Needless to say, I hope, the antidote is not the view that 'Astrology is true, now and always, and unfairly maligned'. It is rather to be historically self-conscious and self-critical. After all, historians too have a history; awareness is needed of the extent to which our own tools-of-the-trade, both as concepts and values, have been shaped by the past – including, in this case, a suspicion of both astrology, as a specific subject, and of theoretical maturity, as part of a general historical approach. Only then will we be able to refrain from the damaging anachronism of irresponsibly-applied hindsight.

It may be objected that anachronism is unavoidable; for example, I have used concepts of social class which (let's say, for purposes of argument) didn't exist in the late seventeenth century. Certainly; but what is illegitimate is the use of such concepts primarily for the purpose of either blaming or praising, as distinct from understanding, your historical subjects. The price of that practice is an agenda which restricts its questions to narrow dimensions that are inhumane when not simply boring: truth-values and implicit teleology, in history of science, and important men and parties, in history. This impoverishes both historical subjects and present-day students, and along the way – to return to the example of astrology – invaluable resources are squandered. Early modern astrology was continually used as an 'other', a foil: as 'superstition' by divines, in order to define true religious belief; as 'common' by literati, to define true 'morals and manners'; as 'ignorance' by natural philosophers, to define true knowledge. In so doing, the identity and meaning of astrology itself changed; but throughout, it provides a unique window on its mainstream opponents.

In the last instance especially, as we have seen, hitherto astrological discourse was itself used to construct the body of experimental knowledge – which was therefore, purportedly, uniquely socially disinterested – as against the sectarian and seditious machinations of astrologers themselves (judicial astrologers were almost invariably described as 'pretending' – that is, merely claiming – to cast nativities: in other words, the latter were not permitted to retain any meaning even in their own terms). Indeed, the construction of astrology as a useful antagonism survives to this day, thanks to modern debunkers of 'pseudo-science', oblivious of the extent to which early modern science was itself responsible for the creation of a plebeian and occult astrology. The history of astrology therefore provides a unique insight into scientific, as well as religious and politely literary, received wisdoms.

To return to the question proposed by Thomas that we started with – why did 'intelligent people' believe in astrology? – what he set in motion, and is now well under way, is a project which defies anachronism by asking not 'why did they believe in astrology?' but 'why did they stop believing in it? Why them and not these others?' And it asks without any sense that the answers are simple, or 'obvious'. As I hope I have suggested, the results of such inquiry are highly revealing, both about the social and intellectual worlds of early modern England and about historical processes in general. By implication, and as part of a larger project, they also emphasize the risks of allowing hegemonically common-sensical assumptions to dominate research; and the need for theory when approaching historical data. A healthy social history of ideas doesn't fear ideas; it thrives on them.

Notes

1 Cyprian Blagden, 'The distribution of almanacs in the second half of the seventeenth century', *Studies in Bibliography*, 9, 1958, pp. 107-16, Table 1. For a fuller exposition and set of references to this historical material, see Patrick Curry, *Prophecy and Power: Astrology in Early Modern England*, Cambridge, 1989.

2 Keith Thomas, *Religion and the Decline of Magic*, Harmondsworth, 1973, p. ix.

3 Thomas, *op. cit.*, pp. 418, 424; repeated by Bernard Capp, *Astrology and the Popular Press: English Almanacs 1500-1800*, London, 1979, p. 278.

4 Thomas, *op. cit.*, pp. 774, 786.

5 Hester Thrale, *Thraliana: the Diary of Mrs Hester Lynch Thrale, 1776-1809*, edited by K. C. Balderston, Oxford, 1942, 2 vols., II, p. 786.

6 Ellic Howe, 'The stationers' company almanack: a late eighteenth-century printing and publishing operation', in *Buch und Buchandel in Europa in achtzehnten Jahrhundert*, edited by G. Barber and B. Fabian, Hamburg, 1981, p. 207.

7 See Curry, *op. cit.* (n. 1), pp. 95-100 (on popular astrology) and pp. 121-27 (on judicial astrology).

8 E. P. Thompson, 'Eighteenth-century English society: class struggle without class?', *Social History*, 3, 1978, p. 156.

9 E. P. Thompson, 'Patrician society, plebeian culture', *Journal of Social History*, 7, 1978, p. 156; J. C. D. Clark, *English Society 1688-1832. Ideology, Social Structure and Political Practice During the Ancien Regime*, Cambridge, 1985, pp. 70-1.

10 See H. Medick, 'Plebeian culture in the transition to capitalism', in *Culture, Ideology and Politics*, edited by R. Samuel and G. Stedman Jones, London, 1982, pp. 84-113; and (on 'plebeian') Logie Barrow, *Independent Spirits: Spiritualism and English Plebeians, 1850-1910*, London, 1986, p. 16.

11 Peter Burke, *Popular Culture in Early Modern Europe*, London, 1978.

12 Henry More, *An Explanation of the Grand Mystery of Godliness*, Cambridge, 1661, pp. 53-4, 134; Samuel Parker, *A Free and Impartial Censure of the*

Platonicke Philosophie, Oxford, 1666, pp. 2, 72-73.

13 Thomas Sprat, *The History of the Royal Society of London,* London, 1667, pp. 364-5.

14 John Flamsteed, 'Hecker', quoted by M. Hunter, 'Science and astrology in seventeenth-century England', in *Astrology, Science and Society,* edited by P. Curry, Woodbridge, 1987, pp. 228, 291 (my italics).

15 Christ Church, Oxford MS. 113, ff. 47-50 (my translation).

16 Margaret 'Espinasse, *Robert Hooke,* London, 1956, pp. 112-13.

17 Letter (9 March 1666) reproduced in *The Works of the Honourable Robert Boyle,* edited by Thomas Birch, London, 1744, 5 vols., I, pp. lxxix, 638-9.

18 Royal Society MS. 243, No. 35; Thomas Birch, *The History of the Royal Society of London,* 4 vols., London, 1756-1757, III, pp. 454-5.

19 *The Diary of John Evelyn,* edited by W. Bray, 4 vols., London, 1879, II, p. 144; as quoted in Abel Heywood, *Three Papers on English Printed Almanacks,* London, 1904, p. 77.

20 John Harris, *Lexicon Technicum: or an Universal English Dictionary of Arts and Science,* 2 vols., London, 1704-1710, I, p. i; Ephraim Chambers, *Cyclopaedia: or, an Universal Dictionary of Arts and Sciences,* 3 vols., London, 1728, pp. 162-3, 433, 444.

21 Simon Schaffer, 'Newton's comets and the transformation of astrology', in Curry, ed., *Astrology, cit.* (n. 14), pp. 219-44.

22 Newton quoted from Schaffer, *art. cit.,* pp. 242-3.; Lambert quoted from *Id.,* 'Authorized prophets: comets and astronomers after 1759', *Studies in Eighteenth Century Culture,* 15, pp. 45-74.

23 William Whiston, *A New Theory of the Earth,* London, 1696, p. 373.

24 See note 7, above.

25 Bilbo Baggins from J .R .R. Tolkien's *The Hobbit,* speaking of something rather similar to ideas (namely, adventures).

26 Roger Chartier, *Cultural History: Between Practices and Representations,* Cambridge, 1988, pp. 41, 102.

27 James Younge, *Sidrophel Vapulans: or, The Quack Astrologer Toss'd in a Blanket,* London, 1699, Epistle Dedicatory.

28 See Curry, *op. cit.* (n. 1), pp. 158-62; also Michel Vovelle, *Ideologies and Mentalities,* Cambridge, 1990.

29 Ernesto Laclau and Chantal Mouffe, *Hegemony and Socialist Strategy: Towards a Radical Democratic Politics,* London, 1985, p. 85.

30 [Jonathan Swift], *The Accomplishment of the First of Mr. Bickerstaff's Predictions ...,* London, 1708, p. 3.

31 Lawrence Stone's phrase, in 'The revival of narrative', repr. in his *The Past and Present Revisited,* London, 1987, pp. 74-96.

Select bibliography

The following bibliography has been compiled by the contributors to this volume as a general guide to recent writings in their specialist areas. The majority of texts referred to are in English or available in English translation, though a small number of French, Italian and German works which are not have also been included, where they were deemed to be of particular interest. Books and essays cited in the notes to individual contributions are only included in the present bibliography where they are of more general relevance. Where books are recommended in connection with more than one topic, subsequent references are in abbreviated form.

1 General

In addition to the many specialist readings referred to in Roy Porter's *Introduction* to the present volume, which discuss important questions of method or represent significant recent advances in the study of Renaissance culture, the following may help place in context the issues dealt with in this volume.

Peter Burke, *The Renaissance*, London, 1966, is a good recent survey of the period; J. Stephens, *The Italian Renaissance*, London, 1990, though more narrowly focused, is also useful. Fernand Braudel, *Capitalism and Material Life*, London, 1973 (1st French ed. Paris, 1967), though added to and corrected by much subsequent research, remains a fascinating overview of the material conditions of early modern society, and how these were altered by technological innovation, maritime exploration and the growth of trade and manufacturing. H. Kamen, *European Society 1500-1700*, London, 1984, is a good general introduction to the history of the period, while Peter Burke,

Popular Culture in Early Modern Europe, Aldershot, 1978, deals specifically with the 'world-view' of the 'lower orders', and how it related to those of the social and intellectual elites.

The intellectual changes of the period are charted in *The Impact of Humanism on Western Europe*, edited by A. Goodman and A. Mackay, London, 1990; R. Mandrou, *From Humanism to Science*, Harmondsworth, 1978 (1st French ed. Paris, 1973); C. S. Singleton, *Art, Science and History in the Renaissance*, Baltimore, 1968; A. Hall, *Revolution in Science. 1500-1750*, London, 1983. Michel Foucault, *The Order of Things*, London, 1970 (1st French ed., *Les Mots et les Choses*, Paris, 1966), has revolutionized our understanding of 'pre-scientific' attitudes to describing and classifying the natural world, and challenged our own assumptions concerning the 'truth' of scientific mentalities.

How social and intellectual transformations alike were promoted by the new technology of printing is the subject of Elizabeth L. Eisenstein, *The Printing Press as an Agent of Change*, 2 vols., Cambridge, 1979; Paolo Rossi, *Philosophy, Technology and the Arts in the Early Modern Era*, trans. S. Attanasio, London, 1970, is an invaluable survey of technological developments and the changing relationship between intellectual and craft practices. Artistic change, in its social and intellectual context, is the subject of Michael Levey, *Early Renaissance*, Harmondsworth 1967, and *High Renaissance*, Harmondsworth, 1975; J. Shearman, *Mannerism*, Harmondsworth, 1967; J. R. Martin, *Baroque*, London, 1989. Despite being a specialist monograph, M. Kemp, *Leonardo da Vinci*, London, 1981, has a great deal to say of a more general nature concenring the fine arts' relationship with traditional crafts, new technologies and natural philosophy.

2   The new learning

No general survey is currently available spanning all the issues discussed in Chapter 2. By way of introduction, a well-informed collection of essays dealing with a number of significant questions is *The Renaissance. Essays in Interpretation*, edited by André Chastel *et al.*, London and New York, 1982. Greater detail is to be found in the work of Charles B. Schmitt, particularly two recent collections of his essays, *Studies in Renaissance Philosophy and Science*, London, 1981, and *The Aristotelian Tradition and Renaissance Universities*, London, 1984, as well as the splendidly concise *Aristotle and the Renaissance*, Cambridge, Mass., and London, 1983. To these one must add the collective work edited by Schmitt (with Quentin Skinner), *The Cambridge History of Renaissance Philosophy*, Cambridge, 1988, which brings together a number of excellent specialist readings on subjects ranging from formal logic to the availability of classical texts. A stimulating contrast to this analytical approach is to be found in the synthesis offered by M.Foucault, *The Order of Things, cit.*

Concerning the history of the universities, in the absence of any general study, one must rely on works dealing with individual countries or institutions: *The History of the University of Oxford*, vol. III, edited by James McConica, *The Collegiate University*, Oxford, 1986; *A History of the University of Cambridge*, vol. I, Damian Riehl Leader, *The University to 1546*, Cambridge 1988. Still on the two English universities, Mark H. Curtis, *Oxford and Cambridge in Transition 1558-1642. An Essay on Changing Relations between the English Universities and English Society*, Oxford, 1959. For France, see chs. 3-6 of the *Histoire des universités en France*, edited by Jacques Verger, Toulouse, 1986; L. W. B. Brockliss, *French Higher Education in the 17th and 18th Centuries. A Cultural History*, Oxford, 1987. For a general European perspective, see *Les Universités européennes du XVIe au XVIIIe siècle. Histoire sociale des populations étudiantes*, edited by Dominique Julia and Jacques Revels, 2 vols., Paris, 1986-89. Two specialist periodicals may be usefully consulted, *History of Universities*, Oxford, 1981-, and *Quaderni dell'Università di Padova*, Padua, 1968-, which centres on, but is not restricted to, that institution. The best work on the new colleges remains Gabriel Codina Mir, *Aux Sources de la pédagogie des jésuites: le 'modus parisiensis'*, Rome, 1968.

The profound transformations brought about by the invention of printing are discussed in E. L. Eisenstein, *The Printing Press, cit.* (though the work is not immune from criticism). To this might be added the excellent, and splendidly illustrated, *Histoire de l'édition française*, vol. I, edited by Roger Chartier and Henri-Jean Martin, Paris, 1982. Also by Chartier are *Cultural History: Between Practices and Representations*, Oxford, 1988, and, of particular importance, *The Cultural Uses of Print in Early Modern France*, Princeton, 1988. An example of 'quality' publishing with great intellectual resonance is discussed in Martin Lowry, *The World of Aldus Manutius. Business and Scholarship in Renaissance Venice*, Oxford, 1979. For Ramus and his disciples, see the exemplary study by Walter J. Ong, *Ramus, Method and the Decay of*

Dialogue, 2nd ed., Cambridge, Mass., 1983. The use of illustrations in connection with the various sciences is not the subject of a comprehensive survey, but the following may usefully be consulted: S. K. Heninger Jr, *The Cosmographical Glass. Renaissance Diagrams of the Universe*, San Marino, Cal., 1977, and William B. Ashworth Jr, 'Iconography of a new Physics', *History and Technology*, 4, 1987, pp. 267-97. Similarly, there is no satisfactory single work dealing with the ways in which the status and form of vernacular languages alters in the Renaissance. See, for English, Richard Foster Jones, *The Triumph of the English Language. A Survey of Opinions Concerning the Vernacular from the Introduction of Printing to the Restauration*, Stanford, Cal., 1953; for French, Peter Rickard, *La Langue française au seizième siècle. Étude suivie de textes*, Cambridge, 1968 (with a notable introduction); for Italian, Armand L. De Gaetano, *Giambattista Gelli and the Florentine Academy. The Rebellion Against Latin*, Florence, 1976.

An excellent introduction to the medieval philosophical tradition, its forms and styles, is John Marenbon, *Later Medieval Philosophy (1150-1350). An Introduction*, London and New York, 1987, chs 1-2. Concerning the changes in philosophical practice in the Renaissance, see *The Cambridge History*, cit., and particularly the contributions by Brian P. Copenhaver, 'Translation, terminology and style in philosophical discourse', pp. 77-110; E. J. Ashworth, 'Traditional logic', pp. 143-72; Lisa Jardine, 'Humanist logic', pp. 173-98; Charles B. Schmitt, 'The Rise of the Philosophical Textbook', pp. 792-804. The importance of dialogue in Italy is discussed by David Marsh, *The Quattrocento Dialogue: Classical Tradition and Humanist Innovation*, Cambridge, Mass., 1980, and the more recent Jon R. Snyder, *Writing the Scene of Speaking. Theories of Dialogue in the Late Italian Renaissance*, Stanford, Cal., 1989 (but what is said of Italy also applies to the rest of Europe).

The role of the Academies was perhaps first underlined by Frances A. Yates, *The French Academies of the 16th Century*, London, 1947. In addition to De Gaetano's study of the Accademia Fiorentina referred to in the previous paragraph, John F. D'Amico, *Renaissance Humanism in Papal Rome. Humanists and Churchmen on the Eve of the Reformation*, Baltimore and London, 1985 (and particularly ch. 4) also deals with the Italian academic milieu. *Marsilio Ficino e il ritorno di Platone*, edited by Gian Carlo Garfagnini, 2 vols., Florence, 1986, recapitulates several decades of research, and opens with a long essay (in English) by Paul Oscar Kristeller. Arthur Field, *The Origins of the Platonic Academy of Florence*, Princeton, 1988, may also be usefully consulted.

Finally, on the subject of the 'longevity' of Aristotle, the writings of C. B. Schmitt referred to at the beginning of this section are indispensable, and to them may be added *L'Aristotelisme au XVIe siècle*, special issue of *Les Études Philosophiques*, 41, 1986, n. 3, edited by Luce Giard.

3   Natural philosophy and the science of history

An invaluable starting-point for further study is *The Cambridge History*, cit., and especially the chapters by Donald R. Kelley, 'The theory of history', pp. 746-61; William A. Wallace, 'Traditional natural philosophy', pp. 201-35; Alfonso Ingegno, 'The new philosophy of nature', pp. 236-63; Richard H. Popkin, 'Theories of knowledge', pp. 668-84; Cesare Vasoli, 'The Renaissance concept of philosophy', pp. 57-74; and B. Copenhaver, cit. No less useful is the excellent reappraisal by J. Schuster, 'The Scientific Revolution', in the *Companion to the History of Modern Science*, edited by R. C. Olby, G. N. Cantor, J. R. R. Christie and M. S. J. Hodge, London, 1990. Edward Grant, *Physical Science in the Middle Ages*, Cambridge, 1977, is a starting-point for scholastic Aristotelianism, and late Renaissance natural philosophy is sympathetically described in John L. Heilbron, *Elements of Early Modern Physics*, Berkeley, Cal., 1982. The full range of Renaissance Aristotelianism is portrayed in C. B. Schmitt, *Aristotle in the Renaissance*, cit.

What has been described in this volume as 'reactionary' Neoplatonism and Hermeticism has been treated by D. P. Walker, *The Ancient Theology. Studies in Christian Platonism from the Fifteenth to the Eighteenth Century*, London, 1972. Less cautious claims about the Hermetic tradition were made by Frances A. Yates, *Giordano Bruno and the Hermetic Tradition*, London, 1964, but see also her more synthetic contribution in the interdisciplinary *Art, Science and History in the Renaissance*, edited by C. S. Singleton, Baltimore, 1967, which also contains a brief survey of Renaissance historiography by F. Gilbert. The role of Hermeticism in the development of natural philosophy was questioned in R. S. Westman and J. E. McGuire, *Hermeticism and the Scientific Revolution*, Berkeley, Cal., 1977.

A. Chalmers, *What Is This Thing Called Science?*, Milton Keynes, 1978, contains a simple introduction to the problems of scientific empiricism. The historical problem of making experimentalism persuasive is the subject of S. Shapin and S. Schaffer, *Leviathan and the Air*

Pump. Hobbes, Boyle and the Experimental Life, Princeton, 1985.

The ideological function of traditions is discussed by J. G. A. Pocock, 'Time, Institutions and Action; An Essay on Traditions and their Understanding' in his *Politics, Language and Time: Essays on Political Thought and History*, London, 1971, pp. 232-72. On Renaissance historiography and senses of the past, Peter Burke, *The Renaissance Sense of the Past*, London, 1969, remains a clear, authoritative introduction with copious illustrations from primary sources. D. R. Kelley, *Foundations of Modern Historical Scholarship: Language, Law and History in the French Renaissance*, New York, 1970, deals thoroughly with the scholarly techniques of late Renaissance historicism. He places French historiography in a broader context in his already cited contribution to *The Cambridge History*. Similar ground – but with a wider focus – is covered by George Huppert, *The Idea of Perfect History: Historical Erudition and Historical Philosophy in Renaissance France*, Chicago, 1970. Both these scholars were developing J. G. A. Pocock, *The Ancient Constitution*, 1st ed., Cambridge, 1957. F. Smith Fussner, *The Historical Revolution: English Historical Writing and Thought 1580-1640*, London, 1962, is informative but Whiggish. Out of many overlapping treatments, E. Cochrane, *Historians and Historiography in the Italian Renaissance*, Chicago, 1981, is an extensive survey. Renaissance senses of the past are compared with other periods in Ernst Breisach, *Historiography, Ancient, Medieval, Modern*, Chicago, 1983, and G. W. Trompf, *The Idea of Historical Recurrence in Western Thought: from Antiquity to the Reformation*, Berkeley, Cal., 1979. For senses of past ideas there is Lucien Braun, *Histoire de l'Histoire de la Philosophie*, Paris, 1973, and the explanatory context of history writing is reviewed by D. E. Brown, *Hierarchy, History and Human Nature: The Social Origins of Historical Consciousness*, Tucson, Ariz., 1988, which contains a useful survey of secondary literature. Relevant articles by Kinser, Huppert, Kelley and Gundesheimer appear in *Renaissance Studies in Honour of Hans Baron*, edited by Anthony Molho and John A. Tedeschi, Florence, 1971.

Huppert and Kelley (*op. cit.*) both deal with individual French legal humanists. In addition the best English introduction to Jean Bodin is still J. Franklin, *Jean Bodin and the Sixteenth-Century Revolution in the Methodology of Law and History*, London, 1963; Bodin's *Methodus* (1566) has been translated as Jean Bodin, *Method for the Easy Comprehension of History*, translated by B. Reynolds, New York, 1945; Louis Le Roy, *La Vicissitude ou variété des choses de l'univers*, Paris, 1584, appeared in English as Loys le Roy called Regius, *Of the Interchangeable Course, or Variety of Things in the Whole World*, translated by R. Ashley, London, 1594; he is discussed in Werner L. Gundesheimer, *The Life and Works of Louis le Roy*, Geneva, 1966.

The relationship between new history and new science in the Renaissance has been little explored. For Francis Bacon dubious methodological links were raised by Fussner (*op. cit.*) and developed by G. Wylie Sypher, 'Similarities between the Scientific and the Historical Revolutions at the end of the Renaissance', *Journal for the History of Ideas*, 26, 1965, pp. 353-68. A deconstructive role for history in Bacon was identified by Paolo Rossi, *Francis Bacon: From Magic to Science*, London, 1968, ch. 3, and a similarity of outlook between Bacon and Gilbert alluded to in R. F. Jones, *Ancients and Moderns*, revised ed., St Louis, 1961. Recently 'Historiography and validation' in Renaissance astronomy has been discussed in N. Jardine, *The Birth of History and Philosophy of Science: Kepler's A Defence of Tycho against Ursus with essays on its provenance and significance*, Cambridge, 1984, and interconnections in the work of Gassendi receive detailed treatment in L. Sumida Joy, *Gassendi the Atomist: Advocate of history in an age of science*, Cambridge, 1987, Gilbert's historiography has yet to receive attention in print.

4   Natural philosophy and rhetoric

There is no substitute for reading the major texts of classical rhetoric. In addition to the *Topics* and *Rhetoric*, in volumes I and XI respectively of *The Works of Aristotle translated into English*, edited by D. W. Ross, Oxford, 1928, Plato's dialogues, and in particular *Gorgias* and *Protagoras*, may usefully be studied (a number of translations are available). For the Roman rhetoricians, see the translations in the Loeb Classical Library: Cicero, *De Oratore*, trans. E. W. Sutton and H. Rackham, 2 vols., London, 1942-48, *Brutus and Orator*, trans. G. L. Hendrickson and H. M. Hubbell, rev. ed., London, 1962; Quintilian, *Institutio Oratoria*, trans. H. E. Butler, 4 vols., London 1920-22.

An excellent recent overview of the whole subject of rhetoric is Brian Vickers, *In Defence of Rhetoric*, Oxford, 1988, which also contains in appendix a useful glossary of terms. On classical rhetoric the most comprehensive modern studies are those by George Kennedy, *The Art of Persuasion in Greece*, London, 1963, and *The Art of Persuasion in the Roman World*, Princeton, NJ, 1972. Roland Barthes, 'L'Ancienne

Rhétorique: Aide-mémoire', *Communications*, 16, 1966, pp. 172-229 is brief and to the point. There are no equivalent comprehensive studies of Renaissance rhetoric, but a good summary is again provided by Brian Vickers, 'Rhetoric and poetics', in *The Cambridge History*, cit.; in the same volume, Lisa Jardine, 'Humanist logic' is also relevant. For the crucial role which rhetoric played in Renaissance education see Anthony Grafton and Lisa Jardine, *From Humanism to the Humanities: Education and the Liberal Arts in Fifteenth- and Sixteenth-Century Europe*, Cambridge, Mass., 1986. There is no really satisfactory study of the attempt by late Renaissance and Baroque rhetoricians to describe and redefine the heuristic function of metaphor, but on the chief of them, Emanuele Tesauro, see M. Zanardi, 'La metafora e la sua dinamica di significazione nel "Cannocchiale Aristotelico" di Emanuele Tesauro', *Giornale Storico della Letteratura Italiana*, 157, pp. 321-68, and Giuseppe Conte, *La metafora barocca*, Milan, 1975.

On the rhetorical-mnemonic aspects of encyclopaedism, the major text in English is, of course, Frances Yates, *The Art of Memory*, London, 1966, though the earlier Paolo Rossi, *Clavis Universalis. Arti mnemoniche e logica combinatoria da Lullo a Leibniz*, Milan–Naples, 1960, is much to be preferred, for its greater clarity and balance (an English translation is forthcoming from Athlone Press). On Giulio Camillo, by far the best work is that by Lina Bolzoni, *Il teatro della memoria: studi su Giulio Camillo*, Padua, 1984, whose *L'universo dei poemi possibili: studi su Francesco Patrizi da Cherso*, Rome, 1980, is also of interest.

Historians and philosophers of science are only gradually becoming aware of the role of the 'arts of discourse', and rhetoric in particular, in the development of scientific method. L. Jardine, *Francis Bacon. Discovery and the Art of Discourse*, Cambridge, 1974, remains fundamental (and opens with a lucid account of how dialectic was taught in the Renaissance). M. A. Finocchiaro, *Galileo and the Art of Reasoning*, Dordrecht, Boston and London, 1980, is a massive study of the forms of argumentation in the *Dialogue Concerning the Two Chief Systems of the World*. Despite its sub-title, 'Rhetorical Foundations of Logic and Scientific Method', it does not involve a systematic comparison with rhetorical theory and practice, but nevertheless uncovers a great deal of important material. Also on Galileo's rhetoric, Andrea Battistini, 'Scienza e Retorica. L'esempio di Galileo', in *Come si legge un testo*, edited by M. L. Altieri Biagi, Milan, 1989, pp. 79-100; see also, in the same volume, Bruno Basile, 'Argomentazione e Scienza:

due esempi secenteschi', pp. 103-11, which examines the rhetoric of two short excerpts from the Jesuit Daniello Bartoli and the 'Galilean' Francesco Redi. Other interesting material may be found in P. France, *Rhetoric and Truth in France. Descartes to Diderot*, Oxford, 1972. The general question of rhetoric's relation to science, both in the making of the 'scientific revolution' and since, is now discussed in *The Politics and Rhetoric of Scientific Method*, edited by John A. Schuster and Richard R. Yeo, Dordrecht–Boston–Lancaster–Tokyo, 1986, which includes an important contribution on Descartes by Schuster, 'Cartesian Method and Mythic Speech: A Diachronic and Structural Analysis', pp. 33-95. Of more general interest are Richard Boyd, 'Metaphor and Theory Change: What is "Metaphor" a Metaphor for?', and Thomas S. Kuhn, 'Metaphor in Science', both in *Metaphor and Thought*, edited by Andrew Ortony, Cambridge, 1979, pp. 356-419. These do not deal specifically with our period, but attempt to discuss the status and functions of metaphor in modern scientific discourse.

5   Natural philosophy and its public concerns

An adequate general survey of the social history of the sciences throughout this period remains to be written. In its absence, several texts provide excellent introductory orientations to various elements of such a social history: P. Rossi, *Philosophy, Technology and the Arts*, cit.; T. K. Rabb, *The Struggle for Stability in Early Modern Europe*, Oxford, 1975; R. S. Westman, 'The Astronomer's role in the sixteenth century: a preliminary study', *History of Science*, 18, 1980, pp. 105-47; Charles Webster, *From Paracelsus To Newton: Magic and the Making of Modern Science*, Cambridge, 1982; Michael A. R. Graves and Robin H. Silcox, *Revolution, Reaction and The Triumph of Conservatism: English History, 1558-1700*, Auckland, 1984, a useful general survey of English political and religious history; J. A. Bennett, 'The Mechanics' Philosophy and the Mechanical philosophy', *History of Science*, 24, 1986, pp. 1-28; Anthony Grafton and Lisa Jardine, *From Humanism to the Humanities*, cit.; Mario Biagioli, 'The Social Status of Italian Mathematicians, 1450-1600', *History of Science*, 27, 1989, pp. 41-95.

On Francis Bacon, the following may be usefully consulted: Paolo Rossi, *Francis Bacon. From Magic To Science*, translated by S. Rabinovitch, Chicago, 1968; Lisa Jardine, *Francis Bacon: Discovery and the Art of Discourse*, cit.; Julian Martin, *Francis Bacon: The State, the Common Law and the Reform of Natural Philosophy*, Cambridge, forthcoming.

Seventeenth-century natural philosophy is viewed in its social context by Charles Webster, *The Great Instauration: Science, Medicine and Reform, 1626-1660*, London, 1975; Robert G. Frank, Jr, *Harvey and the Oxford Physiologists: A Study of Scientific Ideas and Social Interaction*, Berkeley, 1980; Paul Wood, 'Methodology and apologetics: Thomas Sprat's "History of the Royal Society"', *British Journal for the History of Science*, 13, 1980, pp. 1-26; Michael Hunter, *Science and Society in Restoration England*, Cambridge, 1981; Steven Shapin, 'Of Gods and Kings: Natural Philosophy and politics in the Leibniz-Clark Disputes', *Isis*, 72, 1981, pp. 187-215; S. Shapin and S. Schaffer, *Leviathan and the Air-Pump*, cit.; L. W. B. Brockliss, *French Higher Education*, cit.

6   The role of the church

Various ecclesiastical aspects of the Church's preoccupation with the 'New Science' (as distinct from specifically theological issues) are treated by William B. Ashworth, Jr, 'Catholicism and Early Modern Science,' in *God and Nature: Historical Essays on the Encounter Between Christianity and Science*, edited by David C. Lindberg and Ronald L. Numbers, Berkeley, Los Angeles and London, 1986, pp. 136-66. Properly theological as opposed to ecclesiastical issues may be approached through Reijer Hooykaas, *Religion and the Rise of Modern Science*, Grand Rapids, Mich., 1972; Eugene M. Klaaren, *Religious Origins of Modern Science: Belief in Creation in Seventeenth-Century Thought*, Grand Rapids, Mich., 1977; Francis Oakley, *Omnipotence, Covenant and Order: An Excursion in the History of Ideas from Abelard to Leibniz*, Ithaca, NY, 1984; Amos Funkenstein, *Theology and the Scientific Imagination from the Middle Ages to the Seventeenth Century*, Princeton, NJ, 1986. The latter, together with Edward Grant, 'Science and Theology in the Middle Ages', in *God and Nature*, cit., pp. 49-75, provides a useful treatment of later medieval theology and philosophy.

On medieval cosmology, Edward Grant, 'Cosmology', in *Science in the Middle Ages*, edited by David C. Lindberg, Chicago, 1978, pp. 265-302, is a good overview, as is Thomas S. Kuhn, *The Copernican Revolution*, Cambridge, Mass., 1957, which also provides a valuable introduction to Copernicus. The issue of censorship and the Index is discussed in Paul F. Grendler, 'Printing and Censorship', in *The Cambridge History*, cit., pp. 25-53. On implications of the Council of Trent's actions for natural philosophy, see Robert S. Westman, 'The Copernicans and the Churches', in *God and Nature*, cit., pp. 75-113, and

Rivka Feldhay, 'Knowledge and Salvation in Jesuit Culture', *Science in Context*, 1, 1987, pp. 195-213; Westman's essay also considers Melanchthon's Lutheran reforms, as do his 'The Melanchthon Circle, Rheticus, and the Wittenberg Interpretation of the Copernican Theory', *Isis*, 66, 1975, pp. 165-93; Hans Blumenberg, *The Genesis of the Copernican World*, Cambridge, Mass., and London, 1987 (especially Part III, chap. 3); Bruce T. Moran, 'The Universe of Philip Melanchthon: Criticism and Use of the Copernican Theory', *Comitatus* 4, 1973, pp. 1-23. Edward Rosen, 'Kepler and the Lutheran Attitude toward Copernicanism', *Vistas in Astronomy*, 18, 1975, pp. 317-37, looks especially at the censorship of Kepler's *Mysterium cosmographicum*, including the excision of the material on scriptural exegesis that later appeared in the *Astronomia nova*. The heresies of Servetus and Bruno are discussed in C. D. O'Malley, *Michael Servetus: Translation of his Geographical, Medical and Astrological Writings*, Philadelphia, 1953, and E. A. Gosselin and L. A. Lerner, 'Galileo and the Long Shadow of Bruno', *Archives internationales d'histoire des sciences*, 25, 1975, pp. 223-46, which looks at the aftershocks of the Bruno affair. The Counter-Reformation may be approached through Eugene F. Rice, Jr, *The Foundations of Early Modern Europe, 1460-1559*, New York, 1970, ch. 5, and for greater depth, Heiko Oberman, *Masters of the Reformation: The Emergence of a New Intellectual Climate in Europe*, Cambridge, 1981; the importance of the Council of Trent's rulings on Scriptural exegesis for the Galileo affair is well considered in William R. Shea, 'Galileo and the Church', in *God and Nature*, cit., pp. 114-35, and Olaf Pedersen, 'Galileo and the Council of Trent: The Galileo Affair Revisited', *Journal for the History of Astronomy*, 14, 1983, pp. 1-29. Other important treatments of Galileo's troubles are Giorgio de Santillana's still indispensable *The Crime of Galileo*, Chicago, 1955; Jerome J. Langford, *Galileo, Science, and the Church*, rev. ed., Ann Arbor, 1971; Richard S. Westfall, 'The Trial of Galileo: Bellarmino, Galileo, and the Clash of Two Worlds', *Journal for the History of Astronomy*, 20, 1989, pp. 1-23. Maurice A. Finocchiaro, *The Galileo Affair: A Documentary History*, Berkeley, Los Angeles and London, 1989, is a comprehensive presentation of relevant documents in translation together with narrative linkage and notes. Another useful source of Galileo translations with commentary and narrative is Stillman Drake, *Discoveries and Opinions of Galileo*, Garden City, NY, 1957. On the fate of Copernicanism among Catholics after 1633, see Christine Jones Schofield, *Tychonic and Semi-Tychonic World Systems*, New York, 1981; John L. Russell,

'Catholic Astronomers and the Copernican System after the Condemnation of Galileo', *Annals of Science*, 46, 1989, pp. 365-86. On atomism and the Eucharist see, for background, Marie Boas Hall 'The Establishment of the Mechanical Philosophy', *Osiris*, 10, 1952, pp. 412-541, and Edith Sylla, 'Autonomous and Handmaiden Science: St. Thomas Aquinas and William of Ockham on the Physics of the Eucharist', in *The Cultural Context of Medieval Learning*, edited by John E. Murdoch and Edith D. Sylla, *Boston Studies in the Philosophy of Science*, 26, Dordrecht, 1975, pp. 361-72; for detail on the controversies, Richard A. Watson, 'Transubstantiation among the Cartesians' and Ronald Laymon, 'Transubstantiation: Test Case for Descartes's Theory of Space', both in *Problems of Cartesianism*, edited by Thomas M. Lennon, John M. Nicholas and John W. Davis, Kingston and Montreal, 1982, pp. 127-48, 149-70; also Pietro Redondi's controversial *Galileo: Heretic*, Princeton, 1987, esp. ch. 9. Redondi argues that Galileo's atomism, and the implied challenge to the doctrine of transubstantiation, were the real issues behind Galileo's trial, the eventual charge of Copernicanism being just a smokescreen; for convincing rebuttals of his claims, see Vincenzo Ferrone and Massimo Firpo, 'From Inquisitors to Microhistorians: A Critique of Pietro Redondi's *Galileo eretico*', *Journal of Modern History*, 58, 1986, pp. 485-524, and Richard S. Westfall, 'Galileo Heretic: Problems, as They Appear to Me, with Redondi's Thesis', *History of Science*, 26, 1988, pp. 399-415. On the nature of Jesuit education, natural philosophy and mathematical sciences, see, in addition to Feldhay, 'Knowledge and Salvation', cit., John L. Heilbron, *Electricity in the 17th and 18th Centuries: A Study of Early Modern Physics*, Berkeley, Los Angeles and London, 1979, relevant parts of which are reprinted in the same author's *Elements of Early Modern Physics*, Berkeley, Los Angeles and London, 1982, and Peter Dear, 'Jesuit Mathematical Science and the Reconstitution of Experience in the Early Seventeenth Century', *Studies in History and Philosophy of Science*, 18, 1987, pp. 133-75. W. B. Ashworth, 'Catholicism and Early Modern Science', cit., considers general priestly involvement in science. Catholic reactions against the new philosophical departures in the second half of the seventeenth century in France are discussed in Trevor McClaughlin, 'Censorship and Defenders of the Cartesian Faith in Mid-Seventeenth Century France', *Journal of the History of Ideas*, 40, 1979, pp. 563-81, while the role of Puritanism in mid-seventeenth-century England is considered by Charles Webster, 'Puritanism, Separatism, and Science', in *God and*

Nature, cit., pp. 192-217.

7 The dissemination of knowledge

For the problems of reconstructing the beliefs of popular culture see the following works by C. Ginsburg, *The Cheese and the Worms, the Cosmos of a Sixteenth-Century Miller*, London, 1980; *The Night Battles*, London, 1983; see also P. Burke, *Popular Culture*, cit., and by the same author *The Historical Anthropology of Early Modern Italy*, Cambridge, 1989, contains an important collection of essays on perception and communication.

On prophecy see *Magia, Astrologia e Religione nel Rinascimento*, Warsaw, 1974; *Astrologi hallucinati*, edited by P. Zambelli, Berlin and New York, 1986; *Scienze, credenze occulte, livelli di cultura*, Florence 1982.

For censorship see J. Hilgers, 'Bücherverbot und Bucherzenzur des sechzehnten Jahrhunderts in Italien', *Zentralblatt für Bibliothekswesen*, 28, 1911, pp. 108-22; idem, *Der Index der Verboten Bücher*, Freiburg, 1905; F. H. Reusch, *Die Index Librorum Prohibitorum des Sechzehnten Jahrhunderts*, Tübingen, 1886; idem, *Der Index der Verboten Bücher: Ein Beitrag zur Kirchen-und literaturgeschichte*, 2 vols., Bonn, 1883-85. P. F. Grendler, *The Roman Inquisition and the Venetian Press*, Princeton, NJ, 1977; *Le Pouvoir et la plume. Incitation, contrôle et repression dans l'Italie du XVIe siècle*, Paris, 1982. For France see F. M. Higman, *Censorship and the Sorbonne*, Geneva, 1979. Censorship in England is surveyed in F. S. Siebert, *Freedom of the Press in England 1476-1776*, Urbana, 1965. Early decrees are examined by R. Hirsch, 'Pre-Reformation censorship of printed books', *The Library Chronicle*, vol. XXI, 1955, pp. 100-5. For the Protestant reaction see G. Bonnant, 'Les Index prohibitifs et expurgatoires contrefaits par des protestants au XVIe et au XVIIe', *Bibliothèque d'Humanisme et Renaissance*, 31, 1969, pp. 611-40. The European sixteenth-century Indices are currently being published by Librairie Droz, Geneva. The volumes for Paris, Louvain, Venice, Spain and Antwerp have already appeared.

On Neo-Platonism, Hermeticism, magic, and talismans see F. Yates, *Giordano Bruno and the Hermetic Tradition*, London, 1964; D. P. Walker, *Spiritual and Demonic Magic from Ficino to Campanella*, London, 1958; On the Yates thesis see R. S. Westman, J. E. McGuire, *Hermeticism and the Scientific Revolution*, Los Angeles, 1977. Illustrations of magical objects are reproduced in the excellent catalogue *La*

Corte il Mare i Mercanti, La Rinascita della Scienza, Editoria e Societa, Astrologia, Magia e Alchimia, Florence, 1980; a stimulating account of less well known aspects of Renaissance culture is E. Battisti, *L'Antirinascimento,* Milan, 1989.

The structure of patronage is dealt with in *Patronage, Art and Society in Renaissance Italy,* edited by F .W. Kent, P. Simons with J. C. Eade, Oxford, 1897. On the European Courts see S. Bertelli, *Italian Renaissance Courts,* London, 1986; *The Courts of Europe,* edited by A. G. Dickens, London, 1977. A good review of the debate concerning the interpretation of the Court is P. Merlin, 'Il tema della Corte nella storiografia italiana ed europea', *Studi Storici,* 1986, pp. 203-44.

For the theatre see R. Strong, *Art and Power,* Woodbridge, 1984; S. Orgel, *The Illusion of Power, Political Theatre in the English Renaissance,* Berkeley, 1975, and the catalogue *Il Luogo Teatrale a Firenze,* Milan, 1975.

On astrology and art see the excellent studies by J. Cox-Rearick, *Dynasty and Destiny in Medici Art,* Princeton, 1984, and C. Rousseau, *Cosimo I de' Medici, Astrology and the Symbolism of Prophecy,* PhD thesis, Ann Arbor, Mich., 1933. The astrological interests of the Medici in the seventeenth century are treated by M. Campbell, *Pietro Cortona at the Pitti Palace,* Princeton, 1977; for Florence in this period see E. Cochrane, *Florence in the Forgotten Centuries,* Chicago and London, 1973.

Medici interests in the natural world and alchemy are examined by M. Dezzi Bardeschi, *Lo Stanzino del Principe in Palazzo Vecchio,* Florence, 1980; G. Lensi Orlandi, *Cosimo e Francesco de' Medici Alchimisti,* Florence, 1978; P. Galluzzi, 'Il mecenatismo mediceo e le scienze', in *Idee Istituzioni Scienza ed Arti nella Firenze dei Medici,* cit., pp. 189-215. On Italian collections see the excellent survey in *La Scienza a Corte. Collezionismo eclettico, natura e immagine a Mantova fra Rinascimento e Manierismo,* Rome, 1979.

The exploration of the natural world is the subject of *Science and the Arts in the Renaissance,* edited by J. W. Shirley and F. David Hoeniger, London, 1985; W. Eamon, 'Arcana disclosed: the advent of printing, the books of secrets tradition and the development of experimental science in the sixteenth century', *History of Science,* 22, 1984, pp. 111-50; W. George, 'Sources and background to discoveries of new animals in the sixteenth and seventeenth centuries', *History of Science,* 18, 1980, pp. 79-104; D. N. Livingstone 'Science, Magic and Religion: a contextual reassessment of Geography in the sixteenth and seventeenth centuries', *History of Science*, 26, 1988, pp. 269-94.

For Piccolomini see M. Celse-Blanc, 'Alessandro Piccolomini disciple d'Aristote, ou les detours de la reécriture', in *Scitture di Scritture*, edited by G. Mazzacurati, Rome, 1987, pp. 109-46; on Della Porta see L. Muraro, *Giambattista della Porta mago e scienziato*, Milan, 1978. For Kircher see J. Godwin, *Athanasius Kircher*, London,1979; on the influence of Hermeticism in the seventeenth century see R. Taylor, 'Hermeticism and mystical architecture in the Society of Jesus', in *Baroque Art: The Jesuit Contribution*, edited by R. Wittkower and I. B. Jaffe, New York, 1972, pp. 63-97; R. J. Van Pelt, 'The utopian exit of the Hermetic temple', in *Hermeticism and the Renaissance: Intellectual History and the Occult in Early Modern Europe*, edited by I. Merkel and A. G. Debus, London and Washington, 1988, cit., pp. 400-23.

8   Artisans, instruments and practical mathematics

On the development of instrumentation in the early modern period, see J. A. Bennett, *The Divided Circle: a History of Instruments for Astronomy, Navigation and Surveying*, Oxford, 1987; Penelope Gouk, *The Ivory Sundials of Nuremberg, 1500-1700*, Cambridge, 1988; A. Turner, *Early Scientific Instruments, Europe 1400-1800*, London, 1987.

On exploration and navigation, see two books by J. H. Parry, *The Age of Reconnaissance*, Berkeley, Los Angeles and London, 1981, and *The Discovery of the Sea*, San Diego, 1981. On navigational technique, E. G. R. Taylor, *The Haven-Finding Art*, London, 1971, is still useful, and D. W. Waters, *The Art of Navigation in England in Elizabethan and Early Stuart Times*, London, 1978, is not so restricted to England as its title suggests.

The modern literature on the history of surveying is very unsatisfactory, but there are valuable contributions in *English Map-making 1500-1650*, edited by S. Tyacke, London, 1983. The literature on the history of cartography, by comparison, is vast; a valuable general account is L. Barlow, *History of Cartography*, edited by R. A. Skelton, London, 1964.

There is little to recommend on the influence of practical mathematics because, as explained in the essay above, the thesis has not been popular, but see J. A. Bennett, 'The mechanics' philosophy', *History of Science*, 24, 1986, pp. 1-28, and P. Rossi, *Philosophy and the Arts*, cit.

9   Medicine and popular healing

There are a number of standard histories of medicine, of which one of the most convenient, though rather brief, is E. H. Ackerknecht, *A Short History of Medicine*, rev. ed., Baltimore and London, 1982. Equally brief but still extremely useful is Roy Porter, *Disease, Medicine and Society in England 1550-1860*, London, 1987. The notion of the 'medical marketplace' derives from two studies of eighteenth-century medicine by N. Jewson, 'Medical Knowledge and the Patronage System in Eighteenth-Century England', *Sociology*, 8, 1974, pp. 369-85, and 'The Disappearance of the Sick Man from Medical Cosmology 1770-1870', *Sociology*, 10, 1976, pp. 225-44. This useful notion has been extended back to the seventeenth century in Harold J. Cook, *The Decline of the Old Medical Regime in Stuart London*, Ithaca and London, 1986, pp. 28-69, and Lucinda Beier, *Sufferers and Healers: The Experience of Illness in Seventeenth-Century England*, London, 1987, pp. 8-50. These both provide a good survey of the variety of options open to one in need of healing in the late Renaissance period. In addition, Cook's book provides an excellent account of the licensing activities of the Royal College of Physicians of London. For a superb account of the reality behind the rhetoric of the threefold partition of medicine see Margaret Pelling and Charles Webster, 'Medical Practitioners', in *Health, Medicine and Mortality in the Sixteenth Century*, edited by Charles Webster, Cambridge, 1979, pp. 165-236.

Little work has been done on the popular knowledge of medical theories and practices. On Shakespeare's knowledge of medicine see R. R. Simpson, *Shakespeare and Medicine*, Edinburgh and London, 1959, and F. N. L. Poynter, 'Medicine and Public Health', in *Shakespeare in His Own Age*, edited by Allardyce Nicoll, Cambridge, 1976, pp. 152-66. For more general treatments see Beier's book cited above and (although it is mostly concerned with eighteenth-century issues) *Patients and Practitioners: Lay Perceptions of Medicine in Pre-Industrial Society*, edited by Roy Porter, Cambridge, 1985. For a cautious but nonetheless informative account of the popular medical literature in the Renaissance see Paul Slack, 'Mirrors of Health and Treasures of Poor Men: The Uses of the Vernacular Medical Literature of Tudor England', in *Health, Medicine and Mortality*, cit., pp. 237-73.

The role of astrology in medicine is discussed in Allan Chapman, 'Astrological Medicine', in *Health, Medicine and Mortality*, cit., pp. 275-300. For further details on the notion of the degree of a drug see

Michael R. McVaugh, 'The Two Faces of a Medical Career: Jordanus de Turre of Montpellier', in *Mathematics and Its Application to Science and Natural Philosophy in the Middle Ages*, edited by Edward Grant and John E. Murdoch, Cambridge, 1987, pp. 301-24.

The social and intellectual processes which led to the development of iatromechanism are superbly outlined, at least with regard to England, in Robert G. Frank, *Harvey and the Oxford Physiologists: Scientific Ideas and Social Interaction*, Berkeley, Los Angeles and London, 1980, and Theodore M. Brown, *The Mechanical Philosophy and the 'Animal Oeconomy': A Study in the Development of English Physiology in the Seventeenth and Early Eighteenth Century*, New York, 1981. Further examples of cases from Thomas Willis's practice can be seen in *Willis's Oxford Casebook (1650-52)*, edited by Kenneth Dewhurst, Oxford, 1981.

For those who wish to know more about the details of Paracelsianism, a good simple introduction to his medical ideas can be found in Lester S. King, *The Growth of Medical Thought*, Chicago, 1963, pp. 86-138. The more ambitious might like to proceed to Walter Pagel, *Paracelsus: An Introduction to Philosophical Medicine in the Era of the Renaissance*, New York, 1958, and the collection of Pagel's essays *From Paracelsus to Van Helmont: Studies in Renaissance Medicine and Science*, edited by Marianne Winder, London, 1986. The best account of the endeavours of Puritan and Paracelsian medical reformers in Interregnum England is Charles Webster, *The Great Instauration: Science, Medicine and Reform 1626-1660*, London, 1975. Paracelsianism in England in the sixteenth century is covered by the same author in an essay on 'Alchemical and Paracelsian Medicine' in *Health, Medicine and Mortality*, cit., pp. 301-34. For a European-wide survey of Paracelsian reformers see Hugh Trevor-Roper, 'The Paracelsian Movement', in his *Renaissance Essays*, London, 1985, pp. 149-99. The best study of Théophraste Renaudot in English is Howard M. Solomon, *Public Welfare, Science, and Propaganda in Seventeenth Century France: The Innovations of Théophraste Renaudot*, Princeton, NJ, 1972.

Magical healing is discussed in the context of a wider study on the nature of popular magic in the excellent Keith Thomas, *Religion and the Decline of Magic*, London, 1973, pp. 209-51. The survival of magical healing into the eighteenth century can be seen in N. C. Hultin, 'Medicine and Magic in the Eighteenth Century: The Diaries of James Woodford', *Journal of the History of Medicine and Applied Sciences*, 30, 1975, pp. 349-66; and, more comprehensively for France, in Matthew Ramsey, *Professional and Popular Medicine in France, 1770-*

1830, Cambridge, 1988. Keith Thomas, op. cit., pp. 335-458, also has a great deal to say about astrology, but for a briefer survey of astrology in medicine see the article by Allan Chapman, already cited. For an understanding of intellectual views on the nature of the 'occult' qualities of substances see Keith Hutchison, 'What Happened to Occult Qualities in the Scientific Revolution?', *Isis*, 73, 1982, pp. 233-53; and Ron Millen, 'The Manifestation of Occult Qualities in the Scientific Revolution', in *Religion, Science and Worldview: Essays in Honor of Richard S. Westfall*, edited by Margaret J. Osler and Paul L. Farber, Cambridge, 1985, pp. 185-216. On the medical dimension see Linda Deer Richardson, 'The Generation of Disease: Occult Causes and Diseases of the Total Substance', in *The Medical Renaissance of the Sixteenth Century* edited by A. Wear, R. K. French and I. M. Lonie, Cambridge, 1985, pp. 175-94, and M. L. Bianchi, 'Occulto e manifesto nella medicina del Rinascimento: Jean Fernel e Pietro Severino', *Atti e memorie dell'Accademia toscana di scienze e lettere 'La Colombaria'*, 33, 1982, pp. 185-248. For fuller accounts of magic, showing its importance to the development of the modern worldview and its intellectual appeal even to the most highly educated classes, see *Occult and Scientific Mentalities in the Renaissance*, edited by Brian Vickers, Cambridge, 1984, and Charles Webster, *From Paracelsus to Newton*, cit..

Inevitably, there are a large number of issues relevant to the story of medicine in popular culture which it has not been possible to tackle in the present volume. Three spring immediately to mind: the impact of plague; female health, including obstetrics and the role of women in the institutions of medicine; and popular ideas about the human body. I can do no more here than point the interested reader to suitable starting-points in what are extensive bodies of literature.

Plague was a continual presence in medieval and Renaissance Europe and, since learned medicine was completely inadequate for dealing with it, it served to undermine elitist medical movements and promote, if only through desperation, popular medicine. For a superb comprehensive survey of the English scene see Paul Slack, *The Impact of Plague in Tudor and Stuart England*, London, 1985. The situation in Renaissance Italy has been detailed by Carlo M. Cipolla in three excellent books: *Cristofano and the Plague: A Study in the History of Public Health in the Age of Galileo*, London, 1973; *Faith, Reason and the Plague: A Tuscan Story of the Seventeenth Century*, Brighton, 1977; *Public Health and the Medical Profession in the Renaissance*, Cambridge, 1976.

The important role of women in the pre-modern domestic art of medicine has been implicit in this essay but this is a multi-faceted and wide-ranging topic in its own right. There is a chapter on women and medicine in Lucinda Beier, *Sufferers and Healers*, cit., pp. 211-41, which is a good starting-point. The standard works are now rather old-fashioned but nobody will be disappointed by the wealth of information they provide: Kate Campbell Hurd-Mead, *A History of Women in Medicine from the Earliest Times to the Beginning of the Nineteenth Century*, Haddam, Conn., 1938 (reprinted New York, 1977) and Muriel Joy Hughes, *Women Healers in Medieval Life and Literature*, New York, 1943. Historians have paid considerable attention to the history of obstetrics, once an exclusively female preserve. Two good introductions to this topic are: Jean Towler and Joan Bramall, *Midwives in History and Society*, London, 1986 and Jean Donnison, *Midwives and Medical Men: A History of the Struggle for the Control of Childbirth*, New Barnet, 1988. Another important aspect of the history of women in medicine is, of course, the witchcraft connection. There is much on this in Keith Thomas, *Religion and the Decline of Magic*, cit., but see also T. R. Forbes, *The Midwife and the Witch*, New Haven, Conn., 1966.

The medical thinking of any given age is always inextricably bound up with wider notions on the nature of man, woman and society. This has given rise to a number of studies in the history of medicine in which a given society's concept of the human body can be shown to reflect the nature of that society and the role of medicine within it. Two recent books by Pietro Camporesi present the reader with extremely rich and fascinating materials to illustrate these interconnections, even if it is not clear what conclusion the author wishes to draw from them. These are: *The Incorruptible Flesh: Bodily Mutation and Mortification in Religion and Folklore*, Cambridge, 1988, and *The Body in the Cosmos: Natural Symbols in Medieval and Early Modern Italy*, Cambridge, 1989. For a study which focuses more on learned medicine see Marie-Christine Pouchelle, *The Body and Surgery in the Middle Ages*, Cambridge, 1989. But even these extra suggestions do not exhaust the topic covered by the title of this essay. A full survey would surely have to include popular attitudes towards death, the effect of syphilis and other epidemic diseases, mental health, and much more.

10   Witchcraft

There is no proper study of the relationship between demonology and science in early modern Europe, but the naturalistic basis of learned witchcraft beliefs is apparent from the survey by Wayne Shumaker, *The Occult Sciences in the Renaissance*, London, 1972, pp. 70-85, and it is stressed by Stuart Clark, 'The Scientific Status of Demonology', in *Occult and Scientific Mentalities in the Renaissance*, edited by Brian Vickers, Cambridge, 1984, pp. 351-74. Evidence of a general affinity between demonology and the 'new science' is offered by Irving Kirsch, 'Demonology and science during the scientific revolution', *Journal of the History of the Behavioural Sciences*, 16, 1980, pp. 359-68. The technical aspects of occult causation in pre-modern science are best approached in Keith Hutchison, 'What happened to occult qualities in the scientific revolution?', *Isis*, 73, 1982, pp. 233-53, and John Henry, 'Occult qualities and the experimental philosophy: Active principles in pre-Newtonian matter theory', *History of Science*, 24, 1986, pp. 335-81.

Abundant evidence of the beliefs and practices associated with popular techniques for securing good fortune, healing, divination, etc., is available in the classic studies by K. Thomas, *Religion and the Decline of Magic*, cit., and Robert Muchembled, *Popular Culture and Elite Culture in France, 1400-1750*, translated by Lydia Cochrane, London, 1985. An especially exciting source is the records of the various Mediterranean Inquisitional tribunals. These have been looked at both for evidence of popular culture and for the Inquisitors' construction of the notion of 'superstition' in, for example, *Inquisition and Society in Early Modern Europe*, edited and translated by Stephen Haliczer, London, 1987; Ruth Martin, *Witchcraft and the Inquisition in Venice, 1550-1650*, Oxford, 1989, and Stephen Haliczer, *Inquisition and Society in the Kingdom of Valencia, 1478-1834*, London, 1990. Mary O'Neil's forthcoming book on the Inquisition at Modena will be especially important for the themes discussed in the present volume.

The best introduction to the cultural history of conscience, also neglected, is now *Conscience and Casuistry in Early Modern Europe*, edited by Edmund Leites, Cambridge, 1988. Indispensable for the general relationship between religious reformation, the concept of 'superstition' and the belief in witchcraft is William Monter, *Ritual, Myth and Magic in Early Modern Europe*, Brighton, 1983. The particularly Protestant aspects of this relationship are dealt with by Stuart Clark, 'Protestant demonology: Sin, superstition and society, c.1520-1630' in *Early Modern European Witchcraft Centres and Peripheries*, edited by Bengt Ankarloo and Gustav Henningsen, London, 1990, pp. 45-81.

The classic essay on the magical powers accorded to language in traditional society is S. J. Tambiah, 'The magical power of words', *Man*, 3, 1968, pp. 175-208, reprinted in his *Culture, Thought and Social Action: An Anthropological Perspective*, London, 1985, pp. 17-59. Apart from the studies cited in notes 37 and 38 (chapter 10), Hans Aarsleff's *From Locke to Saussure: Essays on the Study of Language and Intellectual History*, Minneapolis, 1982, is reliable on the history of early modern and modern attitudes to language.

Help with understanding the significance of the works of Michel Foucault cited in connection with the study of witchcraft beliefs is available in J. G. Merquior, *Foucault*, London, 1985, and, at a more advanced (but still lucid) level, Gary Gutting, *Michel Foucault's Archaeology of Scientific Reason*, Cambridge, 1989. Merquior reports (p. 137) that in an interview of 1970, Foucault was asked if he intended to work on the history of religion, and, in reply, said that his real interest in this field was witchcraft. Foucault had just published an essay on the subject: 'Médecins, juges et sorciers au xviie siècle', *Médecine de France*, 200, 1969, pp. 121-8.

11 Astrology

Renaissance astrological literature is a vast and complex topic which has not yet been properly investigated, even if in the last decades there has been a notable increase in works on this subject, ignored for centuries as irrational and superstitious. Some fundamental pioneering studies have pointed out the importance of astrology in the history of Renaissance ideas, and its connection with philosophy, religion, magic, medicine, politics, prophecy and iconology. One can cite here the works of E. Cassirer, *The Individual and the Cosmos in Renaissance Philosophy*, Philadelphia, 1972; F. Saxl, *Lectures*, 2 vols., London, 1957, a collection of papers which include an abundance of material on the role of astrological iconography in the arts; E. Klibansky, E. Panofsky and F. Saxl, *Saturn and Melancholy*, London, 1964, and more recently the works of F. A. Yates and K. Thomas, cit..

L. Thorndike, *A History of Magic and Experimental Science*, New York, 1923-58, remains a fundamental point of reference for astrology, magic, alchemy and medicine; it contains a wealth of information on authors and texts from antiquity to the eighteenth century. Volumes V and VI are dedicated to the Renaissance. Eugenio Garin has written many works on Renaissance astrology, and edited with an Italian parallel translation G. Pico della Mirandola's *Disputationes adversus astrologiam divinatricem*, 2 vols., Florence, 1946-52. Other works by Garin include *Astrology in the Renaissance: the Zodiac of Life*, London, 1983.

There are a number of very useful collections of essays, including *Magia, astrologia e religione*, cit.; *Scienze, credenze occulte*, cit.; for an overview of the literature on these themes, see P. Zambelli, 'Teorie su astrologia, magia e alchimia (1348-1586) nelle interpretazioni recenti', *Rinascimento*, 27, 1987, pp. 95-119. For a well-balanced account of astrology and magic see B. Copenhaver, art. cit., in *The Cambridge History*, cit., pp. 264-300, which deals mainly with 'natural' magic. A fascinating, well-documented panorama of the history of astrology is L. Aurigemma, *Le Signe zodiacal du Scorpion dans les traditions occidentales de l' antiquité greco-latine à la Renaissance*, Paris, 1976. For a lively history of astrology dealing particularly with the medieval period see S. Caroti, *L'Astrologia in Italia*, Rome, 1983. For studies of Renaissance astrology see D. C. Allen, *The Star-Crossed Renaissance*, New York, 1973. A more technical work is J. D. North, *Horoscopes and History*, London, 1986.

The recent volume *Astrology, Science and Society*, edited by Patrick Curry, Woodbridge, 1987, is a collection of essays which deal particularly with the medieval period, and with the relationship between astrology and the scientific revolution in seventeenth-century England.

For works more specifically related to Germana Ernst's essay in the present volume: on Cardano and astrology, see J. Ochman, 'Il determinismo astrologico di G. Cardano', in *Magia, astrologia e religione*, cit., pp. 123-9. See also J. C. Margolin, 'Rationalisme et irrationalisme dans la pensée de Jerome Cardan', in *Revue de l'Université de Bruxelles*, 21, 1969, pp. 89-128, and G. Canziani, 'Une encyclopédie naturaliste de la Renaissance devant la critique du XVIIe Siècle: le "Theophrastus redivivus" lecteur de Cardan', XVIIe Siècle, 26, 1985, pp. 379-406. A conference on Cardano was held in Wolfenbuttel in 1989, and the proceedings are forthcoming. An English translation of the horoscope of Christ can be found in W. Shumaker, *Renaissance Curiosa*, New York, 1982, p. 53 ff..

On the life and works of Giuntini see L. Thorndike, *A History of Magic*, cit., vol. VI, pp. 129-33; D. Hellman discusses Giuntini, with particular reference to the comet of 1577 in *The comet of 1577: its Place in the History of Astronomy*, New York, 1971, pp. 377-9, and 'The gradual abandonment of the Aristotelian Universe: a preliminary note on some sidelights', *L' Aventure de la science. Mélanges A. Koyr,* Vol. I, Paris, 1964, pp. 283-93.

12 The decline of astrological beliefs

The classic book on English astrology in the sixteenth and seventeenth centuries is Keith Thomas, *Religion and the Decline of Magic,* cit.. This should be supplemented by Bernard Capp, *Astrology and the Popular Press: English Almanacs 1500-1800,* London, 1979. The most recent book-length study of the subject is Patrick Curry, *Prophecy and Power: Astrology in Early Modern England,* Cambridge, 1989 (which covers 1642-1800).

Other valuable work on early modern English astrology includes: An *Astrological Diary of the Seventeenth Century: Samuel Jeake of Rye, 1625-1699,* edited by Michael Hunter and Annabel Gregory, including the useful introduction; parts of Charles Webster, *From Paracelsus to Newton,* cit.; Michael MacDonald, *Mystical Bedlam: Madness, Anxiety and Healing in Seventeenth-Century England,* Cambridge, 1982. There are also several relevant essays in *Astrology, Science and Society,* edited by Patrick Curry, Woodbridge, 1987, including Michael Hunter, 'Science and Astrology in Seventeenth-Century England: An Unpublished Polemic by John Flamsteed', pp. 261-300; Simon Schaffer, 'Newton's Comets and the Transformation of Astrology', pp. 219-44; Jacques Halbronn, 'The Revealing Process of Translation and Criticism in the History of Astrology', pp. 197-218. This collection is also of interest in connection with the historiographical problems posed by the study of astrology.

The gap in the study of judicial astrology is soon to be remedied by the publication of a book based on Ann Geneva, 'England's Propheticall Merline Decoded: A Study of the Symbolic Art of Astrology in Seventeenth Century England', PhD thesis, State University of New York, 1988. Readers may also consult Patrick Curry, 'Afterword' and 'Bibliographical Appendix', and Geoffrey Cornelius, 'A Modern Astrological Perspective', in William Lilley, *Christian Astrology,* London 1985, pp. 856-71. A recent and accessible European overview (which however stops in mid-seventeenth century) is Jim Tester, *A History of Western Astrology,* Woodbridge, 1987.

There are some excellent essays on the 'occult' generally in Renaissance and early modern Europe in *Hermeticism and the Renaissance,* cit., and *Occult and Scientific Mentalities,* cit..

Material that is useful and suggestive for the place and period as a whole includes E. P. Thompson, 'Patrician Society, Plebeian Culture', *Journal of Social History*, 7, 1974, pp. 382-405; idem, 'Eighteenth-Century English Society: Class Struggle Without Class?', *Social History*, 3, 1978, pp. 133-65; Harry C. Payne, 'Elite vs. Popular Mentality in the Eighteenth Century', *Studies in Eighteenth-Century Culture*, 8, 1979, pp. 3-32.

Exciting recent work in the history and sociology of science includes (selected out of a large literature): *Natural Order: Historical Studies of Scientific Culture*, edited by Barry Barnes and Steven Shapin; *The Ferment of Knowledge: Studies on the Historiography of Eighteenth-Century Science*, edited by G. S. Rousseau and Roy Porter, Cambridge, 1980, which includes Steven Shapin, 'The Social Uses of Science', pp. 93-142; Simon Schaffer, 'Godly Men and Mechanical Philosophers: Souls and Spirits in Restoration Natural Philosophy', *Science in Context*, 1, 1987, pp. 55-86.

Valuable discussions of the historiography of the occult include Roger Cooter, 'Deploying "Pseudoscience": Then and Now', in *Science, Pseudo-Science and Society*, edited by M. P. Hanen, M. J. Osler and R. G. Weyant, Waterloo, Ont., 1980, pp. 237-72; and Simon Schaffer, 'Occultism and Reason', in *Philosophy: its History and Historiography*, edited by A. J. Holland, Dordrecht, 1985, pp. 115-41.

For the best discussion and analysis of popular/elite cultures and mentalities and ideologies, and a resolution of these antinomies, see Roger Chartier, *Cultural History: Between Practices and Representations*, Cambridge, 1988; by the same author, 'Culture as Appropriation: Popular Cultural Uses in Early Modern France', in *Understanding Popular Culture: Europe from the Middle Ages to the Nineteenth Century*, edited by Steven L. Kaplan, Amsterdam, 1984, pp. 229-54. Also excellent are Michel Vovelle, *Ideologies and Mentalities*, Cambridge, 1989 (first French edition, Paris, 1982); *Constructing the Past: Essays in Historical Methodology*, edited by Jacques Le Goff and Pierre Nora, Cambridge, 1985 (first French edition, Paris, 1974). An early discussion of such ideas in English-speaking countries was chiefly initiated by Peter Burke, *Popular Culture in Early Modern Europe*, London, 1978, and more recently in 'Revolution in Popular Culture', in *Revolution in History*, edited by Roy Porter and Mikuláš Teich, Cambridge, 1987, pp. 206-25.

The literature on Antonio Gramsci is vast. On hegemony as a possible resolution of structuralism/culturalism, see *Popular Culture and Social Relations,* edited by Tony Bennett, Colin Mercer and Jane Woollacott, especially pp. xi-xix. The 'post-Marxist' idea of hegemony is developed in Ernesto Laclau and Chantal Mouffe, *Hegemony and Socialist Strategy: Towards a Radical Democratic Politics,* London, 1985; and their 'Post-Marxism Without Apologies', *New Left Review,* 156, 1987, pp. 79-106.

For good discussions of theory in relation to the practice of history, see E. P. Thompson's polemical *The Poverty of Theory,* London, 1978; Gregor McLennan, 'E. P. Thompson and the Discipline of Historical Context', in *Making Histories: Studies in History-Writing,* edited by Richard Johnson, Gregor McLennan, Bill Schwarz and David Sutton, London, 1982, pp. 96-132; Gareth Stedman Jones, 'From Historical Sociology to Theoretical History', *British Journal of Sociology,* 27, 1976, pp. 295-305; Philip Abrams, *Historical Sociology,* Somerset, 1982.

Finally, for an informed discussion of relativism, see Paul Feyerabend, *Farewell to Reason,* London, 1987; Richard Rorty, *Contingency, Solidariety and Irony,* Cambridge, 1987; T. S. Kuhn's forthcoming book, incorporating his 1987 Sherman Lectures at the University of London.

Index

Abbreviations: *fl.* floruit; *r.* reigned

Medici, Giovanni de' (1475-1521) *see* Leo X

Medici, Lorenzo de' (the Magnificent) (1449-92), 155, 157-9, 161

Medici, Virginia de' (*d.* 1615), 162

medicine, **191-221**
  and Arabs, 56, 203, 208
  and astrology, 207-8
  attempts at change, 205-6
  Company of Barber-Surgeons, 192
  diagnosis, 200
  drugs, 208-9
  education, 206-7
  Faculty of Physicians and Surgeons, 193
  hierarchies, 192-3, 209
  historiography, 195
  humours, 199
  Incorporation of Surgeons and Barbers, 193
  licensing, 191-3, 196
  in literature *see* Marlowe; Shakespeare
  and magical healing, 215-18
  and mechanical philosophy, 210-11
  Paracelsianism, 212-15, 218
  popular books of, 201
  and popular knowledge, 197-205, 213
  practical experience, 195
  practitioners of, 191, 197, 209-10
  and reformist movements, 213-14
  Royal College of Physicians, 192, 196, 213-14
  temperaments, 200
  textual tradition, 194-5
  tripartite division of, 193, 195-7
  uroscopy, 203-5
  Worshipful Society of Apothecaries, 192
  see also Index Librorum Prohibitorum; popular culture

Melanchthon, Philipp (1497-1560), 37, 43, 61, 124-5, 137, 147, 149

Menocchio (Domenico Scandella) (1532-1600?), 102

Mercator, Gerard (1512-94), 178, 182, 184

Mersenne, Marin (1588-1648), 104, 130, 135-6, 243-4

Merton, Robert, 2

Mesland, Denis, 132-3

Mohammed (*c.* 570-632), 258

Molière (Jean Baptiste Poquelin) (1622-1733), 209

Mondino (Raimondino dei Liucci) (*c.* 1275-*c.* 1326), 56

Montaigne, Michel Eyquem de (1533-92), 38, 154, 171

Montefeltro, Guidobaldo da (Duke of Urbino) (1478-1508), 158

Moore, 277-8

Morandi, Orazio (Abbot of Santa Prassede) (*c.* 1570-1630), 263, 267-70

More, Henry (1614-87), 279-80

More (Sir) Thomas (1478-1535), 38, 42

Moses, 58-9, 126

Mouffe, Chantal, 286-7

Murner, Thomas (1475-*c.* 1533), 29

Nabod, 264

nationalism *see* history

natural philosophy
  and authority, 48-9, 52-7, 66
  and Catholics, 122-3
  and churches, 136-7
  at Courts, 102-3
  and history, 48
  and law, 109-13
  and printing, 101
  and protestants, 122, 124-5
  and ruling elite, 101-9
  scientia/ars, 65-6, 68
  and socio-political concerns, **100-18**
  and theology, 121, 125-30
  and transubstantiation, 124, 130-3
  see also astrology in England; demonology; history; mathematics; rhetoric

Navarro, Vincente, 239

neo-Platonism, 35, 41-2, 57-61, 85, 155-72 *passim*, 243, 260

Newton, Isaac (1642-1727), 6, 51, 60, 69, 88, 116, 211, 276, 280, 282